Process Control
A Practical Approach

Process Control
A Practical Approach

Myke King

Whitehouse Consulting, Isle of Wight, UK

A John Wiley and Sons, Ltd., Publication

This edition first published 2011
2011 © John Wiley & Sons Ltd

Registered office
John Wiley & Sons Ltd, The Atrium, Southern Gate, Chichester, West Sussex, PO19 8SQ, United Kingdom

For details of our global editorial offices, for customer services and for information about how to apply for permission to reuse the copyright material in this book please see our website at www.wiley.com.

The right of the author to be identified as the author of this work has been asserted in accordance with the Copyright, Designs and Patents Act 1988.

2   2011

All rights reserved. No part of this publication may be reproduced, stored in a retrieval system, or transmitted, in any form or by any means, electronic, mechanical, photocopying, recording or otherwise, except as permitted by the UK Copyright, Designs and Patents Act 1988, without the prior permission of the publisher.

Wiley also publishes its books in a variety of electronic formats. Some content that appears in print may not be available in electronic books.

Designations used by companies to distinguish their products are often claimed as trademarks. All brand names and product names used in this book are trade names, service marks, trademarks or registered trademarks of their respective owners. The publisher is not associated with any product or vendor mentioned in this book. This publication is designed to provide accurate and authoritative information in regard to the subject matter covered. It is sold on the understanding that the publisher is not engaged in rendering professional services. If professional advice or other expert assistance is required, the services of a competent professional should be sought.

Library of Congress Cataloging-in-Publication Data

King, Michael, 1951-
  Process control : a practical approach / Michael King.
       p. cm.
  Includes bibliographical references and index.
  ISBN 978-0-470-97587-9 (cloth)
1. Chemical process control. I. Title.
  TP155.75.K56 2011
  660'.2815–dc22
                                2010034824

A catalogue record for this book is available from the British Library.

Print ISBN: 9780470975879
e-PDF ISBN: 9780470976555
o-Book ISBN: 9780470976562
e-Pub ISBN: 9780470976661

Set in 10/12 Times Roman by Thomson Digital, Noida, India
Printed in Singapore by Fabulous Printers Pte Ltd.

Contents

Preface ix
About the Author xv

1. **Introduction** 1

2. **Process Dynamics** 3
   2.1 Definition 3
   2.2 Cascade Control 9
   2.3 Model Identification 11
   2.4 Integrating Processes 20
   2.5 Other Types of Process 22
   2.6 Robustness 24
   2.7 Laplace Transforms for Processes 27
   References 28

3. **PID Algorithm** 29
   3.1 Definitions 29
   3.2 Proportional Action 30
   3.3 Integral Action 33
   3.4 Derivative Action 35
   3.5 Versions of Control Algorithm 39
   3.6 Interactive PID Controller 41
   3.7 Proportional-on-PV Controller 43
   3.8 Nonstandard Algorithms 50
   3.9 Tuning 51
   3.10 Ziegler-Nichols Tuning Method 52
   3.11 Cohen-Coon Tuning Method 56
   3.12 Tuning Based on Penalty Functions 57
   3.13 Manipulated Variable Overshoot 60
   3.14 Lambda Tuning Method 61
   3.15 IMC Tuning Method 63
   3.16 Choice of Tuning Method 65
   3.17 Suggested Tuning Method for Self-Regulating Processes 66
   3.18 Tuning for Load Changes 66
   3.19 Tuning for Unconstrained MV Overshoot 71
   3.20 PI Tuning Compared to PID Tuning 72
   3.21 Tuning for Large Scan Interval 74
   3.22 Suggested Tuning Method for Integrating Processes 76

3.23	Implementation of Tuning	78
3.24	Loop Gain	79
3.25	Adaptive Tuning	79
3.26	Initialisation	80
3.27	Anti-Reset Windup	81
3.28	On-Off Control	81
3.29	Laplace Transforms for Controllers	83
3.30	Direct Synthesis	85
References		88

4. Level Control — 91
4.1	Use of Cascade Control	91
4.2	Parameters Required for Tuning Calculations	93
4.3	Tight Level Control	97
4.4	Averaging Level Control	100
4.5	Error-Squared Controller	105
4.6	Gap Controller	108
4.7	Impact of Noise on Averaging Control	111
4.8	General Approach to Tuning	113
4.9	Three-Element Level Control	114

5. Signal Conditioning — 117
5.1	Instrument Linearisation	117
5.2	Process Linearisation	119
5.3	Constraint Conditioning	122
5.4	Pressure Compensation of Distillation Tray Temperature	124
5.5	Pressure Compensation of Gas Flow Measurement	125
5.6	Filtering	126
5.7	Exponential Filter	127
5.8	Higher Order Filters	129
5.9	Nonlinear Exponential Filter	130
5.10	Averaging Filter	131
5.11	Least Squares Filter	132
5.12	Control Valve Characterisation	136
5.13	Equal Percentage Valve	137
5.14	Split-Range Valves	140

6. Feedforward Control — 147
6.1	Ratio Algorithm	147
6.2	Bias Algorithm	151
6.3	Deadtime and Lead-Lag Algorithms	152
6.4	Tuning	155
6.5	Laplace Derivation of Dynamic Compensation	161

7. Deadtime Compensation — 163
7.1	Smith Predictor	163

	7.2	Internal Model Control	166
	7.3	Dahlin Algorithm	167
	References		168

8. Multivariable Control — 169
- 8.1 Constraint Control — 169
- 8.2 SISO Constraint Control — 170
- 8.3 Signal Selectors — 171
- 8.4 Relative Gain Analysis — 174
- 8.5 Steady State Decoupling — 177
- 8.6 Dynamic Decoupling — 180
- 8.7 MVC Principles — 184
- 8.8 Parallel Coordinates — 187
- 8.9 Enhanced Operator Displays — 188
- 8.10 MVC Performance Monitoring — 189
- References — 195

9. Inferentials and Analysers — 197
- 9.1 Inferential Properties — 197
- 9.2 Assessing Accuracy — 203
- 9.3 Laboratory Update of Inferential — 208
- 9.4 Analyser Update of Inferential — 210
- 9.5 Monitoring On-stream Analysers — 212
- Reference — 214

10. Combustion Control — 215
- 10.1 Fuel Gas Flow Correction — 215
- 10.2 Measuring NHV — 220
- 10.3 Dual Firing — 222
- 10.4 Inlet Temperature Feedforward — 223
- 10.5 Fuel Pressure Control — 225
- 10.6 Combustion Air Control — 227
- 10.7 Boiler Control — 236
- 10.8 Fired Heater Pass Balancing — 237

11. Compressor Control — 243
- 11.1 Polytropic Head — 243
- 11.2 Flow Control (Turbo-Machines) — 246
- 11.3 Flow Control (Reciprocating Machines) — 251
- 11.4 Anti-Surge Control — 252

12. Distillation Control — 259
- 12.1 Key Components — 262
- 12.2 Relative Volatility — 263
- 12.3 McCabe-Thiele Diagram — 266
- 12.4 Cut and Separation — 271

	12.5	Effect of Process Design	281
	12.6	Basic Controls	285
	12.7	Pressure Control	285
	12.8	Level Control	299
	12.9	Tray Temperature Control	315
	12.10	Pressure Compensated Temperature	325
	12.11	Inferentials	335
	12.12	First-Principle Inferentials	342
	12.13	Feedforward on Feed Rate	344
	12.14	Feed Composition Feedforward	348
	12.15	Feed Enthalpy Feedforward	349
	12.16	Decoupling	350
	12.17	Multivariable Control	352
	12.18	On-stream Analysers	360
	12.19	Towers with Sidestreams	361
	12.20	Column Optimisation	364
	12.21	Optimisation of Column Pressure	366
	12.22	Energy/Yield Optimisation	368
	References		370
13.	**APC Project Execution**		**371**
	13.1	Benefits Study	371
	13.2	Benefit Estimation for Improved Regulatory Control	373
	13.3	Benefits of Closed-Loop Real-Time Optimisation	380
	13.4	Basic Controls	382
	13.5	Inferentials	384
	13.6	Organisation	385
	13.7	Vendor Selection	389
	13.8	Safety in APC Design	391
	13.9	Alarms	392
	References		393

Index **395**

Preface

So why write yet another book on process control? There are already many published, but they are largely written by academics and intended mainly to support courses taught at universities. Excellent as some of these books are in meeting that aim, the content of many academic courses has only limited relevance to control design in the process industry. There are a few books that take a more practical approach but these usually provide only an introduction to the technologies. They contain enough detail if used as part of a wider engineering course but not enough for the practitioner. This book aims more to meet the needs of industry.

Most engineers responsible for the design and maintenance of control applications find daunting much of the theoretical mathematics that is common in the academic world. In this book we have aimed to keep the mathematics to a minimum. For example, Laplace transforms are only included so that the reader may relate what is in this book to what will be found in most theoretical texts and in the documentation provided by many DCS (distributed control system) vendors. They are not used in any of the control design techniques. And while we present the mathematical derivation of these techniques, to show that they have a sound engineering basis, the reader can skip these if too daunting and simply apply the end result.

The book aims to present techniques that have an immediate practical application. In addition to the design methods it describes any shortcuts that can be taken and how to avoid common pitfalls. The methods have been applied on many processes on a wide range of controllers. They should work!

In addition to providing effective design methods, this book should improve the working practices of many control engineers. For example, the majority still prefer to tune PID (proportional, integral, derivative) controllers by trial and error. This is time-consuming and rarely leads to controllers performing as well as they should. This might be because of a justified mistrust of published tuning methods. Most do have serious limitations. This book addresses this and offers a method proven to be effective in terms of both controller performance and engineering effort.

DCS include a wide array of control algorithms with many additional engineer-definable parameters. The DCS vendors are poor at explaining the purpose of these algorithms with the result that the industry is rife with misinterpretation of their advantages and disadvantages. These algorithms were included in the original system specification by engineers who knew their value, but this knowledge has not passed to the industry. The result is that there are substantial improvements that can be made on almost every process unit, surpassing what the control engineer is even aware of – let alone knows how to implement. This book addresses all the common enhancements.

This book takes a back-to-basics approach. The use of MVC (multivariable controllers) is widespread in industry. Control engineering staff and their contractors have invested thousands of man-hours in the necessary plant testing and commissioning. Improving the basic controls is not usually an option once the MVC is in place. Improvements are likely to change the process dynamics and would thus involve substantial re-engineering of the MVC. Thus poor basic control remains the status quo and becomes the accepted standard to the point where it is not addressed even when the opportunity presents itself. This book raises the standard of what might be expected from the performance of basic controls.

Before MVC, ARC (advanced regulatory control) was commonplace. MVC has rightly replaced many of the more complex ARC techniques, but it has been used by too many as the panacea to any control problem. There remain many applications where ARC outperforms MVC; but appreciation of its advantages is now hard to find in industry. The expertise to apply it is even rarer. This book aims to get the engineer to reconsider where ARC should be applied and to help develop the necessary implementation skills.

However due credit must be given to MVC as a major step forward in the development of APC (advanced process control) techniques. This book focuses on how to get the best out of its application, rather than replicate the technical details that appear in many text books, papers and product documentation.

The layout of the book has been designed so that the reader can progress from relatively straightforward concepts through to more complex techniques applied to more complex processes. It is assumed that the new reader is comfortable with mathematics up to a little beyond high school level. As the techniques become more specific some basic knowledge of the process is assumed, but introductory information is included – particularly where it is important to control design. Heavily mathematical material, daunting to novices and not essential to successful implementation, has been relegated to the end of each chapter.

SI units have been mainly used throughout but, where important and practical, conversion to imperial units is given in the text. Methods published in non-SI units have been included without change if doing so would make them too complex.

The book is targeted primarily for use in the continuous process industry, but even predominantly batch plants have continuous controllers and often have sections of the process which are continuous. My experience is mainly in the oil and petrochemicals industries and, despite every effort being taken to make the process examples as generic as possible, it is inevitable that this will show through. However this should not be seen as a reason for not applying the techniques in other industries. Many started there and have been applied by others to a wide range of processes.

It is hoped that the academic world will take note of the content. While some institutions have tried to make their courses more relevant to the process industry, practitioners still perceive a huge gulf between theory and practice. Of course there is a place for the theory. Many of the modern control technologies now applied in the process industry are developed from it. And there are other industries, such as aerospace, where it is essential.

The debate is what should be taught as part of chemical engineering. Very few chemical engineers benefit from the theory currently included. Indeed the risk is that many potentially excellent control engineers do not enter the profession because of the poor image that theoretical courses create. Further, those that do follow a career in process control, can find themselves working in an organisation managed by a chemical engineering graduate who has no appreciation of what process control technology can do and its importance to the business.

It is the nature of almost any engineering subject that the real gems of useful information get buried in amongst the background detail. Listed here are the main items worthy of special attention by the engineer because of the impact they can have on the effectiveness of control design.

- Understanding the process dynamics is essential to the success of almost every process control technique. These days there is very little excuse for not obtaining these by plant testing or from historically collected data. There are a wide range of model identification products available plus enough information is given in Chapter 2 for a competent engineer to develop a simple spreadsheet-based application.

- Often overlooked is the impact that apparently unrelated controllers can have on process dynamics. Their tuning and whether they are in service or not, will affect the result of steptests and hence the design of the controller. Any changes made later can then severely disrupt controller performance. How to identify such controllers, and how to handle their effect, is described in Chapters 2 and 8.

- Modern DCS include a number of versions of the PID controller. Of particular importance in the proportional-on-PV algorithm. It is probably the most misunderstood option and is frequently dismissed as too slow compared to the more conventional proportional-on-error version. In fact, if properly tuned, it can make a substantial improvement to the way that process disturbances are dealt with – often shortening threefold the time it takes the process to recover. This is fully explained in Chapter 3.

- Controller tuning by trial and error should be seen as an admission of failure to follow proper design procedures, rather than the first choice of technique. To be fair to the engineer, every published tuning technique and most proprietary packages have serious limitations. Chapter 3 presents a new technique that is well proven in industry and gives sufficient information for the engineer to extend it as required to accommodate special circumstances.

- Derivative action is too often excluded from controllers. Understandably introducing a third parameter to tune by trial and error might seem an unnecessary addition to workload. It also has a poor reputation in the way that it amplifies measurement noise, but, engineered using the methods in Chapter 3, it has the potential to substantially lessen the impact of process disturbances.

- Tuning level controllers to exploit surge capacity in the process can dramatically improve the stability of the process. However the ability to achieve this is often restricted by poor instrument design, and, often it is not implemented because of difficulty in convincing the plant operator that the level should be allowed to deviate from SP (set-point) for long periods. Chapter 4 describes the important aspects in sizing and locating the level transmitter and how the conventional linear PID algorithm can be tuned – without the need even to perform any plant testing. It also shows how nonlinear algorithms, particularly gap control, can be set up to handle the situation where the size of the flow disturbances can vary greatly.

- While many will appreciate how signal conditioning can be applied to measurements and controller outputs to help linearise the behaviour, not so commonly understood is how it can be applied to constraint controllers. Doing so can enable constraints to be approached more closely and any violation dealt with more quickly. Full details are given in Chapter 5.

- Many engineers are guilty of installing excessive filtering to deal with noisy measurements. Often implemented only to make trends look better they introduce additional lag and can have a detrimental impact on controller performance. Chapter 5 gives guidance on when to install a filter and offers a new type that actually reduces the overall process lag.

- Split-ranging is commonly used to allow two or more valves to be moved sequentially by the same controller. While successful in some cases the technique is prone to problems with linearity and discontinuity. A more reliable alternative is offered in Chapter 5.

- Feedforward control is often undervalued or left to the MVC. Chapter 6 shows how simple techniques, applied to few key variables, can improve process stability far more effectively than MVC.

- A commonly accepted problem with MVC is that, if not properly monitored, they become over-constrained. In fact, if completely neglected, they are effectively fully disabled – even though they may show 100 % up-time. Chapter 8 offers a range of monitoring tools, supplementary to those provided by the MVC vendor, which can be readily configured by the engineer.

- There are many examples of MVC better achieving the wrong operating objective; unbeknown to the implementer they are reducing process profitability. Rather than attempt to base the cost coefficients on real economics they are often adjusted to force the MVC to follow the historically accepted operating strategy. Some MVC are extremely complex and it is unlikely that even the most competent plant manager will have considered every opportunity for adopting a different strategy. Chapter 12 shows how properly setting up the MVC can reveal such opportunities.

- There are literally thousands of inferential properties, so called 'soft sensors', in use today that are ineffective. Indeed many of them are so inaccurate that process profitability would be improved by decommissioning them. Chapter 9 shows how many of the statistical techniques that are used to assess their accuracy are flawed and can lead the engineer into believing that their performance is adequate. It also demonstrates that automatically updating the inferential bias with laboratory results will generally aggravate the problem.

- Simple monitoring of on-stream analysers, described in Chapter 9, ensures that measurement failure does not disrupt the process and that the associated reporting tools can do much to improve their reliability and use.

- Compensating fuel gas flow measurement for variations in pressure, temperature and molecular weight requires careful attention. Done for accounting purposes, it can seriously degrade the performance of fired heater and boiler control schemes. Chapter 10 presents full details on how it should be done.

- Manipulating fired heater and boiler duty by control of fuel pressure, rather than fuel flow, is common practice. However it restricts what improvements can be made to the controller to better handle process disturbances. Chapter 10 shows how the benefits of both approaches can be captured.

- Fired heater pass balancing is often installed to equalise pass temperatures in order to improve efficiency. Chapter 10 shows that the fuel saving is negligible and that, in some cases, the balancing may accelerate coking. However there may be much larger benefits available from the potential to debottleneck the heater.

- Compressor control packages are often supplied as 'black boxes' and many compressor manufacturers insist on them being installed in special control systems on the basis that DCS-based schemes would be too slow. Chapter 11 describes how these schemes work and, using the tuning method in Chapter 3, how they might be implemented in the DCS.

- A common failing in many distillation column control strategies is the way in which they cope with changes in feed rate and composition. Often only either the reboiler duty or the reflux flow is adjusted to compensate – usually under tray temperature control. Chapter 12 shows that failing to adjust both is worse than making no compensation. Other common misconceptions include the belief that column pressure should always be minimised and that the most economic strategy is to always exactly meet all product specifications.

- There are many pitfalls in executing an advanced control project. Significant profit improvement opportunities are often overlooked because of the decision to go with a single supplier for the benefits study, MVC, inferentials and implementation. Basic controls, inferentials and advanced regulatory controls are not given sufficient attention before awarding the implementation contract. The need for long-term application support is often underestimated and poor management commitment will jeopardise the capture of benefits. Chapter 13 describes how these and many other issues can be addressed.

Gaining the knowledge and experience now contained in this book would have been impossible if it were not for the enthusiasm and cooperation of my clients. I am exceedingly grateful to them and indeed would welcome any further suggestions on how to improve or add to the content.

Myke King
July 2010, Isle of Wight

About the Author

Myke King is the founder and director of Whitehouse Consulting, an independent consulting organisation specialising in process control. He has over 35 years experience working with over 100 clients from more than 30 countries. As part of his consulting activities Myke has developed training courses covering all aspects of process control. To date, around 2000 delegates have attended these courses. To support his consulting activities he has developed a range of software to streamline the design of controllers and to simulate their use for learning exercises.

Myke graduated from Cambridge University in the UK with a Master's degree in chemical engineering. His course included process control taught as part of both mechanical engineering and chemical engineering. At the time he understood neither! On graduating he joined, by chance, the process control section at Exxon's refinery at Fawley in the UK. Fortunately he quickly discovered that the practical application of process control bore little resemblance to the theory he had covered at university. He later became head of the process control section and then moved to operations department as a plant manager. This was followed by a short period running the IT section.

Myke left Exxon to co-found KBC Process Automation, a subsidiary of KBC Process Technology, later becoming its managing director. The company was sold to Honeywell where it became their European centre of excellence for process control. It was at this time Myke set up Whitehouse Consulting.

Myke is a Fellow of the Institute of Chemical Engineers in the UK.

1
Introduction

In common with many introductions to the subject, process control is described here in terms of layers. At the lowest level is the process itself. Understanding the process is fundamental to good control design. While the control engineer does not need the level of knowledge of a process designer, an appreciation of how the process works, its key operating objectives and basic economics is vital. In one crucial area his or her knowledge must exceed that of the process engineer, who needs primarily an understanding of the *steady-state* behaviour. The control engineer must also understand the *process dynamics*, i.e. how process parameters move between steady states.

Next up is the field instrumentation layer, comprising measurement transmitters, control valves and other actuators. This layer is the domain of instrument engineers and technicians. However the control engineer needs an appreciation of some of the hardware involved in control. He or she needs to be able to recognise a measurement problem or a control valve working incorrectly and must be aware of the accuracy and the dynamic behaviour of instrumentation.

Above the field instrumentation is the DCS and process computer. These will be supported by a system engineer. It is normally the control engineer's responsibility to configure the control applications, and their supporting graphics, in the DCS. So he or she needs to be well-trained in this area. In some sites only the system engineer is permitted to make changes to the system. However this does not mean that the control engineer does not need a detailed understanding of how it is done. Close cooperation between control engineer and system engineer is essential.

The lowest layer of process control applications is described as *regulatory control*. This includes all the basic controllers for flow, temperature, pressure and level. But it also includes control of product quality. Regulatory is not synonymous with basic. Regulatory controls are those which maintain the process at a desired condition, or SP, but that does not mean they are simple. They can involve complex instrumentation such as on-stream analysers. They can employ 'advanced' techniques such as signal conditioning, feedforward, dynamic compensation, overrides, inferential properties etc. Such techniques are often described as *advanced regulatory control (ARC)*. Generally they are implemented

---

Process Control: A Practical Approach   Myke King
© 2011 John Wiley & Sons, Ltd

within the DCS block structure, with perhaps some custom code, and are therefore sometimes called 'traditional' advanced control. This is the domain of the control engineer.

There will be somewhere a division of what falls into the responsibilities between the control engineer and others working on the instrumentation and system. The simplistic approach is to assign all hardware to these staff and all configuration work to the control engineer. But areas such as algorithm selection and controller tuning need a more flexible approach. Many basic controllers, providing the tuning is reasonable, do not justify particular attention. Work on those that do requires the skill more associated with a control engineer. Sites that assign all tuning to the instrument department risk overlooking important opportunities to improve process performance.

Moving up the hierarchy, the next level is *constraint control*. This comprises control strategies that drive the process towards operating limits, where closer approach to these limits is known to be profitable. Indeed, on continuous processes, this level typically captures the large majority of the available process control benefits. The main technology applied here is the *multivariable controller* (*MVC*). Because of its relative ease of use and its potential impact on profitability it has become the focus of what is generally known as *advanced process control* (*APC*). In fact, as a result, basic control and ARC have become somewhat neglected. Many sites (and many APC vendors) no longer have personnel that appreciate the value of these technologies or have the know-how to implement them.

The topmost layer, in terms of closed loop applications, is *optimisation*. This is based on key economic information such as feed price and availability, product prices and demand, energy costs etc. Optimisation means different things to different people. The planning group would claim they optimise the process, as would a process support engineer determining the best operating conditions. MVC includes some limited optimisation capabilities. It supports objective coefficients which can be set up to be consistent with process economics. Changing the coefficients can cause the controller to adopt a different strategy in terms of which constraints it approaches. However those MVC based on linear process models cannot identify an unconstrained optimum. This requires a higher fidelity process representation, possibly a rigorous simulation. This we describe as *closed-loop real-time optimisation* (*CLRTO*) or more usually just *RTO*.

Implementation should begin at the base of the hierarchy and work up. Any problems with process equipment or instrumentation will affect the ability of the control applications to work properly. MVC performance will be restricted and RTO usually needs to work in conjunction with the MVC. While all this may be obvious, it is not necessarily reflected in the approach that some sites have towards process control. There are sites investing heavily in MVC but which give low priority to maintaining basic instrumentation. And **most** give only cursory attention to regulatory control before embarking on implementation of MVC.

2
Process Dynamics

Understanding process dynamics is essential to effective control design. Indeed, as will become apparent in later chapters, most design involves performing simple calculations based solely on a few dynamic parameters. While control engineers will commit several weeks of round-the-clock effort to obtaining the process dynamics for MVC packages, most will take a much less analytical approach to regulatory controls. This chapter aims to demonstrate that process dynamics can be identified easily and that, when combined with the design techniques described in later chapters, it will result in controllers that perform well without the need for time-consuming tuning by trial-and-error.

2.1 Definition

To explore dynamic behaviour, as an example, we will use a simple fired heater as shown in Figure 2.1. It has no automatic controls in place and the minimum of instrumentation – a temperature indicator (TI) and a fuel control valve. The aim is to ultimately commission a temperature controller which will use the temperature as its *process variable* (*PV*) and the fuel valve position as it *manipulated variable* (*MV*).

Figure 2.2 shows the effect of manually increasing the opening of the valve. While the temperature clearly rises as the valve is opened, the temperature trend is somewhat different from that of the valve. We use a number of parameters to quantify these differences.

The test was begun with the process steady and sufficient time was given for the process to reach a new steady state. We observed that the steady state change in temperature was different from that of the valve. This difference is quantified by the *steady state process gain* and is defined by the expression

$$\text{process gain} = \frac{\text{change in temperature}}{\text{change in valve position}} \tag{2.1}$$

Process gain is given the symbol K_p. If we are designing controls to be installed in the DCS, as opposed to a computer-based MVC, K_p should generally have no dimensions. This

---

Process Control: A Practical Approach   Myke King
© 2011 John Wiley & Sons, Ltd

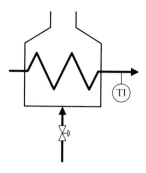

Figure 2.1 Process diagram

is because the DCS works internally with measurements represented as fractions (or percentages) of instrument range.

$$K_p = \frac{\Delta PV}{\Delta MV} \quad (2.2)$$

where

$$\Delta PV = \frac{\text{change in temperature}}{\text{range of temperature transmitter}} \quad (2.3)$$

and

$$\Delta MV = \frac{\text{change in valve position}}{\text{range of valve positioner}} \quad (2.4)$$

Instrument ranges are defined when the system is first configured and generally remain constant. However it is often overlooked that the process gain changes if an instrument is

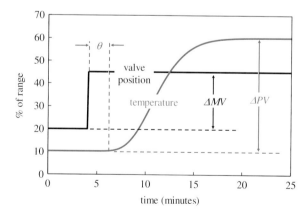

Figure 2.2 Process response

later re-ranged and, if that instrument is either a PV or MV of a controller, then the controller should be re-tuned to retain the same performance.

Numerically K_p may be positive or negative. In our example temperature rises as the valve is opened. If we were to increase heater feed rate (and keep fuel rate constant) then the temperature would fall. K_p, with respect to changes in feed rate, would therefore be negative. Nor is there is any constraint on the absolute value of K_p. Very large and very small values are commonplace. In unusual circumstances K_p may be zero; there will be a transient disturbance to the PV but it will return to its starting point.

The other differences, in Figure 2.2, between the trends of temperature and valve position are to do with timing. We can see that the temperature begins moving some time after the valve is opened. This delay is known as the *process deadtime*; until we develop a better definition, it is the time difference between the change in MV and the first perceptible change in PV. It is usually given the symbol θ. Deadtime is caused by *transport delays*. In this case the prime cause of the delay is the time it takes for the heated fluid to move from the firebox to the temperature instrument. The DCS will generate a small delay, on average equal to half the controller scan interval (ts). While this is usually insignificant compared to any delay in the process it is a factor in the design of controllers operating on processes with very fast dynamics – such as compressors. The field instrumentation can also add to the deadtime; for example on-stream analysers may have sample delays or may be discontinuous.

Clearly the value of θ must be positive but otherwise there is no constraint on its value. Many processes will exhibit virtually no delay; there are some where the delay can be measured in hours or even in days.

Finally the shape of the temperature trend is very different from that of the valve position. This is caused by the 'inertia' of the system. The heater coil will comprise a large mass of steel. Burning more fuel will cause the temperature in the firebox to rise quickly and hence raise the temperature of the external surface of the steel. But it will take longer for this to have an impact on the internal surface of the steel in contact with the fluid. Similarly the coil will contain a large quantity of fluid and it will take time for the bulk temperature to increase. The field instrumentation can add to the lag. For example the temperature is likely to be a thermocouple located in a steel thermowell. The thermowell may have thick walls which cause a lag in the detection of an increase in temperature. Lag is quite different from deadtime. Lag does not delay the **start** of the change in PV. Without deadtime the PV will begin changing immediately but, because of lag, takes time to reach a new steady state. We normally use the symbol τ to represent lag.

To help distinguish between deadtime and lag, consider liquid flowing at a constant rate (F) into a vessel of volume (V). The process is at steady state. The fraction (x) of a component in the incoming liquid is changed at time zero ($t=0$) to x_{new}. By mass balance the change in the quantity of the component in the vessel is the difference between what has entered less what has left. Assuming the liquid is perfectly mixed then

$$V.dx = F.dt.x_{new} - F.dt.x \qquad (2.5)$$

Rearranging

$$\frac{V}{F}\frac{dx}{dt} + x = x_{new} \qquad (2.6)$$

solving gives

$$x = x_{new}\left(1 - e^{-t/\tau}\right) \quad \text{where} \quad \tau = \frac{V}{F} \tag{2.7}$$

In the well-mixed case the delay (θ) would be zero. The outlet composition would begin changing immediately, with a lag determined by V/F. However, if absolutely no mixing took place in the vessel, the change in composition would pass through as a step change – delayed by the residence time of the vessel, i.e.

$$\theta = \frac{V}{F} \tag{2.8}$$

In this case the lag would be zero. In practice, neither perfect mixing nor no mixing is likely and the process will exhibit a combination of deadtime and lag.

When trying to characterise the shape of the PV trend we also have to consider the *order* (*n*) of the process. While processes in theory can have very high orders, in practice we can usually assume that they are first order. However there are occasions where this assumption can cause problems, so it is important to understand how to recognise this situation.

Conceptually order can be thought of as the number of sources of lag. In our example the overall lag will be dictated by the lag of the valve positioner, the mass of combustion products in the firebox, the mass of the heater casing and its coil, the mass of the fluid in the coil and the steel in the thermowell. Figure 2.3 shows a process contrived to demonstrate the effect of combining lags. It comprises two identical vessels, both open to the atmosphere and both draining through identical valves. Both valves are simultaneously opened fully. The flow through each valve is determined by the head of liquid in the vessel so, as this falls, the flow through the valve reduces and the level falls more slowly.

We will use A as the cross-sectional area of the vessel and h as the height of liquid (starting at 100 %). If we assume for simplicity that flow is related linearly to h with k as the constant of proportionality, then

$$A\frac{dh}{dt} = -kh \tag{2.9}$$

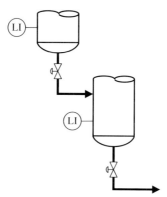

Figure 2.3   Illustration of order

Thus

$$A \int_{100}^{h} \frac{dh}{h} = -k \int_{0}^{t} dt \qquad (2.10)$$

Integrating gives

$$A[\ln(h)]_{100}^{h} = -k[t]_{0}^{t} \qquad (2.11)$$

$$h = 100e^{-kt/A} \qquad (2.12)$$

$$h = 100e^{-t/\tau} \quad \text{where} \quad \tau = \frac{A}{k} \qquad (2.13)$$

The shape of the resulting trend is governed by Equation (2.13). Trend A in Figure 2.4 shows the level in the upper vessel. It shows the characteristic of a first order response in that the rate of change of PV is greatest at the start of the change. Trend B shows the level in the lower vessel – a second order process. Since this vessel is receiving liquid from the first then, immediately after the valves are opened, the inlet and outlet flows are equal. The level therefore does not change immediately. This *apparent deadtime* is a characteristic of higher order systems and is additive to any real deadtime caused by transport delays. Thus by introducing additional deadtime we can approximate a high order process to first order. This approximation is shown as the dashed line close to trend B.

The accuracy of the approximation is dependent on the combination of process lags. While trend B was drawn with both vessels identical, trend C arises if we increase the lag for the top vessel (e.g. by reducing the size of the valve). We know that the system is still second order but visually the trend could be first order. Our approximation will therefore be very accurate. However, if we reduce the lag of the top vessel below that of the bottom one then we obtain trend D. This arises because, on opening both valves, the flow entering

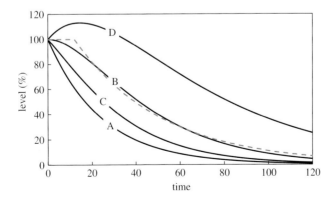

Figure 2.4 Effect of combination of process lags

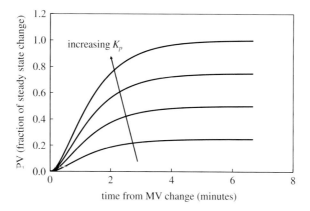

Figure 2.5 Effect of K_p

the bottom vessel is greater than that leaving and so the level initially rises. This is *inverse response*; the PV initially moves in a direction opposite to the steady-state change. Fitting a first order model to this response would be extremely inaccurate. Examples of processes prone to this type of response include steam drum levels, described in Chapter 4, and some schemes for controlling pressure and level in distillation columns, as described in Chapter 12.

Figures 2.5 to 2.8 show the effect of changing each of these dynamic parameters. Each response is to the same change in MV. Changing K_p has no effect on the behaviour of the process over time. The time taken to reach steady state is unaffected; only the actual steady state changes. Changing θ, τ or n has no effect on actual steady state; only the time taken to reach it is affected. The similarity of the family of curves in Figures 2.7 and 2.8 again shows the principle behind our approximation of first order behaviour – increasing θ has an effect very similar to that of increasing n.

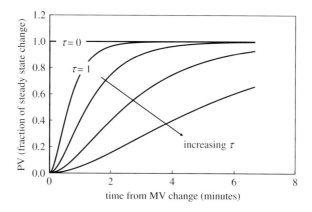

Figure 2.6 Effect of τ

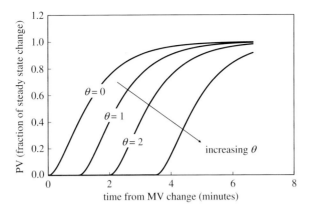

Figure 2.7 Effect of θ

2.2 Cascade Control

Before attempting to determine the process dynamics we must first explore how they might be affected by the presence of other controllers. One such situation is the use of *cascade control*, where one controller (the *primary* or *master*) adjusts the SP of another (the *secondary* or *slave*). The technique is applied where the process dynamics are such that the secondary controller can detect and compensate for a disturbance much faster than the primary. Consider the two schemes shown in Figure 2.9. If there is a disturbance to the pressure of the fuel header, for example because of an increase in consumption on another process, the flow controller will respond quickly and maintain the flow close to SP. As a result the disturbance to the temperature will be negligible. Without the flow controller, correction will be left to the temperature controller. But, because of the process dynamics, the temperature will not change as quickly as the flow and nor can it correct as quickly once

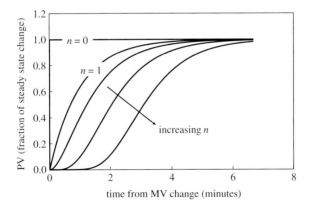

Figure 2.8 Effect of n (by adding additional lags equal to τ)

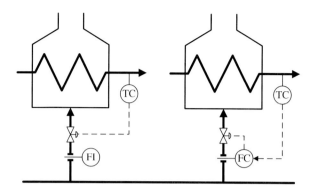

Figure 2.9 Direct versus cascade control

it has detected the disturbance. As a result the temperature will deviate from SP for some significant time.

Cascade control also removes any control valve issues from the primary controller. If the valve characteristic is nonlinear, the positioner poorly calibrated or subject to minor mechanical problems, all will be dealt with by the secondary controller. This helps considerably when tuning the primary controller.

Cascade control should not normally be employed if the secondary cannot act more quickly than the primary. Imagine there is a problem with the flow meter in that it does not detect the change in flow for some time. If, during this period, the temperature controller has dealt with the upset then the flow controller will make an unnecessary correction when its measurement does change. This can make the scheme unstable.

Tuning controllers in cascade should always be completed from the bottom up. Firstly the secondary controller will on occasions be in use without the primary. There may, for example, be a problem with the primary or its measurement may be out of range during start-up or shutdown of the process. We want the secondary to perform as effectively as possible and so it should be optimally tuned as a standalone controller. The second reason is that the MV of the primary controller is the SP of the secondary. When performing step tests to tune the primary we will make changes to this SP. The secondary controller is now effectively part of the process and its tuning will affect the dynamic relationship between the primary PV and MV. If, after tuning the primary, we were to change the tuning in the secondary then the tuning in the primary would no longer be optimum.

Cascade control, however, is not the only case where the sequence of controller tuning is important. In general, before performing a plant test, the engineer should identify any controllers that will take corrective action during the test itself. Any such controller should be tuned first. In the case of cascade control, clearly the secondary controller takes corrective action when its SP is changed. But consider the example shown in Figure 2.10. The heater has a simple flue gas oxygen control which adjusts a damper to maintain the required excess air. When the downward step is made to the fuel flow SP the oxygen controller, if in automatic mode, will take corrective action to reduce the air rate and return the oxygen content to SP. However, if this controller is in manual mode then no corrective action is taken, the oxygen level will rise and the heater efficiency will fall. As a result the heater outlet temperature will fall by more than it did in the first test. Clearly this affects

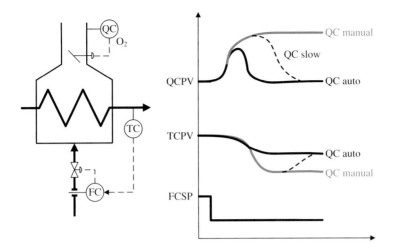

Figure 2.10   Effect of other controllers

the process gain between temperature and fuel. Imagine now that the oxygen control is retuned to act more slowly. The dynamic behaviour of the temperature with respect to fuel changes will be quite different. So we have the situation where an apparently unrelated controller takes corrective action during the step test. It is important therefore that this controller is properly tuned before conducting the test.

In the case of testing to support the design of a MVC, the MVs are likely to be mainly basic controllers and it is clear that these controllers should be well-tuned before starting the step tests. However, imagine that one of the MVs is the feed flow controller. When its SP is stepped there is likely to be a large number of regulatory controllers that will take corrective action during the test. Many of these will not be MVs but nevertheless need to be well tuned before testing begins.

2.3   Model Identification

Model identification is the process of quantifying process dynamics. The techniques available fall into one of two approaches – *open loop* and *closed loop* testing. Open loop tests are performed with either no controller in place or, if existing, with the controller in manual mode. A disturbance is injected into the process by directly changing the MV. Closed loop tests may be used if a controller exists and already provides some level of stable control. Under these circumstances the MV is changed indirectly by making a change to the SP of the controller.

Such plant testing should be well organised. While it is clear that the process operator must agree to the test there needs to be discussion about the size and duration of the steps. It is in the engineer's interest to make these as large as possible. The operator of course would prefer that no disturbance be made! The operator also needs to appreciate that other changes to the process should not be made during the test. While it is possible to determine the dynamics of simultaneous changes to several variables, the analysis is complex and more prone to error.

It seems too obvious to state that the process instrumentation should be fully operational.

Many data historians included a compression algorithm to reduce the storage requirement. When later used to recover the original data some distortion will occur. While this is not noticeable in most applications, such as process performance monitoring and accounting, it can affect the apparent process dynamics. Any compression should therefore be disabled prior to the plant tests.

It is advisable to collect more than just the PV and MV. If the testing is to be done closed loop then the SP should also be recorded. Any other process parameter which can cause changes in the PV should also be collected. This is primarily to ensure that they have not changed during the testing, or to help diagnose a poor model fit. While such disturbances usually invalidate the test, it may be possible to account for them and so still identify an accurate model.

Ideally, testing should be planned for when there are no other scheduled disturbances. It can be a good idea to avoid shift changeovers – partly to avoid having to persuade another crew to accept the process disturbances but also to avoid the changes to process conditions that operators often make when returning from lengthy absences. If ambient conditions can affect the process then it is helpful to avoid testing when these are changing rapidly, for example at dawn or dusk and during rainstorms. Testing should also be scheduled to avoid any foreseen changes in feed composition or operating mode.

Laboratory samples are often collected during plant tests. These are usually to support the development of inferential properties (as described in Chapter 9). Indeed steady operation, under conditions away from normal operation, can provide valuable data 'scatter'. Occasionally a series of samples are collected to obtain dynamic behaviour, for example if an onstream analyser is temporarily out of service or its installation delayed. The additional laboratory testing generated may be substantial compared to the normal workload. If the laboratory is not expecting this, then analysis may be delayed for several days with the risk that the samples may degrade.

The most accurate way of determining the dynamic constants is by a computer-based curve fitting technique which uses the values of the MV and PV collected frequently throughout the test. If we assume that the process can be modelled as first order plus deadtime, then in principle this involves fitting the following equation to the collected data.

$$PV_n = aPV_{n-1} + bMV_{n-\theta/ts} + bias \quad (2.14)$$

$$a = e^{-ts/\tau} \quad \text{and} \quad b = K_p\left(1 - e^{-ts/\tau}\right) \quad (2.15)$$

Or, if we make the first order Taylor approximation

$$e^{-ts/\tau} = 1 - \frac{ts}{\tau} \quad (2.16)$$

then

$$a = \frac{\tau - ts}{\tau} \quad \text{and} \quad b = K_p\frac{ts}{\tau} \quad (2.17)$$

Or, with the more accurate first order Padé approximation

$$e^{-ts/\tau} = \frac{2 - \dfrac{ts}{\tau}}{2 + \dfrac{ts}{\tau}} \quad (2.18)$$

then

$$a = \frac{2\tau - ts}{2\tau + ts} \quad \text{and} \quad b = K_p \frac{2ts}{2\tau + ts} \quad (2.19)$$

The values of K_p, θ, τ and *bias* are adjusted to minimise the sum of the squares of the error between the predicted PV and the actual PV. When θ is not an exact multiple of the data collection interval (ts), then the MV is interpolated between the two values either side of the required value.

$$MV_{n-\theta/ts} = MV_{n-\text{int}(\theta/ts)} - \left(\frac{\theta}{ts} - \text{int}\left(\frac{\theta}{ts}\right)\right)\left(MV_{n-\text{int}(\theta/ts)} - MV_{n-\text{int}(\theta/ts)-1}\right) \quad (2.20)$$

An alternative approach is to apply linear regression to identify a, b and *bias* for a chosen value of θ. An iterative approach is then followed to find the best value for θ. Once the coefficients are known then K_p can be derived as

$$K_p = \frac{b}{1-a} \quad (2.21)$$

The derivation of τ can be from any of the following depending on the approximation made.

$$\tau = \frac{-ts}{\ln(a)} \approx \frac{ts}{1-a} \approx \frac{ts(1+a)}{2(1-a)} \quad (2.22)$$

More complex equations can be used to identify higher order models. For example Equation (2.23) was developed by applying a z-transform to the Laplace form of a second order process (with lags τ_1 and τ_2, and *lead* τ_3) as shown in Equation (2.68). Lead is required if there is PV overshoot ($\tau_3 > 0$) or inverse response ($\tau_3 < 0$).

$$PV_n = a_1 PV_{n-1} + a_2 PV_{n-2} + b_1 MV_{n-\theta/ts} + b_2 MV_{n-\theta/ts-1} + bias \quad (2.23)$$

$$a_1 = e^{-ts/\tau_1} + e^{-ts/\tau_2} \quad \text{and} \quad a_2 = -e^{-ts/\tau_1} e^{-ts/\tau_2} \quad (2.24)$$

$$b_1 = K_p\left[1 + \frac{\tau_3 - \tau_1}{\tau_1 - \tau_2} e^{-ts/\tau_1} + \frac{\tau_2 - \tau_3}{\tau_1 - \tau_2} e^{-ts/\tau_2}\right] \quad (2.25)$$

$$b_2 = K_p\left[e^{-ts/\tau_1} e^{-ts/\tau_2} + \frac{\tau_3 - \tau_1}{\tau_1 - \tau_2} e^{-ts/\tau_2} + \frac{\tau_2 - \tau_3}{\tau_1 - \tau_2} e^{-ts/\tau_1}\right] \quad (2.26)$$

$$MV_{n-\theta/ts} = MV_{n-\text{int}(\theta/ts)} - \left(\frac{\theta}{ts} - \text{int}\left(\frac{\theta}{ts}\right)\right)\left(MV_{n-\text{int}(\theta/ts)} - MV_{n-\text{int}(\theta/ts)-1}\right) \quad (2.27)$$

$$MV_{n-\theta/ts-1} = MV_{n-\text{int}(\theta/ts)-1} - \left(\frac{\theta}{ts} - \text{int}\left(\frac{\theta}{ts}\right)\right)\left(MV_{n-\text{int}(\theta/ts)-1} - MV_{n-\text{int}(\theta/ts)-2}\right) \quad (2.28)$$

A similar approach can be taken fitting this model. K_p, θ, τ_1, τ_2, τ_3 and *bias* can be fitted directly. Or a_1, a_2, b_1, b_2 can be identified by linear regression for the best value of θ. K_p can then be derived from

$$K_p = \frac{b_1 + b_2}{1 - a_1 - a_2} \quad (2.29)$$

The values for τ_1 and τ_2 are interchangeable. Arbitrarily we select $\tau_1 > \tau_2$, and so they can be derived from

$$\tau_1 = \frac{-ts}{\ln\left(\dfrac{a_1 + \sqrt{a_1^2 + 4a_2}}{2}\right)} \quad (2.30)$$

$$\tau_2 = \frac{-ts}{\ln\left(\dfrac{a_1 - \sqrt{a_1^2 + 4a_2}}{2}\right)} \quad (2.31)$$

The condition for τ_1 and τ_2 to be real is that

$$a_1^2 + 4a_2 > 0 \quad (2.32)$$

With regression there is no guarantee that this will be the case. If not, then the process cannot be described by the second order model.

The value for τ_3 is obtained by substituting the results for K_p, τ_1 and τ_2 into either Equation (2.25) or (2.26). For example, from Equation (2.25)

$$\tau_3 = \frac{\left(\dfrac{b_1}{K_p} - 1\right)(\tau_1 - \tau_2) + \tau_1 e^{-ts/\tau_1} - \tau_2 e^{-ts/\tau_2}}{e^{-ts/\tau_1} - e^{-ts/\tau_2}} \quad (2.33)$$

Note that Equations (2.25) and (2.26) cannot be applied if

$$a_1^2 + 4a_2 = 0 \quad (2.34)$$

This means that $\tau_1 = \tau_2$ and b_1 and b_2 would therefore be indeterminate. If $\tau_1 = \tau_2 = \tau$ then the coefficients become

$$a_1 = 2e^{-ts/\tau} \quad \text{and} \quad a_2 = -e^{-2ts/\tau} \quad (2.35)$$

$$b_1 = K_p\left[1 - e^{-ts/\tau} + \frac{(\tau_3 - \tau)ts}{\tau^2} e^{-ts/\tau}\right] \quad (2.36)$$

$$b_2 = K_p\left[e^{-2ts/\tau} - e^{-ts/\tau} - \frac{(\tau_3 - \tau)ts}{\tau^2} e^{-ts/\tau}\right] \quad (2.37)$$

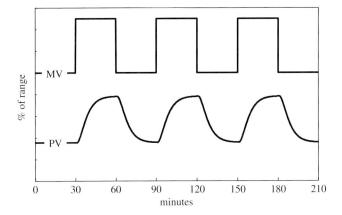

Figure 2.11 Correlated steps

Equation (2.29) can still be used to obtain K_p but the lags (τ) are obtained from

$$\tau = \frac{-ts}{\ln\left(\frac{a_1}{2}\right)} = \frac{-2ts}{\ln(-a_2)} \qquad (2.38)$$

The value for τ_3 is obtained by substituting the results for K_p and τ into either Equation (2.36) or (2.37). For example, from Equation (2.36)

$$\tau_3 = \tau + \frac{\left(\frac{b_1}{K_p} - 1 + e^{-ts/\tau}\right)\tau^2}{e^{-ts/\tau}} \qquad (2.39)$$

The data collection interval can be quite long. We will show later that steady state is virtually reached within $\theta + 5\tau$. Assuming we need around 30 points to achieve a reasonably accurate fit and that we make both an increase and a decrease in the MV, then collecting data at a one-minute interval would be adequate for a process which has time constants of around two or three minutes.

This model identification technique can be applied to both open and closed loop tests. Multiple disturbances can be made in order to check the repeatability of the results and to check linearity. However it is important to avoid *correlated steps*. Consider the series of steps shown in Figure 2.11. There is clearly a strong correlation between the PV and the MV, with K_p of 1.0 and θ of around 3.0 minutes. However, there is an equally accurate model with K_p of -1.0 and θ of around 33.0 minutes.

Performing a series of steps of varying size and duration, as in Figure 2.12, would avoid this problem. While not necessary for every step made, model identification will be more reliable if the test is started with the process as steady as possible and allowed to reach steady state after at least some of the steps.

Model identification software packages will generally report some measure of confidence in the model identified. A low value may have several causes. Noise in either the MV or PV, if of a similar order of magnitude to the changes made, can disguise the model.

16   Process Control

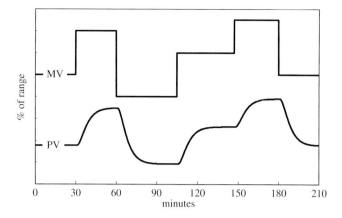

Figure 2.12   Noncorrelated steps

If the MV is a valve or similar actuator, then problems such as *stiction* and *hysteresis* will reduce model accuracy. These are shown in Figure 2.13. Stiction, caused by excessive friction, requires that the signal change to **start** the valve moving is greater than the signal to **keep** it moving. Thus a small change in the signal may have no affect on the PV, whereas a subsequent change will affect it as expected. Hysteresis is usually caused by wear in couplings and bearings resulting in some 'play' in the mechanism. As the signal is increased this 'play' is first overcome before the valve begins to move. It will then behave normally until the signal is reversed, when the 'play' must again be overcome.

If suspected, these faults can usually be diagnosed by making a series of steps in one direction followed by a series in the opposite direction. If the change in PV at each step is not in constant proportion to the change in MV, the valve should be overhauled.

The relationship between PV and MV may be inherently nonlinear. Some model identification packages can analyse this. If not, then plotting the steady-state values of PV against MV will permit linearity to be checked and possibly a linearising function developed.

While computer-based packages are readily available, there may be circumstances where they cannot be applied. For example, if no facility exists to collect process data in numerical

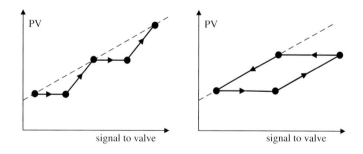

Figure 2.13   Stiction and hysteresis in a control valve

form at regular intervals, then there are graphical techniques that can be applied to process trends. They can also only be used to identify first order plus deadtime models and the MV must be changed as a single step, starting and ending at steady state.

This is not always possible.

- Any existing controller will need to be switched to manual mode. This may be undesirable on an inherently unstable process.
- There are many processes which rarely reach true steady state and so it would be optimistic to start and finish the test under these conditions.
- The size of the step must be large enough to have a noticeable effect on the process. If the PV is subject to noise, small disturbances will be difficult to analyse accurately. The change in PV needs to be at least five times larger than the noise amplitude. This may cause an unacceptable process disturbance.
- Dynamics, as we shall see later in Chapter 6, are not only required for changes in the MV but also for *disturbance variables* (*DV*). It may be that these cannot be changed as steps.

If a single step is practical it will still be necessary to conduct multiple tests, analysing each separately, to confirm repeatability and to check for linearity.

The most widely published method is based on the principle that a process with zero deadtime will complete 63.2 % of the steady state change within one process lag. If, in Equation (2.7), we set t equal to τ, we get

$$x = 0.632 x_{new} \qquad (2.40)$$

This calculation can be repeated for multiples of τ, resulting in the graph shown in Figure 2.14.

While, in theory, the process will never truly reach steady state, within five time constants it will be very close – having completed 99.3 % of the change.

In general, however, we have to accommodate deadtime in our calculation of dynamics. Ziegler and Nichols (Reference 1) proposed the method using the *tangent of steepest slope*. Shown in Figure 2.15 it involves identifying the point at which the PV is changing most

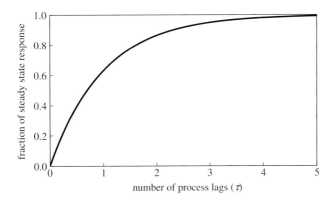

Figure 2.14 Time to reach steady state

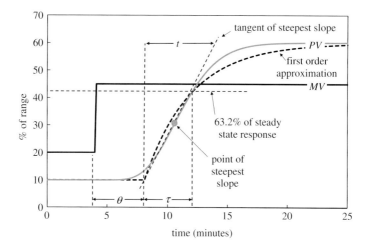

Figure 2.15 Ziegler-Nichols steepest slope method

rapidly and then drawing a tangent to the curve at this point. Where it crosses the value of the PV at the start of the test gives the process deadtime (θ). There are two methods for determining the process lag (τ). While not mentioned by Ziegler and Nichols, the time taken to reach 63.2 % of the steady state response is $\theta + \tau$, so once θ is known τ can be derived. Ziegler and Nichols, as we shall see later when looking at their controller tuning method, instead characterised the process by determining the slope of the tangent (R). This is equivalent to defining τ as the distance labelled t in Figure 2.15. For a truly first order process with deadtime this will give the same result. For higher order systems this approach is inaccurate. K_p is determined from Equation (2.2).

The resulting first order approximation is included in Figure 2.15. The method forces it to pass through three points – the intersection of the tangent with the starting PV, the 63.2% response point and the steady state PV. The method is practical but may be prone to error. Correctly placing the line of steepest slope may be difficult – particularly if there is measurement noise. Drawing it too steeply will result in an overestimate of θ and an underestimate of τ. The ratio θ/τ, used by most controller tuning methods, could thus be very different from the true value.

An alternative approach is to identify two points on the response curve. A first order response is then forced through these two points and the steady-state values of the PV. Defining t_a as the time taken to reach a % of the steady-state response and t_b as the time taken to reach b %, the process dynamics can be derived from the formulae

$$\tau = \frac{t_b - t_a}{\ln\left(1 - \frac{a}{100}\right) - \ln\left(1 - \frac{b}{100}\right)} \qquad (2.41)$$

$$\theta = t_a + \tau.\ln\left(1 - \frac{a}{100}\right) \qquad (2.42)$$

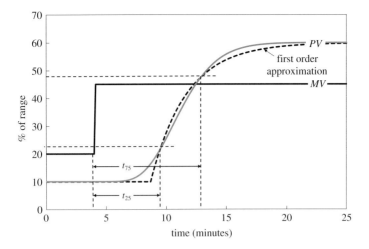

Figure 2.16 Two-point method

The values of *a* and *b* need not be symmetrical but for maximum accuracy they should not be close together nor too close to the start and finish steady-state conditions. Choosing values of 25 % and 75 % reduces Equations (2.41) and (2.42) to

$$\tau = 0.910(t_{75}-t_{25}) \tag{2.43}$$

$$\theta = t_{25}-0.288\tau \tag{2.44}$$

Figure 2.16 shows the application of this method and the resulting first order approximation. Another approach is to use more points from the curve and apply a least squares technique to the estimation of θ and τ. Rearranging Equation (2.42) we get

$$t_a = \theta + \tau \ln\left(\frac{100}{100-a}\right) \tag{2.45}$$

So, by choosing points at 10 % intervals

$$t_{10} = \theta + 0.1054\tau \tag{2.46}$$

$$t_{20} = \theta + 0.2231\tau \tag{2.47}$$

$$t_{30} = \theta + 0.3567\tau \tag{2.48}$$

$$t_{40} = \theta + 0.5108\tau \tag{2.49}$$

$$t_{50} = \theta + 0.6931\tau \tag{2.50}$$

$$t_{60} = \theta + 0.9163\tau \tag{2.51}$$

$$t_{70} = \theta + 1.2040\tau \tag{2.52}$$

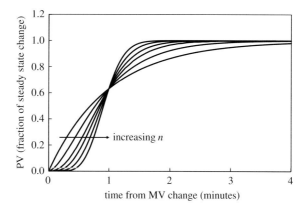

Figure 2.17 Effect of n (by keeping 63 % response time equal)

$$t_{80} = \theta + 1.6094\tau \qquad (2.53)$$

$$t_{90} = \theta + 2.3026\tau \qquad (2.54)$$

Using a spreadsheet package θ and τ would be adjusted to minimise the sum of the square of the errors between the actual time to reach each percentage point of steady-state and the time predicted by each of the Equations (2.46) to (2.54).

With any model identification technique care should be taken with units. As described earlier in this chapter K_p should be dimensionless if the value is to be used in tuning a DCS-based controller. For computer-based MVC K_p would usually be required in engineering units. θ and τ should be in units consistent with the tuning constants. It is common for the integral time (T_i) and the derivative time (T_d) to be in minutes, in which case the process dynamics should be in minutes; but this is not universally the case.

Figure 2.17 shows the effect of increasing order, but unlike Figure 2.8, by adjusting the time constants so that the overall lag remains the same, i.e. all the responses reach 63 % of the steady state change after one minute. It shows that, for large values of n, the response becomes closer to a step change. This confirms that a series of lags can be approximated by deadtime. But it also means that deadtime can be approximated by a large number of small lags. We will cover, in Chapters 6, 7 and 8, control schemes that require a deadtime algorithm. If this is not available in the DCS then this approximation would be useful.

2.4 Integrating Processes

The fired heater that we have worked with is an example of a *self-regulating* process. Following the disturbance to the fuel valve the temperature will reach a new steady state without any manual intervention. Not all processes behave this way. For example, if we were trying to obtain the dynamics for a future level controller we would make a step change to the manipulated flow. The level would not reach a new steady state unless some intervention is made. This *non-self-regulating* process can also be described as an *integrating* process.

While level is the most common example there are many others. For example, many pressure controllers show a similar behaviour. Pressure is a measure of the inventory of gas in a system, much like a level is a measure of liquid inventory. An imbalance between the gas flow into and out of the system will cause the pressure to ramp without reaching a new steady state. However, not all pressures show pure integrating behaviour. For example if the flow in or out of the system is manipulated purely by valve position, i.e. no flow control, then the resulting change in pressure will cause the flow through the valve to change until a new equilibrium is reached. Even with flow controllers in place, if flow is measured by an uncompensated orifice type meter, the error created in the flow measurement by the change in pressure will also cause the process to be self-regulating.

Some temperatures can show integrating behaviour. If increasing heater outlet temperature also causes heater inlet temperature to rise, through some recycle or heat integration, then the increase in energy input will cause the outlet temperature to ramp up.

The response of a typical integrating process is shown as Figure 2.18. Since it does not reach steady state we cannot immediately apply the same method of determining the process gain from the steady-state change in PV. Nor can we use any technique which relies on a percentage approach to steady state.

By including a bias (because it is not true that the PV is zero when the MV is zero) we can modify Equation (2.2) for a self-regulating process to

$$PV = K_p.MV + bias \tag{2.55}$$

In the case of an integrating process, the PV also varies with time, so we can describe it by

$$PV = K_p \int MV.dt + bias \tag{2.56}$$

or

$$\frac{dPV}{dt} = K_p.MV + bias \tag{2.57}$$

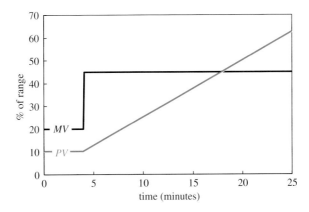

Figure 2.18 Integrating process

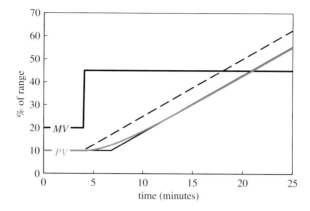

Figure 2.19 Effect of lag on an integrating process

By replacing PV with its derivative we can therefore apply the same model identification techniques used for self-regulating processes. While for DCS-based controllers, PV and MV remain dimensionless, K_p must now have the units of reciprocal time. The units will depend on whether rate of change of PV is expressed in $\sec^{-1}$, $\min^{-1}$ or hr^{-1}. Any may be used provided consistency is maintained. We will use $\min^{-1}$ throughout this book.

We can omit the lag term when characterising the process dynamics of an integrating process. Although the process is just as likely to include a lag, this manifests itself as deadtime. Figure 2.19 illustrates the effect of adding lag to the PV. In this case a lag of 3 minutes has caused the apparent deadtime to increase by about the same amount. After the initial response the PV trend is still a linear ramp. We can thus characterise the response using only K_p and θ.

2.5 Other Types of Process

In addition to self-regulating and integrating processes there are a range of others. There are processes which show a combination of these two types of behaviour. For example steam header pressure generally shows integrating behaviour if boiler firing is changed. If there is a flow imbalance between steam production and steam demand the header pressure will not reach a new steady state without intervention. However, as header pressure rises, more energy is required to generate a given mass of steam and the imbalance reduces. While the effect is not enough for the process to be self-regulating, the response will include some self-regulating behaviour.

Figure 2.20 shows another example. Instead of the planned temperature controller being mounted on a tray in the distillation column it has been installed on the reboiler outlet. As the reboiler duty is increased, by increasing the flow of the heating fluid, the outlet temperature will increase. This will in turn cause the reboiler inlet temperature to increase – further increasing the outlet temperature which will then show integrating behaviour. However the higher outlet temperature will result in increased vaporisation in the base of the column, removing some of the sensible heat as heat of vaporisation. This self-regulating

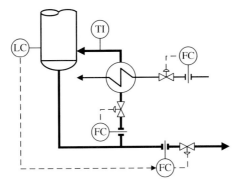

Figure 2.20   Mixed integrating and self-regulating process

effect will usually override the integrating behaviour and the process will reach a new steady state.

The term *open-loop unstable* is also used to describe process behaviour. Some would apply it to any integrating process. But others would reserve it to describe inherently unstable processes such as exothermic reactors. Figure 2.21 shows the impact that increasing the reactor inlet temperature has on reactor outlet temperature. The additional conversion caused by the temperature increase generates additional heat which increases conversion further. It differs from most non-self-regulating processes in that the rate of change of PV increases over time. It often described as a *runaway* response. Of course, the outlet temperature will eventually reach a new steady state when all the reactants are consumed; however this may be well above the maximum permitted.

The term open-loop unstable can also be applied to controllers that have *saturated*. This means that the controller output has reached either its minimum or maximum output but not eliminated the deviation between PV and SP. It can also be applied to a controller using a discontinuous on-stream analyser that fails. Such analysers continue to transmit the last measurement until a new one is obtained. If, as a result of analyser failure, no new measurement is transmitted then the controller no longer has *feedback*.

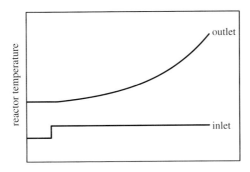

Figure 2.21   Exothermic reactor

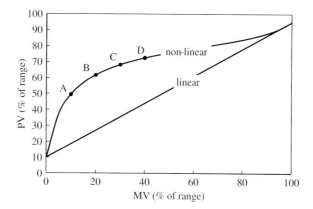

Figure 2.22   Nonlinear process

2.6   Robustness

For a controller to be *robust* it must perform well over the normal variation of process dynamics. Dynamics are rarely constant and it is important to assess how much they might vary before finalising controller design.

Dynamics vary due to a number of reasons. The process may be inherently nonlinear so that, as process conditions vary, a controller tuned for one set of conditions may not work well under others. This is illustrated by Figure 2.22. A step test performed between points A and B would give a process gain of about 1.2, while one performed between points C and D would give a value of about 0.4. As a guideline, linear controllers are reasonably robust provided the process dynamics stay within $\pm 20\,\%$ of the values used to design the controller. In our example an average gain of 0.8 could be used but the variation would be $\pm 50\,\%$. This would require a modified approach to controller design, such as the inclusion of some linearising function, so it is important that we conduct plant tests over the whole range of conditions under which the controller will be expected to operate.

A common oversight is not taking account of the fact that process dynamics vary with feed rate. Consider our example of a fired heater. If it is in a nonvaporising service we can write the heat balance

$$F_{feed} \cdot c_p (T - T_{inlet}) = F \cdot NHV \cdot \eta \qquad (2.58)$$

On the feed side F_{feed} is the flow rate to the heater, c_p is the specific heat, T is the outlet temperature and T_{inlet} is the inlet temperature. On the fuel side F is the flow of fuel, NHV the net heating value (calorific value) and η the heater efficiency. Rearranging we get

$$T = \frac{F \cdot NHV \cdot \eta}{F_{feed} \cdot c_p} + T_{inlet} \qquad (2.59)$$

Differentiating we get

$$K_p = \frac{dT}{dF} = \frac{NHV \cdot \eta}{F_{feed} \cdot c_p} \qquad (2.60)$$

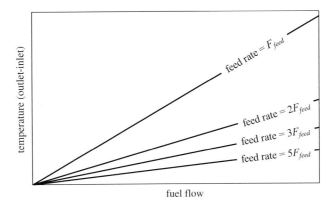

Figure 2.23 Variation of process gain with feed rate

While the process gain is sensitive to operating conditions, such as *NHV*, η and c_p, of most concern is its sensitivity to feed flow rate. In fact it is inversely proportional to feed rate. A little thought would have predicted this. Making the same increase in fuel at a higher feed rate would result in a smaller temperature increase because there is more feed to heat. Figure 2.23 shows how the relationship, between the rise in temperature across the heater and fuel flow, varies with feed rate. So, for example, doubling the feed rate halves the gradient of the line. Some might describe the behaviour as nonlinear, using the term for any process in which the process gain is variable. Strictly this is a linear process; changing feed rate clearly affects the process gain but behaviour remains linear at a given feed rate.

This effect is not unique to fired heaters; almost all process gains on a plant will vary with feed rate. Given that we tolerate $\pm 20\%$ variation in process gain, we can therefore tolerate $\pm 20\%$ variation in feed rate. Assuming a reference feed rate of 100, our controller will work reasonably well for feed rates between 80 and 120. The *turndown ratio* of a process is defined as the maximum feed rate divided by the minimum. We can see that if this value exceeds 1.5 (120/80) then the performance of almost all the controllers on the process will degrade noticeably as the minimum or maximum feed rate is approached. Fortunately most processes have turndown ratios less than 1.5, so providing the controllers are tuned for the average feed rate their performance should be acceptable. The technique used, if this is not the case, is covered in Chapter 6.

Feed flow rate may also affect process deadtime. If the prime cause of deadtime is transport delay than an increase in feed will cause the residence time to fall and a reduction in deadtime. At worst, deadtime may be inversely proportional to feed rate. If so then the maximum turndown limit of 1.5 will apply. In fact controllers are more sensitive to increases in deadtime than decreases. Rather than design for the average deadtime, a value should be chosen so that it varies between -30% and $+10\%$. Techniques for accommodating excessive variation in deadtime are covered in Chapter 7.

Feed rate generally has little effect on process lag- although Equation (2.7) would appear to suggest otherwise. However, this only applies when there is perfect mixing. In general, only in relatively small sections of most processes does this occur. But lag is often sensitive

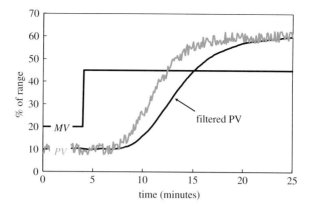

Figure 2.24 Effect of filter on process dynamics

to the inventory of the process. For example, the lag caused by a vessel will change depending on the level of liquid in the vessel – as shown Equation (2.7). Changes in vessel inlet temperature or composition will be more slowly detected at the vessel outlet if the level is high. Whether this is significant will depend on a number of factors. There are likely to be other sources of lag which, when added to that caused by the vessel, reduce the impact of inventory changes. Similarly although the indicated level in the vessel may appear to change a great deal, it is unlikely that the level gauge operates over the full height of the vessel. A change in level from an indicated 10 % to 90 % would not mean that there is a ninefold increase in liquid volume. However a check should be made if averaging level control, as described in Chapter 4, is used – since this can permit large sustained changes in inventory.

The addition of filtering, to deal with measurement noise, can also affect the process dynamics. Figure 2.24 shows the same plant test but with noise added to the PV. This noise has then been removed by the additional of a filter (as described in Chapter 5).

The filter adds lag and, because it increases the order of the system, also increases the apparent deadtime. Adding a filter after a controller has been tuned is therefore inadvisable. Either the plant test should be repeated to identify the new dynamics or, if the model identification package permits it, the original test data may be used with the filter simulated in the package.

It is very common for filters to be implemented unnecessarily. They are often added visually to smooth the trended measurement. But the main concern should be the impact they have on the final control element, for example the control valve. This is a function not only of the amplitude of measurement noise but also the gains through which it passes. These may be less than one and so attenuate the noise.

Not all filtering is implemented in the DCS. Most transmitters include filters. Provided the filter constant is not changed then model identification will include the effect of the transmitter filter in the overall dynamics. However, if the filter in the transmitter is changed by a well-intentioned instrument technician unaware of its implications, this can cause degradation in controller performance.

We will show later that controllers can be tuned to respond more quickly as K_p and θ/τ reduce. If dynamics can vary from those obtained by plant testing, it is better that the controller becomes more sluggish than more oscillatory. It is therefore safer to base controller tuning on higher values of K_p and θ and on a lower value of τ.

2.7 Laplace Transforms for Processes

While this book only uses Laplace transforms when the alternative would be overly complex, they are used in many text books and by control system vendors. So that they can be recognised, the transforms for the common types of process are listed here.

a. Self-regulating first order plus deadtime (FOPDT)

$$PV = \frac{K_p e^{-\theta s}}{1 + \tau s} MV \tag{2.61}$$

b. Integrating with deadtime

$$PV = \frac{K_p e^{-\theta s}}{s} MV \tag{2.62}$$

c. Self-regulating second order plus deadtime (SOPDT)

$$PV = \frac{K_p e^{-\theta s}}{(1 + \tau_1 s)(1 + \tau_2 s)} MV \tag{2.63}$$

d. Self-regulating second order with inverse response

As shown in the example in this chapter, inverse response is caused by two competing processes – the faster of which takes the process first in a direction opposite to the steady state. We can approximate this as two first-order processes with gains of opposite sign, so that the combined effect is given by

$$PV = \left[\frac{(K_p)_1}{1 + \tau_1 s} + \frac{(K_p)_2}{1 + \tau_2 s} \right] MV \tag{2.64}$$

Rearranging

$$PV = \frac{((K_p)_1 + (K_p)_2) \left[1 + \frac{(K_p)_1 \tau_2 + (K_p)_2 \tau_1}{(K_p)_1 + (K_p)_2} s \right]}{(1 + \tau_1 s)(1 + \tau_2 s)} MV \tag{2.65}$$

Defining

$$K_p = (K_p)_1 + (K_p)_2 \tag{2.66}$$

and

$$\tau_3 = \frac{(K_p)_1 \tau_2 + (K_p)_2 \tau_1}{(K_p)_1 + (K_p)_2} \tag{2.67}$$

we can write Equation (2.65) as

$$PV = \frac{K_p(1+\tau_3 s)}{(1+\tau_1 s)(1+\tau_2 s)} MV \qquad (2.68)$$

If τ_3 is less than 0 then the process will show inverse response. If τ_3 is greater than 0 then the process will show PV overshoot and τ_3 is said to add lead to the process.

References

1. Ziegler, J.G. and Nichols, N.B. (1942) Optimum settings for automatic controllers. *Transactions of the ASME*, **64**, 759–768.

3
PID Algorithm

The PID (proportional, integral, derivative) algorithm has been around since the 1930s. While many DCS vendors have attempted to introduce other more effective algorithms it remains the foundation of almost all basic control applications.

The basic form of the algorithm is generally well-covered by academic institutions. Its introduction here follows a similar approach but extends it to draw attention to some of the more practical issues. Importantly it also addresses the many modifications on offer in most DCS, many of which are undervalued by industry unaware of their advantages. This chapter also covers controller tuning in detail. Several commonly known published methods are included but mainly to draw attention to their limitations. An alternative, well-proven, technique is offered for the engineer to use.

3.1 Definitions

Before proceeding we must ensure that we define the key terminology. In Chapter 2 we defined *PV* (the process variable that we wish to control) and *MV* (the manipulated variable). The reader should note that some texts use this abbreviation to mean 'measured value', i.e. what we call *PV*. We will also use *M* to represent the controller output, which will normally be the same as *MV*. To these definitions we have also added *SP* (i.e. the target for *PV*).

The error (*E*) is defined as the deviation from *SP* but its definition varies between DCS. Our definition is

$$E = PV - SP \qquad (3.1)$$

Many texts and some systems define error as *SP* – *PV*. Misinterpreting the definition will result in the controller taking corrective action in the direction opposite to that it should, worsening the error and driving the control valve fully closed or fully open.

3.2 Proportional Action

The principle behind proportional control is to keep the controller output (M) in proportion to the error (E).

$$M = K_c.E + C \tag{3.2}$$

K_c is the controller gain and is a tuning constant set by the control engineer. The term C is necessary since it is unlikely to be the case that zero error coincides with zero controller output. In some control systems the value of C may be adjusted by the process operator, in which case it is known as *manual reset*. Its purpose will be explained later in this section.

We have seen that the process gain (K_p) may be positive or negative but the controller gain (K_c) is always entered as an absolute value. The control algorithm includes therefore an additional engineer-defined parameter known as *action*. If set to *direct*, the controller output will increase as the PV increases; if set to *reverse*, output decreases as PV increases. If we consider our fired heater example we would want the controller to reduce the fuel rate if the temperature increases and so we would need to set the action to reverse. In other words if the process gain is positive then the controller should be reverse acting; if the process gain is negative then it should be direct acting. This definition is consistent with that adopted by the ISA (Reference 1) but is not used by all DCS vendors and is not standardised in text books. Some base the action on increasing E, rather than PV. If they also define error as $SP - PV$, then our heater temperature controller would need to be configured as **direct** acting.

Confusion can arise if the controller is manipulating a control valve. Valves are chosen to either fail open or fail closed on loss of signal – depending on which is less hazardous. The signal actually sent to a 'fail open' valve therefore needs to be reverse acting. Some texts take this into account when specifying the action of the controller. However most DCS differentiate between the output from the controller, which is displayed to the operator, and what is sent to the valve. To the operator and the controller all outputs represent the fraction (or percentage) that the valve is **open**. Any reversal required is performed after this. Under these circumstances, valve action need not be taken into account when specifying controller action.

The controller as specified in Equation (3.2) is known as the *full position* form in that it generates the actual controller output. A more useful form is the *incremental* or *velocity* form which generates the change in controller output (ΔM). We can convert the controller to this form by considering two consecutive scans. If E_n is the current error and E_{n-1} is the error at the previous scan then

$$M_n = K_c.E_n + C \tag{3.3}$$

$$M_{n-1} = K_c.E_{n-1} + C \tag{3.4}$$

Subtracting gives

$$\Delta M = K_c(E_n - E_{n-1}) = K_c.\Delta E \tag{3.5}$$

The advantage of this version is that first it eliminates C which is usually not a constant and would require adjustment as process conditions vary. Secondly the controller will have *bumpless initialisation*. When any controller is switched from manual to automatic mode it should cause no disturbance to the process. With the full position version it would be necessary first to calculate C to ensure that M is equal to the current value of the MV. Since the velocity form generates increments it will always start from the current MV and therefore requires no special logic.

Some systems require the *proportional band* (*PB*) rather than gain. It is defined as the percentage change in error required to move the output 100 %. Conversion between the two is straightforward.

$$PB = \frac{100}{K_c} \qquad (3.6)$$

While it will respond to changes in *PV*, the main purpose of proportional action is to generate a *proportional kick* whenever the SP is changed. If we assume *PV* is constant then from Equation (3.1)

$$\Delta E = -\Delta SP \qquad (3.7)$$

And, substituting into Equation (3.5)

$$\Delta M = -K_c.\Delta SP \qquad (3.8)$$

Remembering that for our fired heater example the controller is reverse-acting, the controller will thus make a step **increase** to fuel flow proportional to the change in temperature SP. This is a one-off change because ΔSP will be zero for future scans until another change is made to *SP*. The response is shown in Figure 3.1. In this case K_c has been set to 2.

Of course, increasing the fuel will cause the temperature to rise and reduce the error – so the controller output will only remain at this new value until the process deadtime has expired. The full trend is shown in Figure 3.2. This demonstrates the main limitation of

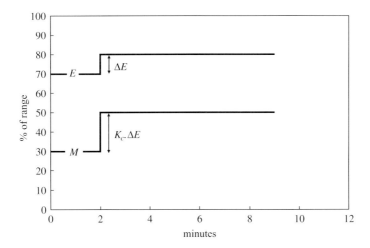

Figure 3.1 Proportional kick

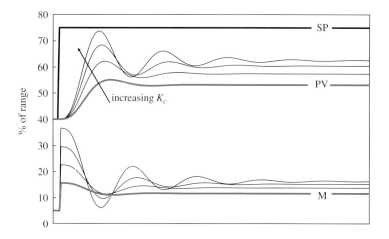

Figure 3.2 Effect of changing K_c

proportional control in that an *offset* will always exist at steady state. The *PV* will never reach *SP* except at initial conditions. Figure 3.2 also shows that the offset can be reduced by increasing K_c but with increasing oscillatory behaviour. We will show that these oscillations, on any process, become unstable before offset can be reduced to zero.

Figure 3.3 represents a control system common to the home. Found in lavatory cisterns and header tanks it provides basic control of level. It operates by a float in effect measuring the deviation from the target level and a valve which is opened in proportion to the position of the float. It is a proportional-only controller. So why does it not exhibit offset? This is because it is not a continuous process. However, should it develop a continuous leak (flow $= f$), in order to maintain steady state the controller would have to maintain an inlet flow of f. The inlet flow can only be nonzero if the error is nonzero.

We can represent this mathematically. Before the leak develops the error is zero. When the process again reaches steady state the controller will have changed the inlet flow by f and the error will be E. By putting these values into Equation (3.5) we get

$$f = K_c.E \qquad \therefore E = \frac{f}{K_c} \tag{3.9}$$

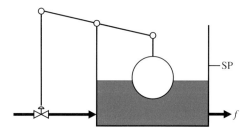

Figure 3.3 Domestic level controller.

This confirms what we know, i.e. that offset can be reduced by increasing K_c but that it cannot be reduced to zero. One might ask why not move the SP to account for the offset? One could, of course, and indeed we have the facility to do the equivalent by adjusting the manual reset term (C), as described in Equation (3.2). However Equation (3.9) shows us that the magnitude of the offset is proportional to the disturbance. Thus we would have to make this correction for virtually every disturbance and automation will have achieved little.

It is not to say, however, that proportional-only control should never be used. There are situations where offset is acceptable (such as in some level controllers as described in Chapter 4). However, in most situations we need the PV always to reach the SP.

3.3 Integral Action

The main purpose of integral action is to eliminate offset. Sometimes called *reset action* it continues to change the controller output for as long as an error exists. It does this by making the rate of change of output proportional to the error, i.e.

$$\frac{dM}{dt} = \frac{K_c}{T_i} E \tag{3.10}$$

T_i is known as the integral time and is the means by which the engineer can dictate how much integral action is taken. Equation (3.10) is already in the velocity form, integrating gives us the form that gives the action its name.

$$M = \frac{K_c}{T_i} \int E.dt \tag{3.11}$$

Converting Equation (3.10) to its discrete form (where ts is the controller scan interval) gives

$$\frac{\Delta M}{ts} = \frac{K_c}{T_i} E_n \tag{3.12}$$

Combining with Equation (3.5) gives proportional plus integral (PI) control

$$\Delta M = K_c \left[(E_n - E_{n-1}) + \frac{ts}{T_i} E_n \right] \tag{3.13}$$

The effect of the addition of integral action is shown in Figure 3.4. K_c has been reduced to a value of 1. The response shows that, for a constant error, the rate of change of output is constant. The change made by integral action will eventually match that of the initial proportional action. The time taken to 'repeat' the proportional action is T_i. In this example T_i is about 5 minutes. In many DCS T_i will have the units of minutes, but some systems use hours or seconds. Others define the tuning constant in *repeats per minute*, i.e. the reciprocal of T_i as we have defined it. The advantage of this is that, should more integral action be required, the engineer would increase the tuning constant. In the form of the algorithm we are using, higher values of T_i give **less** integral action. We therefore have to be careful in the use of zero as a tuning constant. Fortunately most systems recognise this as a special case and disable integral action, rather than attempt to make an infinite change.

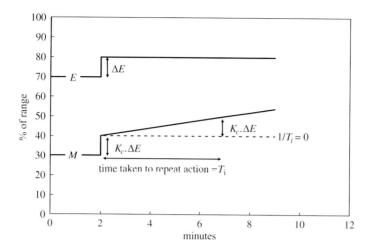

Figure 3.4 Proportional plus integral action

Again the trend in Figure 3.4 is only valid until the deadtime expires, after which the behaviour will be as shown in Figure 3.5. Even a very small amount of integral action will eliminate offset. Attempting to remove it too quickly will, as with any control action, cause oscillatory behaviour. However this can be compensated for by reducing K_c. Optimum controller performance is a trade-off between proportional and integral action.

For most situations a PI controller is adequate. Indeed many engineers will elect not to include derivative action to simplify tuning the controller by trial-and-error. A two-dimensional search for optimum parameters is considerably easier than a three-dimensional one. However in most situations the performance of even an optimally tuned PI controller can be substantially improved.

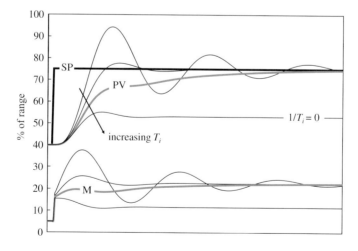

Figure 3.5 Effect of changing T_i

3.4 Derivative Action

Derivative action is intended to be anticipatory in nature; indeed, in older texts, it was called this. It anticipates by taking action if it detects a rapid change in error. The error may be very small (even zero) but, if changing quickly, will surely be large in the future. Derivative action attempts to prevent this by changing the output in proportion to the rate of change of error, i.e.

$$M = K_c . T_d \frac{dE}{dt} \tag{3.14}$$

T_d is known as the *derivative time* and is the means by which the engineer can dictate how much derivative action is taken. Converting Equation (3.14) to its discrete form, gives

$$M_n = K_c . T_d \frac{E_n - E_{n-1}}{ts} \tag{3.15}$$

Writing it for the previous scan interval

$$M_{n-1} = K_c . T_d \frac{E_{n-1} - E_{n-2}}{ts} \tag{3.16}$$

and subtracting

$$\Delta M = \frac{K_c . T_d}{ts} (E_n - 2E_{n-1} + E_{n-2}) \tag{3.17}$$

In order to demonstrate the effect of derivative action we will formulate a proportional plus derivative (PD) controller. This probably has no practical application but including integral action would make the trends very difficult to interpret. Combining Equations (3.5) and (3.17) gives

$$\Delta M = K_c \left[(E_n - E_{n-1}) + \frac{T_d}{ts} (E_n - 2E_{n-1} + E_{n-2}) \right] \tag{3.18}$$

Figure 3.6 shows the response of this controller, again up to the point where the deadtime expires. This time we have not made a step change to the SP, instead it has been ramped. The initial step change in the output is not then the result of proportional action but the derivative action responding to the change, from zero, in the rate of change of error. The subsequent ramping of the output is due to the proportional action responding to the ramping error. The proportional action will eventually change the output by the same amount as the initial derivative action. The time taken for this is T_d which, like T_i, can be expressed in units such as minutes or repeats per minute, depending on the DCS.

Also shown in Figure 3.6 is what the controller response would be without derivative action, i.e. proportional-only. It can be seen that derivative action takes action immediately that the proportional action takes T_d minutes to do. In effect it has anticipated the need for corrective action, even though the error was zero at the time. The anticipatory nature of derivative action is beneficial if the process deadtime is large; it compensates for the delay between the change in PV and the cause of the disturbance.

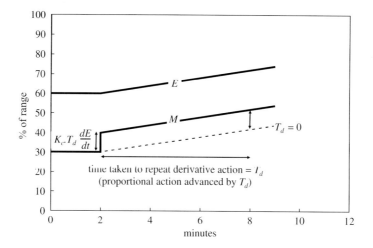

Figure 3.6 Proportional plus derivative action

Processes with a long deadtime, more specifically processes with a large θ/τ ratio, are relatively few. Thus most controllers, when responding to a change in SP, do not obviously benefit from the addition of derivative action. Indeed, if the θ/τ ratio is small, instability can be caused by relatively small amounts of derivative action. However we will demonstrate later that, even as the θ/τ ratio approaches zero, a little derivative action can be very useful in speeding the recovery from a process disturbance.

It is often said that derivative action should only be used in temperature controllers. It is true that temperatures, such as those on the outlet of fired heater and on distillation column trays, will often exhibit significantly more deadtime than measurements such as flow, level and pressure. However this is not universally the case, as illustrated in Figure 3.7. Manipulating the bypass of the stream on which we wish to install a temperature controller, in this case around the tube side of the exchanger, will provide an almost immediate response. Indeed, if accurate control of temperature is a priority, this would be preferred to the alternative configuration of bypassing the shell side.

While there are temperatures with very short deadtimes there will be other measurements that, under certain circumstances, show long deadtimes. In Chapter 4 we include a level control configuration that is likely to benefit from derivative action. In Chapter 7 we describe a composition control strategy with a very large θ/τ ratio.

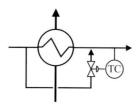

Figure 3.7 Temperature control without deadtime

The full PID equation that we have developed so far is thus

$$\Delta M = K_c \left[(E_n - E_{n-1}) + \frac{ts}{T_i} E_n + \frac{T_d}{ts} (E_n - 2E_{n-1} + E_{n-2}) \right] \quad (3.19)$$

This form of the equation, however, exhibits a problem known as *derivative spike*. Consider how the derivative action responds to a change in SP. If, before the change, the process is at steady state and at SP then

$$E_{n-1} = E_{n-2} = 0 \quad (3.20)$$

The change will introduce an error (E) and so the change in output due to the derivative action will be

$$\Delta M = K_c \frac{T_d}{ts} E \quad (3.21)$$

Assuming the process deadtime is longer than the controller scan interval then, at the next scan, the PV will not yet have responded to this change and so both E_n and E_{n-1} will now have the value E. The derivative action will then be a change of the same magnitude but opposite in direction, i.e.

$$\Delta M = -K_c \frac{T_d}{ts} E \quad (3.22)$$

Until the process deadtime expires the values of E_n, E_{n-1} and E_{n-2} will all be E and so the derivative action will be zero. Bearing in mind that T_d will be of the order of minutes and ts in seconds the magnitude of ΔM is likely to be large, possibly even full scale, and is likely to cause a noticeable process upset. Derivative action is not intended to respond to SP changes. Remembering that we have defined error as $PV - SP$, then

$$E_n - 2E_{n-1} + E_{n-2} = PV_n - 2PV_{n-1} + PV_{n-2} - (SP_n - 2SP_{n-1} + SP_{n-2}) \quad (3.23)$$

If there are no SP changes, then

$$E_n - 2E_{n-1} + E_{n-2} = PV_n - 2PV_{n-1} + PV_{n-2} \quad (3.24)$$

And we can rewrite the PID controller as

$$\Delta M = K_c \left[(E_n - E_{n-1}) + \frac{ts}{T_i} E_n + \frac{T_d}{ts} (PV_n - 2PV_{n-1} + PV_{n-2}) \right] \quad (3.25)$$

This *derivative-on-PV* version will no longer cause a derivative spike when there is a change in SP. But the response of derivative action to process disturbances is unaffected. In many DCS this modification is standard. Others retain both this and the *derivative-on-error* versions as options. It is also common for this algorithm to include some form of filtering to reduce the impact of the spike but, even with this in place, there is no reason why the engineer should ever use the derivative-on-error version if the derivative-on-PV version is available.

While this modified algorithm deals with the problem of a spike resulting from a SP change, it will still produce spikes if there are steps in the PV. These can result if the

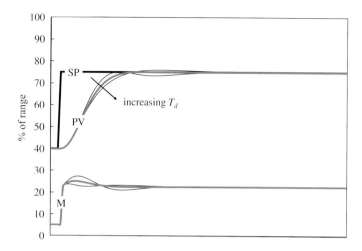

Figure 3.8   Effect of changing T_d

measurement is discontinuous. The most common example is some types of on-stream analysers, such as chromatographs. The *sample-and-hold* technique these employ will exhibit a staircase trend as the PV changes. Each step in the staircase will generate a spike. This is a particular issue because analysers tend to be a significant contributor to deadtime and thus the composition controller would benefit from the use of derivative action. This problem is addressed in Chapter 7.

However the problem can also arise from the use of digital field transmitters. Even if the analog-to-digital conversion is done to a high resolution – say to 0.1 % of range – the resulting 0.1 % steps as the PV changes can be amplified by one or two orders of magnitude by the derivative action. Care should therefore be taken in the selection of such transmitters if they are to be installed in situations where derivative action would be beneficial.

Figure 3.8 shows the performance of the full PID controller as described in Equation (3.25). As might be expected, the response becomes more oscillatory as T_d is increased. Perhaps more surprising is that **reducing** T_d also causes an oscillatory response. This is because the addition of derivative action permits more integral action to be used, so the oscillation observed by removing the derivative action is caused by excessive integral action.

This interdependence means that we cannot simply add derivative action to a well-tuned PI controller. It will only be of benefit if all three tuning constants are optimised. Similarly, if we wished to remove derivative action from a controller, we should re-optimise the proportional and integral tuning.

If measurement noise is present then we need to be cautious with the application of derivative action. While the amplitude of the noise may be very small, it will cause a high rate of change of the PV. Derivative action will therefore amplify this. This is perhaps another reason why there may be a reluctance to use it. However modern DCS provide a range of filtering techniques which permit advantage still to be taken of derivative action. These techniques are covered in Chapter 5.

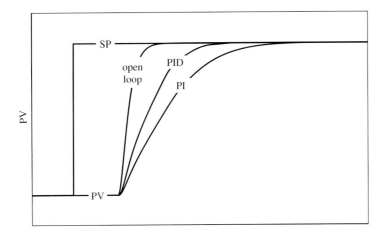

Figure 3.9 Benefit of derivative action

Figure 3.9 shows the benefit of including derivative action. The open loop response was produced by making a step change to the MV of the same magnitude as that ultimately made by the controllers. Applying the methods detailed in earlier in this chapter, the reader can confirm that this is a long deadtime process with a θ/τ ratio of about 2.7. The closed loop responses were placed by overlaying their trends so that change in SP is at the same point in time as the start of the open loop test. With such a process an optimally tuned PID controller will outperform an optimally tuned PI controller by reaching SP in about 30 % less time. Given that it is impossible for any controller to reach SP before the deadtime has elapsed, this is a substantial improvement. As a general rule derivative action will make a noticeable improvement to the response to SP changes if the θ/τ ratio is greater than 0.5. We will show later that it benefits the response to load changes at much lower values of θ/τ.

3.5 Versions of Control Algorithm

Before embarking on tuning the controller it is important that we understand the exact form of the algorithm. Different versions are in common use for two reasons. Firstly, there are a variety of approaches taken by different DCS vendors in converting the equations written in analog form into their discrete version. Secondly vendors have added a range of enhancements to the 'standard' controller.

Addressing the first of these issues we can write, by combining Equations (3.2), (3.11) and (3.14), the conventional analog version of the algorithm.

$$M = K_c \left[E + \frac{1}{T_i} \int E.dt + T_d \frac{dE}{dt} \right] \qquad (3.26)$$

One approach is to rewrite this directly in its discrete form

$$M_n = K_c \left[E_n + \frac{ts}{T_i} \sum_{j=0}^{n} E_j + \frac{T_d}{ts} (E_n - E_{n-1}) \right] \qquad (3.27)$$

In doing so we have applied the *rectangular rule*, i.e. the integral is treated as a series of rectangles of width ts and height E_j. But subtracting from this the equation written for the previous scan $(n-1)$ we would again obtain the algorithm described in Equation (3.19).

An alternative method is to apply the *trapezium rule*, where the integral is treated as series of trapeziums.

$$M_n = K_c' \left[E_n + \frac{ts}{T_i'} \sum_{j=0}^{n} \frac{E_j + E_{j-1}}{2} + \frac{T_d'}{ts}(E_n - E_{n-1}) \right] \tag{3.28}$$

By subtracting the equation for the previous scan we obtain a slightly different version of the velocity form of the controller.

$$\Delta M = K_c' \left[(E_n - E_{n-1}) + \frac{ts}{T_i'} \left(\frac{E_n + E_{n-1}}{2} \right) + \frac{T_d'}{ts}(E_n - 2E_{n-1} + E_{n-2}) \right] \tag{3.29}$$

This uses the average of the last two errors in the integral action rather than the latest value of the error. The algorithm will perform in exactly the same way provided that the tuning is adjusted to take account of the change. By equating coefficients for E_n, E_{n-1} and E_{n-2} we can derive tuning constants for Equation (3.29) from those used in Equation (3.19).

$$K_c' = K_c \left[1 + \frac{ts}{2T_i} \right] \tag{3.30}$$

$$T_i' = T_i + \frac{ts}{2} \tag{3.31}$$

$$T_d' = T_d \left[\frac{2T_i}{2T_i + ts} \right] \tag{3.32}$$

Details of the Laplace form of the control equations are presented at the end of this chapter. Without going into detail, for the 'standard' controller it would be

$$M = K_c \left[1 + \frac{1}{T_i s} + T_d s \right] E \tag{3.33}$$

Again, without going into details, a more rigorous conversion to the z-transform gives

$$\Delta M = K_c' \left[(E_n - E_{n-1}) + \frac{ts}{T_i'} E_{n-1} + \frac{T_d'}{ts}(E_n - 2E_{n-1} + E_{n-2}) \right] \tag{3.34}$$

Again, only the integral term is affected – using E_{n-1} rather than E_n. Again by equating coefficients we can show that.

$$K_c' = K_c \left[1 + \frac{ts}{T_i} \right] \tag{3.35}$$

$$T_i' = T_i + ts \tag{3.36}$$

$$T'_d = T_d \left[\frac{T_i}{T_i + ts} \right] \qquad (3.37)$$

Equations (3.30) to (3.32) and Equations (3.35) to (3.37) all show that, as *ts* tends towards zero, the tuning constants are the same for all three versions of the PID controller. This confirms that all three are good approximations to analog control. It also shows that, provided the tuning constants are large compared to the scan interval, the values required vary little between the algorithms. Tuning constants are generally of the order of minutes, while the scan interval is of the order of seconds, and so it will generally be the case that we do not need to know the precise form of the algorithm. This is somewhat fortunate with many DCS; the vendor will describe the algorithm in its analog form but not always divulge how it has been converted to its discrete form. However, if the process dynamics are very fast, resulting in tuning constants measured in a few seconds, then knowing the precise form of the algorithm becomes important.

3.6 Interactive PID Controller

There is, however, a quite different version of the PID algorithm. For reasons which will become apparent, the controller with which we have been working is known as the *noninteractive* form.

It can also be described as the *parallel* form. This is because Equation (3.33) can be represented diagrammatically, as in Figure 3.10, as number of operations working in parallel. To convert the box diagram to an equation, functions in parallel are additive, those in series are multiplicative.

Figure 3.11 represents the *series* version of the algorithm. This is more representative of the algorithm used by early pneumatic instruments and is retained, usually as an option, in some DCS.

Converted to its Laplace form

$$M = K'_c \left(1 + \frac{1}{T'_i s}\right)(1 + T'_d s)E = K'_c \left[\left(1 + \frac{T'_d}{T'_i}\right) + \frac{1}{T'_i s} + T'_d s\right]E \qquad (3.38)$$

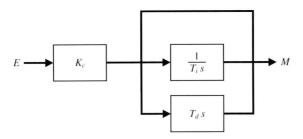

Figure 3.10 Parallel PID

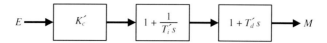

Figure 3.11 Series PID

And then to its digital velocity form

$$\Delta M = K_c' \left[\left(1 + \frac{T_d'}{T_i'}\right)(E_n - E_{n-1}) + \frac{ts}{T_i'} E_n + \frac{T_d'}{ts}(E_n - 2E_{n-1} + E_{n-2}) \right] \quad (3.39)$$

Comparison with our noninteractive controller in Equation (3.19) shows that the proportional action depends not only on K_c' but also on T_i' and T_d'. Changing either the integral action or the derivative action will affect the proportional action – hence its 'interactive' title. Again, by equating coefficients we can develop formulae for modifying the tuning as we change from one algorithm to the other.

$$K_c' = \frac{K_c}{2}\left[1 + \sqrt{1 - 4\frac{T_d}{T_i}}\right] \quad (3.40)$$

$$T_i' = \frac{T_i}{2}\left[1 + \sqrt{1 - 4\frac{T_d}{T_i}}\right] \quad (3.41)$$

$$T_d' = \frac{2T_d}{\left[1 + \sqrt{1 - 4\frac{T_d}{T_i}}\right]} \quad (3.42)$$

While the interactive and noninteractive algorithms can be tuned to give identical performance, this is subject to the restriction that $T_d < 0.25T_i$. This constraint will only be violated if the process has a very large θ/τ ratio. Indeed if no derivative action is required, then the algorithms will perform identically without the need to change tuning.

If T_d/T_i is at the limit of 0.25 then, by combining Equations (3.41) and (3.42) we can see that the value of T_d' will be the same as that for T_i'.

$$\frac{T_d'}{T_i'} = \frac{4T_d}{T_i\left(1 + \sqrt{1 - 4\frac{T_d}{T_i}}\right)^2} = 1 \quad (3.43)$$

In general, the interactive controller requires a much larger derivative tuning constant than the noninteractive equivalent.

So that one algorithm can be adopted as the standard approach for all situations, there are several arguments for choosing the noninteractive algorithm. Firstly, it can be tuned better for processes with a very large θ/τ ratio; these merit much more derivative action with T_d often exceeding $0.25T_i$. Secondly, most DCS use this algorithm – often describing it as *ideal*; others give the option to use either. Finally, most published tuning methods are based on the noninteractive version.

To switch from the interactive to the noninteractive version requires tuning to be changed as follows.

$$K_c = K_c' \frac{T_i' + T_d'}{T_i'} \tag{3.44}$$

$$T_i = T_i' + T_d' \tag{3.45}$$

$$T_d = T_d' \frac{T_i'}{T_i' + T_d'} \tag{3.46}$$

The interactive algorithm is usually modified further to

$$M = K_c \left[1 + \frac{1}{T_i s}\right] \left[\frac{1 + T_d s}{1 + aT_d s}\right] E \tag{3.47}$$

Comparison with Equation (3.33) shows that an additional function has been introduced, i.e. $(1 + aT_d s)$. This introduces a lag into the controller (of time constant aT_d) that is intended to reduce the amplification of measurement noise by the derivative action. Setting a to zero removes this filter, setting it to 1 will completely disable the derivative action. In some systems the value of a is configurable by the engineer. In many it is fixed, often at a value of 0.1. The reciprocal of a is known as the *derivative gain limit*.

Its inclusion is of dubious value. If no noise is present and derivative action is required then we have to modify the controller tuning to account of its presence. Few published tuning methods take this into account. Secondly, the filtering is identical to that provided by the standard DCS filter (see Chapter 5). The DCS filter is generally adjustable by the engineer whereas that in the control algorithm may not be. Secondly, even if a is adjustable, its upper limit means that the filter time constant cannot be increased beyond T_d. If noise is an issue then an engineer-configurable filter is preferred. This strengthens the argument not to use the interactive version of the controller.

Some vendors have also introduced filtering into the noninteractive version as

$$M = K_c \left[1 + \frac{1}{T_i s} + \frac{T_d s}{1 + aT_d s}\right] E \tag{3.48}$$

Comparison with Equation (3.33) shows that the lag is only applied to the derivative action and thus does not force unnecessary filtering of the proportional and integral actions. However a may still not be adjustable by the engineer and is not usually taken account of in published tuning methods.

3.7 Proportional-on-PV Controller

The most misunderstood and most underutilised version of the PID algorithm is the *proportional-on-PV* type. Taking the algorithm developed as Equation (3.25) we modify the proportional action so that it is based on PV rather than error.

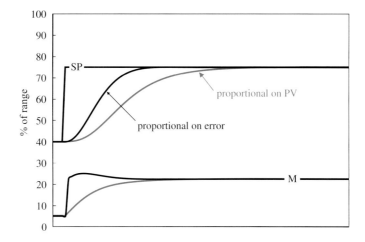

Figure 3.12 Proportional-on-PV algorithm

$$\Delta M = K_c \left[(PV_n - PV_{n-1}) + \frac{ts}{T_i} E_n + \frac{T_d}{ts} (PV_n - 2PV_{n-1} + PV_{n-2}) \right] \qquad (3.49)$$

In the same way that changing the derivative action from using PV instead of error, this change will stop the **proportional** action responding to changes in SP. This would appear to undermine the main purpose of proportional action by eliminating the proportional kick it produces whenever the SP is changed. Indeed, only the integral action will now respond to the SP change, producing a much gentler ramping function. This can be seen in Figure 3.12 where a well-tuned proportional-on-error algorithm has had this modification made. The absence of the initial proportional kick can be seen on the trend of the MV and results in the PV taking much longer to reach its new SP.

Many believe therefore that this algorithm should be applied on processes where the MV should be adjusted slowly. However, if this performance were required, it could be achieved by tuning the more conventional proportional-on-error algorithm. Conversely it is important to recognise that the proportional-on-PV algorithm can be retuned to compensate for lack of the proportional kick and so respond well to SP changes. This is illustrated in Figure 3.13.

Figure 3.14 shows the behaviour of each part of the proportional-on-error control algorithm in response to the SP change above. The proportional kick is clear with the proportional part of the controller returning to zero as the error returns to zero. The derivative action is the greatest as the PV peaks, and so permits more proportional and integral action to be used. It too returns to zero as the rate of change of PV returns to zero.

Figure 3.15 shows the same disturbance but with the proportional-on-PV algorithm. Note that the vertical scale is much larger than that in Figure 3.14. As expected, there is no proportional kick and, since the action is now based on PV, the proportional part does not return to zero. It can be confusing that the proportional part **reduces** as the SP is increased but this is because the controller must be configured as reverse acting. The integral action compensates for this so that there is a net increase in controller output. The derivative action

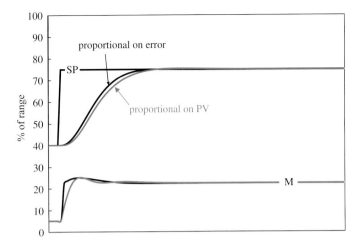

Figure 3.13 Proportional-on-PV algorithm retuned for SP change

behaves in almost the same way as in the proportional-on-error case, but the correction is larger because of the higher controller gain.

Figure 3.13 shows that the performance of a **retuned** proportional-on-PV controller would, on a real process, be indistinguishable from the original proportional-on-error controller. Compensation for the loss of the proportional kick has been achieved mainly by substantially increasing the controller gain. This causes the integral action to ramp the MV much faster. But achieving similar performance begs the questions as to why the proportional-on-PV algorithm is included in most DCS and when it should be used.

Rather than consider SP changes, we should give more attention to **load** changes. Load changes are process disturbances which cause the PV to move away from the SP. On our

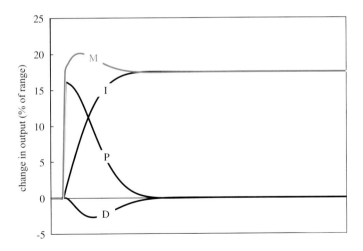

Figure 3.14 Response to SP change (proportional-on-error algorithm)

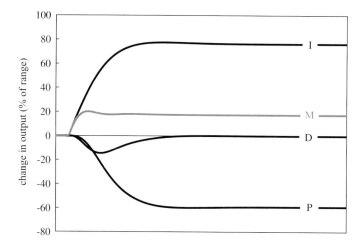

Figure 3.15 Response to SP change (proportional-on-PV algorithm)

fired heater, for example, changes in feed flow rate or heater inlet temperature would cause such a deviation. Most controllers experience many more load changes than SP changes. A heater outlet temperature controller may operate for days or weeks with no change to its SP. But it is likely to experience many process disturbances in the meantime.

Load changes impact the error differently to SP changes since their effect must pass through the process and is subject to the process lag. Rather than in the case of a SP change, when the error changes as a step, it will accumulate more gradually. Figure 3.16 shows the performance of the two controllers, both tuned for SP changes, subjected to a load change. The change could, for example, be an increase in feed flow rate. The open loop trend shows what would happen with no temperature control in place.

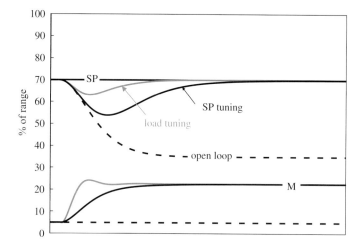

Figure 3.16 Response to a load change

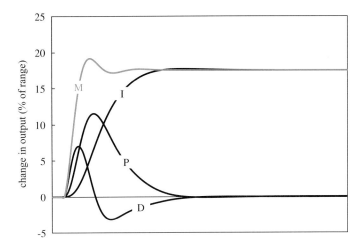

Figure 3.17 Response to a load change (both algorithms)

Switching the algorithm between proportional-on-error and proportional-on-PV has no affect on the way it responds to **load** changes. The difference we see is due to the difference in tuning. The more tightly tuned algorithm deviates from SP by less than half and the duration of the upset is also halved.

This opportunity for substantial improvement is often overlooked by control engineers. Preoccupied with tuning controllers for SP changes they rarely appreciate how much faster the controller can be made to react to process disturbances.

Figure 3.17 shows the breakdown of the control action for the load tuning case above. Since the SP is constant, the response would be the same for both the proportional-on-error and proportional-on-PV algorithm.

It is important to recognise that the process cannot benefit from the tuning change if applied to the proportional-on-error algorithm. The effect of doing so is shown in Figure 3.18. Even if SP changes are rare, when they are made, the controller will now react far too quickly. In our example the MV has overshot its steady-state change by over 200 %. In our fired heater example, this would likely cause a serious upset to fuel combustion.

We have to consider which algorithm and tuning combination should be used if the controller is the secondary of a cascade. Such controllers are subject to SP changes when the SP of the primary is changed but also when the primary takes corrective action during a load change. Unlike a primary, a secondary controller will be subject to frequent SP changes. Theoretically this would suggest that the proportional-on-error algorithm should be used in the secondary, since this will marginally outperform the proportional-on-PV version. One could make the same argument if a MVC is installed, since the controllers it manipulates effectively become secondaries of a cascade.

We will cover later different measures of control performance but the most commonly used is *integral over time of absolute error* (*ITAE*). The higher the value of ITAE, the poorer the controller is at eliminating the error. Figure 3.19 shows the impact that switching from proportional-on-PV to proportional-on-error has on ITAE. Both algorithms have been tuned

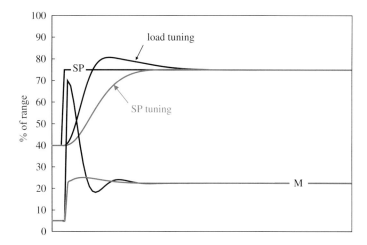

Figure 3.18 Response to a SP change (proportional-on-error algorithm)

for SP changes. As expected the proportional-on-PV algorithm does not perform as well as the proportional-on-error algorithm for SP changes. While this might at first appear significant, it should be compared against the impact it has when the process is subjected to a load change.

Secondaries of cascades generally have a very small θ/τ ratio and so the improvement in the response to load changes, from selecting the proportional-on-PV algorithm, is likely to be around 600 %. The price we pay for this is around a 10 % increase in ITAE when the SP is changed. We must also consider the case when the primary controller is out of service. The secondary will then experience mainly load changes and would benefit from this decision. Further, in the interest of standardisation, universally adopting the same algorithm would be advantageous.

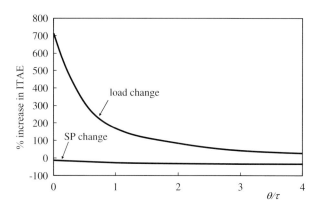

Figure 3.19 Impact of switching from proportional-on-PV to proportional-on-error

Some DCS have the option of automatically switching between the two algorithms as the cascade is switched between auto and manual. While this might seem a good idea it is advisable to disable this feature. The DCS does not make the necessary change to tuning constants when it switches algorithm. Of course it would possible to add this feature as a customisation but that would involve maintaining two sets of tuning. Given that one set is likely to be used rarely, there is a danger of it not being updated to reflect any process or instrumentation changes. This switching also has an impact when step-testing the secondary SP in order to develop tuning constants for the primary. If the switching is enabled then the proportional-on-PV algorithm will be used during step-testing but the proportional-on-error will be used when the cascade is commissioned. This also applies when step-testing for the later addition of a MVC. After the primary (or MVC) is commissioned the dynamics could be quite different from those found in the step-test and may cause performance problems. On systems where this facility cannot be disabled, the proportional-on-error version must be used when the primary (or MVC) is in service and special arrangements have to be made to 'trick' the DCS into using this algorithm during step-testing.

Figure 3.19 does, however, show that, if a controller experiences an equal split of SP and load changes, the advantage of the proportional-on-PV algorithm tends to zero as the θ/τ ratio becomes large. Theoretically here we should use the proportional-on-error algorithm if there are frequent SP changes. However, as we shall see in Chapter 7, there are techniques superior to the PID algorithm for processes with long deadtimes. While these may not always be applicable, the occasions on which the proportional-on-error algorithm is justified will be very rare.

Some DCS include an algorithm described as the *two degrees of freedom controller*. This has the form

$$\Delta M = K_c \left[(x_n - x_{n-1}) + \frac{ts}{T_i} E_n + \frac{T_d}{ts} (y_n - 2y_{n-1} + y_{n-2}) \right] \quad (3.50)$$

where

$$x = PV - \alpha SP \quad \text{and} \quad y = PV - \beta SP \quad (3.51)$$

The algorithm can also be described as having *set point weighting*. The values of α and β can be set by the engineer to a value in the range 0 to 1. Setting them both to 0 will give the controller as described by Equation (3.19), while setting them both to 1 will give the recommended form of the controller as described by Equation (3.49). It is possible to use values between 0 and 1 but there is little benefit in doing so. Optimising the tuning, for both load and SP changes, will always result in a value for α of 1 (and higher if permitted). However this ignores the requirement that the controller should handle both disturbances well with the same tuning constants. For this to be achievable, α must be 0. The optimised value for β will be that which generates the maximum permitted derivative spike, i.e. β will be zero if no spike is permitted.

Another option available in some systems is an *integral only* algorithm. This has the form

$$\Delta M = \frac{ts}{T_i} E_n \quad (3.52)$$

For SP changes this will respond identically to the proportional-on-PV, derivative-on-PV PID algorithm – provided the value used for T_i is that same as K_c/T_i in the PID algorithm. In the same way that derivative action amplifies noise, integral action attenuates it. The full position discrete form of the algorithm, Equation (3.27), shows how integral action is based on the sum of all previous values of the error. This averages out any noise. However, in the absence of proportional and derivative actions, the algorithm will respond more slowly to load changes. So, while it was once commonly used in flow controllers, it is difficult to see what advantage it offers.

3.8 Nonstandard Algorithms

Most of the algorithms presented in this chapter can be represented by the general algorithm

$$\Delta M = a_1 PV_n + a_2 PV_{n-1} + a_3 PV_{n-2} + b_1 SP_n + b_2 SP_{n-1} + b_3 SP_{n-2} \qquad (3.53)$$

So, for example, equating coefficients with the basic PID control defined by Equation (3.19) gives

$$a_1 = K_c \left[1 + \frac{ts}{T_i} + \frac{T_d}{ts} \right] \quad a_2 = -K_c \left[1 + \frac{2T_d}{ts} \right] \quad a_3 = K_c \frac{T_d}{ts} \qquad (3.54)$$

$$b_1 = -K_c \left[1 + \frac{ts}{T_i} + \frac{T_d}{ts} \right] \quad b_2 = K_c \left[1 + \frac{2T_d}{ts} \right] \quad b_3 = -K_c \frac{T_d}{ts} \qquad (3.55)$$

Similarly, equating coefficients with the preferred PID control defined by Equation (3.49) gives

$$a_1 = K_c \left[1 + \frac{ts}{T_i} + \frac{T_d}{ts} \right] \quad a_2 = -K_c \left[1 + \frac{2T_d}{ts} \right] \quad a_3 = K_c \frac{T_d}{ts} \qquad (3.56)$$

$$b_1 = -K_c \frac{ts}{T_i} \quad b_2 = 0 \quad b_3 = 0 \qquad (3.57)$$

However, even in this case, there are more coefficients than conventional tuning constants. So while the coefficients can always be derived from the tuning constants the reverse is not true. It is feasible to optimise the coefficients to obtain the best possible controller performance but it is unlikely that it would be possible to convert these to tuning constants to be used in any of the algorithms available in the DCS.

This begs the question: why retain such algorithms where the general algorithm can be tuned to outperform any of them? Part of the answer is that there would no longer be an obvious connection between observing that the controller response might be improved and knowing which coefficient(s) to adjust to bring about the improvement. Trial-and-error tuning would become far more time-consuming because of this and the increase in the number of parameters to adjust. However, if tuning is derived using some computerised optimiser based on known process dynamics, such algorithms become feasible.

3.9 Tuning

It is probably fair to say that the vast majority of PID controllers in the process industry are not optimally tuned. The majority of tuning is completed using experience and trial-and-error. While this may not adversely affect process performance when the process dynamics are very short it does become an issue otherwise. It is not the intention of this book that rigorous model identification and tuning be applied to **every** controller. Improving a fired heater fuel flow controller so that it reacts to a SP change, say, in 5 seconds as opposed to 10 seconds will have a minor impact on the control of the temperature which has dynamics measured in minutes. However adopting a rigorous approach to the temperature controller is likely to be well worth the effort.

It is unlikely that the control engineer will find a published controller tuning method that will meet the needs of the process. This is despite a considerable amount of research work. In 2000 a survey (Reference 2) identified, for self-regulating processes, 81 published methods for tuning PI controllers and 117 for PID controllers. For integrating processes, it also found 22 methods for PI control and 15 for PID control. **Every one** of these methods has at least one flaw.

The published methods described in this chapter are included primarily to draw the engineer's attention to what limitations might be encountered and permit assessment of any other method offered.

Some tuning methods are based on the *damping ratio*. Damping can describe either open loop or closed loop behaviour. The terminology originates from the analysis of spring/damper combinations as used in vehicle suspension systems. An *overdamped* system, as shown in Figure 3.20, approaches the new steady state gradually without overshooting. An *underdamped* system will exhibit overshoot. A *critically damped* system is one as close as possible to being underdamped without overshoot. An *undamped* system is one which oscillates at constant amplitude.

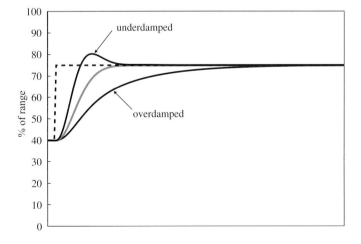

Figure 3.20   Damping

An underdamped system has to be at least second order. In the process industry underdamped **processes** are rare. One example would be the rapid closure of a pressure control valve at the end of a long length or pipework containing liquid travelling quickly. The pressure upstream of the valve would temporarily increase above its steady state value. Known as *fluid hammer*, this can require special attention on pipelines. Certain types of reactor can also exhibit open loop underdamped response. However once a controller is added to an overdamped process then underdamping can be brought about by tightly tuning the controller.

3.10 Ziegler-Nichols Tuning Method

Perhaps the most well known and most frequently published method is that by Ziegler and Nichols (Reference 3). They developed an open loop method in addition to the more frequently published closed loop method. The criterion used by both methods is the *quarter decay ratio*. This is illustrated in Figure 3.21; following a SP change the PV response should be slightly oscillatory with the height of the second peak being one-quarter of that of the first.

The nomenclature used by Ziegler and Nichols is somewhat different from that which we use today. To avoid confusion we will update it to current terminology. If we address first the closed loop method, the technique involves starting with a proportional-only controller and adjusting its gain until the PV oscillates with a constant amplitude. We record the gain at which this is achieved – known as the *ultimate gain* (K_u) and measure the period of oscillation – known as the *ultimate period* (P_u). This is shown in Figure 3.22; the amplitude of both the PV and the MV is constant and they are in exact anti-phase.

One of the problems in applying the method is the practicality of achieving sustained oscillation. Even if triggered by a very small change in SP, the amplitude can be very large

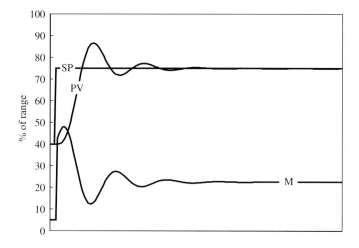

Figure 3.21   Quarter decay ratio

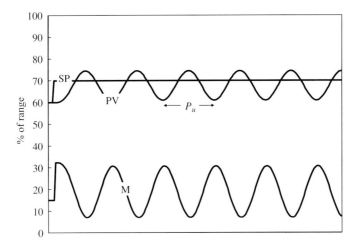

Figure 3.22 Sustained oscillation

and potentially harmful to the process. The preferred proportional-on-PV controller, if configured as proportional-only, will not respond to SP changes and so the oscillation must be triggered by a load change. This is not always straightforward. However it is possible to predict K_u and P_u from the process dynamics.

For a self-regulating process

$$K_u = \frac{-1}{K_p \cos\left(\frac{2\pi\theta}{P_u}\right)} \tag{3.58}$$

$$\frac{2\pi\tau}{P_u} + \tan\left(\frac{2\pi\theta}{P_u}\right) = 0 \tag{3.59}$$

Solving Equations (3.58) and (3.59) is not trivial and requires an iterative approach. P_u must be between 2θ and 4θ. A good starting point for the iteration is given by the formula developed by Lopez, Miller, Smith and Murrill (Reference 4):

$$K_u = \frac{2.133}{K_p}\left[\frac{\theta}{\tau}\right]^{-0.877} \quad \text{for} \quad 0 \leq \theta \leq \tau \tag{3.60}$$

For an integrating process the solution is considerably easier:

$$K_u = \frac{\pi}{2K_p\theta} \tag{3.61}$$

$$P_u = 4\theta \tag{3.62}$$

An alternative approach to adjusting K_p by trial-and-error to attain sustained oscillation is to apply the *Relay Method* (Reference 5). This involves first selecting acceptable low and

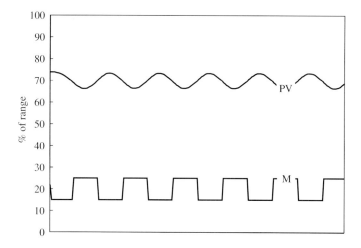

Figure 3.23 Relay method

high values of the MV that are either side of the current operating point (MV_{low} and MV_{high}). The MV is switched automatically between these two values when the PV crosses a target set close to the current operation. This will set up a sustained oscillation in the PV. The period of this oscillation will be P_u. The ultimate gain is derived from the amplitude (A) of the oscillation.

$$K_u = \frac{2(MV_{high} - MV_{low})}{\pi A} \tag{3.63}$$

Figure 3.23 gives an example where the target for the PV is set at 70% and the low and high values for the MV are set at 15% and 25% respectively. The resulting period of oscillation is the same as that in Figure 3.22.

Once K_u and P_u are known, Ziegler and Nichols provide simple calculations for the derivation of tuning constants. These are, for a P only controller

$$K_c = 0.5 K_u \tag{3.64}$$

Because a proportional only controller will never reach SP, the quarter decay is determined with respect to the steady state condition. The reciprocal of the coefficient, in this case the reciprocal of 0.5, is known as the *gain margin*. It is the factor by which the controller gain can be increased before the controller becomes unstable. A proportional only controller tuned according to the Ziegler-Nichols method will therefore have a gain margin of 2.

For a PI controller

$$K_c = 0.45 K_u \quad T_i = \frac{P_u}{1.2} \tag{3.65}$$

For a PID controller

$$K_c = 0.6K_u \quad T_i = \frac{P_u}{2} \quad T_d = \frac{P_u}{8} \quad (3.66)$$

The tuning criterion used by this method is now dated. It is unlikely to be perceived now as a well-tuned controller. It gives substantial overshoot on both the PV and the MV with little advantage to achieving steady state as soon as possible. A number of others have suggested the use of different coefficients in the calculations. But the main problem is that K_u and P_u do not uniquely define the process. For a self-regulating process, they can be derived from K_p, θ and τ but the reverse is not true. There are many combinations of dynamics which would give the same value of K_u and P_u. For the same reason there are many combinations of tuning constants that will give the required quarter decay performance. This restricts the use of the method to integrating processes. Because only two dynamic constants are normally used to characterise such a process, there is a unique relationship, as shown in Equations (3.61) and (3.62).

The Ziegler-Nichols open loop tuning technique is an extension of the steepest slope method that we described in Chapter 2 as a means of obtaining the process dynamics. The method, as originally documented, involves calculating the gradient (R) of the steepest slope and then dividing this by the step change (ΔMV) to give the *unit reaction rate* (R_1). However we can update the method to use the actual process dynamics. Without deadtime a first order self-regulating process is described by

$$\Delta PV = K_p \Delta MV (1 - e^{-t/\tau}) \quad (3.67)$$

Differentiating

$$\frac{d(\Delta PV)}{dt} = \frac{K_p \Delta MV}{\tau} e^{-t/\tau} \quad (3.68)$$

For a first order process, the rate of change of PV is a maximum when $t=0$, so

$$R = \frac{K_p \Delta MV}{\tau} \quad \text{and} \quad R_1 = \frac{K_p}{\tau} \quad (3.69)$$

Adapting the formulae given by Ziegler and Nichols we get, for a P only controller

$$K_c = \frac{\tau}{K_p \theta} \quad (3.70)$$

For a PI controller

$$K_c = \frac{0.9\tau}{K_p \theta} \quad T_i = \frac{\theta}{0.3} \quad (3.71)$$

And for a PID controller

$$K_c = \frac{1.2\tau}{K_p \theta} \quad T_i = \frac{\theta}{0.5} \quad T_d = \frac{\theta}{2} \quad (3.72)$$

For an integrating process, modifying Equation (2.55) gives

$$\frac{d(\Delta PV)}{dt} = K_p \Delta MV \tag{3.73}$$

And so R_1 is the same as K_p and the tuning formulae become, for a P only controller,

$$K_c = \frac{1}{K_p \theta}, \tag{3.74}$$

for a PI controller

$$K_c = \frac{0.9}{K_p \theta} \quad T_i = \frac{\theta}{0.3}, \tag{3.75}$$

and for a PID controller

$$K_c = \frac{1.2}{K_p \theta} \quad T_i = \frac{\theta}{0.5} \quad T_d = \frac{\theta}{2} \tag{3.76}$$

Given that the noninteractive version of the PID algorithm was not available until the advent of electronic controllers, Ziegler and Nichols could not have used this to develop their tuning method. While in theory they would have used the interactive version, they actually used a pneumatic controller which approximates to this. In fact the mathematical definition of the interactive algorithm came after pneumatic controllers were designed. If we use Equations (3.44) to (3.46) to determine the tuning for the noninteractive algorithm, K_c and T_i must be increased by 25 % and T_d reduced by 25 %. For the interactive algorithm Ziegler and Nichols recommend T_d should be set at $0.25T_i$. For the noninteractive controller this reduces to $0.16T_i$.

One could argue that, since the controller is in automatic mode when the *closed loop* test is performed, any difference in control algorithm is taken into account. This is partially true in that it compensates for the change from analog to digital control. But since only the proportional mode is used during testing it does not take account of any other changes to the algorithm.

3.11 Cohen-Coon Tuning Method

Another published method frequently quoted is that by Cohen and Coon (Reference 6). It too uses the quarter decay criterion and is presented as sets of formulae based on process dynamics. For P only control

$$K_c = \frac{1}{K_p} \frac{\tau}{\theta} \left[1 + \frac{\theta}{3\tau} \right] \tag{3.77}$$

For PI control

$$K_c = \frac{1}{K_p} \frac{\tau}{\theta} \left[0.9 + \frac{\theta}{12\tau} \right] \quad T_i = \theta \left[\frac{30 + \frac{3\theta}{\tau}}{9 + \frac{20\theta}{\tau}} \right] \tag{3.78}$$

For PID control

$$K_c = \frac{1}{K_p} \frac{\tau}{\theta} \left[\frac{4}{3} + \frac{\theta}{4\tau} \right] \quad T_i = \theta \left[\frac{32 + \frac{6\theta}{\tau}}{13 + \frac{8\theta}{\tau}} \right] \quad T_d = \theta \left[\frac{4}{11 + \frac{2\theta}{t}} \right] \quad (3.79)$$

3.12 Tuning Based on Penalty Functions

Tables 3.1 to 3.5 give details of the tuning method developed by Smith, Murrill and others (References 4 and 7). It assumes that the θ/τ ratio is between 0 and 1. Tables 3.1 to 3.3 give tuning designed for load changes. With these it is important to use the proportional-on-PV control algorithm so that the controller does not give an excessive response to SP changes. Tables 3.4 and 3.5 giving tuning for SP changes; the method assumes that the proportional-on-error algorithm is used.

The tables use three different tuning criteria – integral of absolute error (IAE), integral of square of error (ISE) and integral over time of the absolute error (ITAE). Not used by Smith and Murrill, another function can be defined – integral over time of the square of the error (ITSE). Each of these is a form of penalty function representing the size and duration of error. The tuning methods aim to minimise the penalty.

Table 3.1 Lopez, Miller, Smith and Murrill (IAE, load change)

	$K_c = \dfrac{A\left[\frac{\theta}{\tau}\right]^B}{K_p}$		$T_i = \dfrac{\tau}{A\left[\frac{\theta}{\tau}\right]^B}$		$T_d = A\left[\frac{\theta}{\tau}\right]^B \tau$	
	A	B	A	B	A	B
P	0.902	−0.985				
PI	0.984	−0.986	0.608	−0.707		
PID	1.435	−0.921	0.878	−0.749	0.482	1.137

Table 3.2 Lopez, Miller, Smith and Murrill (ISE, load change)

	$K_c = \dfrac{A\left[\frac{\theta}{\tau}\right]^B}{K_p}$		$T_i = \dfrac{\tau}{A\left[\frac{\theta}{\tau}\right]^B}$		$T_d = A\left[\frac{\theta}{\tau}\right]^B \tau$	
	A	B	A	B	A	B
P	1.411	−0.917				
PI	1.305	−0.959	0.492	−0.739		
PID	1.495	−0.945	1.101	−0.771	0.560	1.006

Table 3.3 Lopez, Miller, Smith and Murrill (ITAE, load change)

	$K_c = \dfrac{A\left[\frac{\theta}{\tau}\right]^B}{K_p}$		$T_i = \dfrac{\tau}{A\left[\frac{\theta}{\tau}\right]^B}$		$T_d = A\left[\frac{\theta}{\tau}\right]^B \tau$	
	A	B	A	B	A	B
P	0.490	−1.084				
PI	0.859	−0.977	0.674	−0.680		
PID	1.357	−0.947	0.842	−0.738	0.381	0.995

Table 3.4 Smith and Murrill (IAE, SP change)

	$K_c = \dfrac{A\left[\frac{\theta}{\tau}\right]^B}{K_p}$		$T_i = \dfrac{\tau}{A + B\frac{\theta}{\tau}}$		$T_d = A\left[\frac{\theta}{\tau}\right]^B \tau$	
	A	B	A	B	A	B
P						
PI	0.758	−0.861	1.020	−0.323		
PID	1.086	−0.869	0.740	−0.130	0.348	0.914

Table 3.5 Smith and Murrill (ITAE, SP change)

	$K_c = \dfrac{A\left[\frac{\theta}{\tau}\right]^B}{K_p}$		$T_i = \dfrac{\tau}{A + B\frac{\theta}{\tau}}$		$T_d = A\left[\frac{\theta}{\tau}\right]^B \tau$	
	A	B	A	B	A	B
P						
PI	0.586	−0.916	1.030	−0.165		
PID	0.965	−0.855	0.796	−0.147	0.308	0.929

Figure 3.24 shows the IAE. The area between the PV and the SP comprises a series of rectangles of width *ts* (the scan interval) and height |E| (the absolute value of the error). The sum of the areas of these rectangles is the IAE.

We remove the sign of the error when integrating. Otherwise positive errors would be cancelled by negative errors, so that even a sustained oscillation would incur a penalty close to zero. The penalty functions are defined as follows.

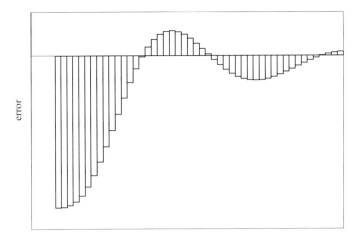

Figure 3.24 Integral of absolute error

$$IAE = \int_0^\infty |E|.dt \qquad (3.80)$$

$$ISE = \int_0^\infty E^2.dt \qquad (3.81)$$

$$ITAE = \int_0^\infty |E|t.dt \qquad (3.82)$$

$$ITSE = \int_0^\infty E^2 t.dt \qquad (3.83)$$

Minimising *ISE* is equivalent to minimising the variance (σ^2) and thus the standard deviation (σ) of the controller error. But the tuning generated by minimising *ISE* (and *IAE*) will result in a controller that eliminates as fast as possible the large error that exists immediately after the deadtime expires, at the expense of causing slightly oscillatory behaviour for some time after the disturbance. The addition of time (t) in Equations (3.82) and (3.83) provides a weighting factor so that small errors existing a long time after the disturbance make a contribution to the penalty function similar in magnitude to a large error at the start of the disturbance. However, since the absolute value of the error never exceeds 1, squaring a small error gives a penalty very close to zero. This undermines the advantage of the time weighting. For these reasons ITAE is generally used. Figure 3.25 illustrates the difference.

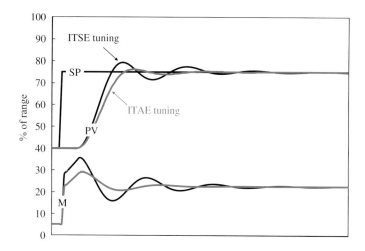

Figure 3.25 Comparison between ITAE and ITSE tuning

3.13 Manipulated Variable Overshoot

The main problem with all four of the tuning methods presented so far is that they all have the sole objective of reaching the SP as soon as possible. With the exception of some special cases, such as averaging level control described in the Chapter 4, this usually is a requirement. But it is not normally the sole requirement. Figure 3.26 shows the performance of a controller tuned to meet this aim. However, depending on the process, this might result in excessive adjustments to the MV. In our fired heater example it is unlikely that the fuel

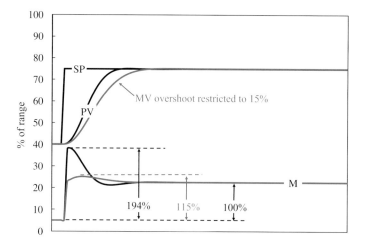

Figure 3.26 Limiting MV overshoot

could be changed as rapidly as shown without causing a problem on the combustion side. For example, it might not be possible to get sufficient air into the firebox as quickly as the rapid increase in fuel would demand. Not doing so would result in incomplete combustion and the controller increasing the fuel even further in an attempt to arrest the resulting fall in temperature. However on another process, the MV might be some minor flow routed to a section of the process that can tolerate very rapid fluctuations.

If we wish to take MV movement into account we must first define some measure of this. Here we use *MV overshoot* and define this as the percentage by which the MV exceeds the steady state change required to meet the new SP. In our example the MV must move by 17 % of range (from 5 % to 22 %) in order for the PV to achieve its SP. If not restricted, the MV temporarily reaches a maximum value of 38 % – giving a maximum change of 33 %. In this case overshoot would be calculated as 100(33/17 – 1) or 94 %. If we were to restrict this to 15 % then the maximum value reached by the MV would be 17 % multiplied by 1.15, plus the starting value of 5 % – i.e. around 25 %. Figure 3.26 shows the effect on the response of the PV if we apply this limit.

It is important we distinguish between the MV overshoot and the PV overshoot. A number of published tuning methods permit definition of the allowable PV overshoot. However this does not satisfy the need to place a defined limit on the movement of the MV. An easy check to determine whether a tuning method takes account of this is to determine what tuning constants would be derived if θ is set to zero. Each of the methods above would give the result

$$K_c \to \infty \quad T_i = 0 \quad T_d = 0 \qquad (3.84)$$

In effect each method suggests that controller gain be set to maximum, the integral action be set to maximum (remembering it uses the reciprocal of T_i) and the derivative action switched off. We might have anticipated the last of these since we have shown that derivative action is only beneficial to SP changes if there is deadtime. The values for K_c and T_i are theoretically correct for analog control. If a process truly has no deadtime (and similarly no scan delay) then increasing controller gain will not cause oscillation. In fact the tuning recommended would ensure the PV follows the SP immediately. The penalty for this type of response is excessive movement of the MV. Figure 3.27 shows just how severe this problem can be (note that the vertical scale is logarithmic!). If we were to tune, with no restriction on MV overshoot, any process with a θ/τ ratio less than 1.8 then the overshoot would exceed our nominal 15% limit. This would therefore apply to the vast majority of controllers.

3.14 Lambda Tuning Method

None of the methods so far described give the engineer any way of explicitly limiting movement of the MV. The only approach would be to start with the calculated tuning and adjust by trial-and-error. One of the approaches to address this issue is the Lambda method first introduced by Dahlin (Reference 8). This includes an additional tuning parameter (λ). This is the desired time constant of the process response to a SP change and gives the engineer the facility to make the controller more or less aggressive.

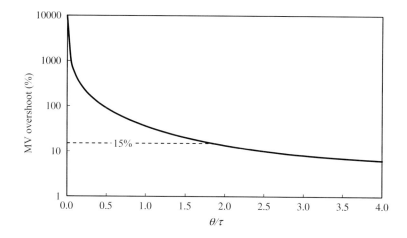

Figure 3.27 Unrestricted MV overshoot

The tuning constants are calculated for a self-regulating process from formulae developed by Chien (Reference 9) as

$$K_c = \frac{\tau}{K_p(\lambda+\theta)} \quad T_i = \tau \quad T_d = 0 \tag{3.85}$$

and for an integrating process as

$$K_c = \frac{2\lambda+\theta}{K_p(\lambda+\theta)^2} \quad T_i = 2\lambda+\theta \quad T_d = 0 \tag{3.86}$$

Figure 3.28 shows the open loop response for a step change in MV. The closed loop responses of three controllers (with λ set at 0.5τ, τ and 2τ) are also shown. With λ set to

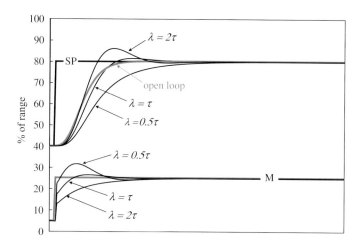

Figure 3.28 Effect of λ on controller response to SP change

the process lag the closed loop response closely follows the open loop response. If λ is increased then the controller will adjust the MV more slowly, reduce the MV overshoot and increase the time taken for the PV to reach its SP. If λ is decreased below τ, the PV will overshoot the SP.

3.15 IMC Tuning Method

Lambda tuning is an example of *internal model control (IMC)* tuning. It is developed using a technique known as *direct synthesis* (Reference 10). It can be applied to higher order processes and to all types of controllers. Some examples are presented in detail at the end of this chapter, but the principle is to synthesise a controller that will respond to a SP change according to a defined trajectory. However the result may not have the form of the PID algorithm and so approximations have to be made. For example, high order terms are neglected. But different developers delete these at different stages in the derivation. Further, if the process has deadtime, this would require the controller to take action before the SP is changed. Again different approximations are made to deal with this.

This results in slight differences between the resulting tuning formulae. Table 3.6 lists the formula commonly required, but many versions will be found in the literature.

It is possible to use the technique to quantify other parameters in the control algorithm. For example Rice and Cooper (Reference 11) developed formula for a – the term used in derivative filtering so that, on those DCS which permit the engineer to change this value, it can also be optimised.

While direct synthesis enables the engineer to decide how aggressive the control should be, he or she has to manipulate λ by trial-and-error. While there are several published techniques for selecting λ, there is no predictable relationship between its value and MV overshoot. Under a different set of process dynamics the relationship between λ and MV overshoot will change. This is illustrated in Figure 3.29. The curves were plotted by testing the tuning

Table 3.6 IMC tuning formulae

	Self-regulating			Integrating		
	K_c	T_i	T_d	K_c	T_i	T_d
PID (non-interactive)	$\dfrac{1}{K_p}\dfrac{\tau+\dfrac{\theta}{2}}{\lambda+\theta}$	$\tau+\dfrac{\theta}{2}$	$\dfrac{\tau\theta}{2\tau+\theta}$	$\dfrac{1}{K_p}\dfrac{2\lambda+\theta}{\left(\lambda+\dfrac{\theta}{2}\right)^2}$	$2\lambda+\theta$	$\dfrac{\lambda\theta+\dfrac{\theta^2}{4}}{2\lambda+\theta}$
PID (interactive)	$\dfrac{1}{K_p}\dfrac{\tau}{\lambda+\theta}$	τ	$\dfrac{\theta}{2}$	$\dfrac{1}{K_p}\dfrac{2\lambda+\dfrac{\theta}{2}}{\left(\lambda+\dfrac{\theta}{2}\right)^2}$	$2\lambda+\dfrac{\theta}{2}$	$\dfrac{\theta}{2}$
PI	$\dfrac{1}{K_p}\dfrac{\tau}{\lambda+\theta}$	τ		$\dfrac{1}{K_p}\dfrac{2\lambda+\theta}{(\lambda+\theta)^2}$	$2\lambda+\theta$	

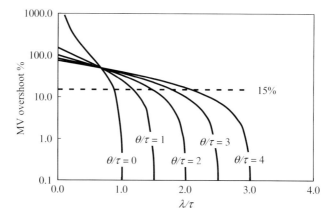

Figure 3.29 Effect of λ on MV overshoot

given in Table 3.6 for the noninteractive PID controller on a self-regulating process. They show that the value of λ required to give a required MV overshoot (say 15 %) varies as the θ/τ ratio varies.

Of course it is possible from this chart to construct another allowing the engineer to choose a value for λ to give the required MV overshoot. Indeed this has been done and the result shown as Figure 3.30.

And, from this chart, simple formulae could be developed. For example for a 15 % MV overshoot.

$$\lambda = 0.31\theta + 0.88\tau \tag{3.87}$$

However for this approach to be adopted, such charts and formulae would have to be developed for every version of the PID algorithm, for every variation of the tuning formulae,

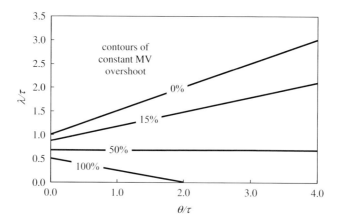

Figure 3.30 Value of λ necessary to give required MV overshoot

for both integrating and self-regulating processes and for both SP and load changes. The number of charts required would be impractically large. Further the formulae in Table 3.6, for self-regulating processes, shows that only K_c changes as λ is changed. We will show later that, to retain optimum tuning, T_i and T_d should also be changed. But the most serious limitation of the method is that, to date, no one has published tuning methods for the preferred proportional-on-PV, derivative-on-PV algorithm.

3.16 Choice of Tuning Method

All the tuning methods we have covered assume analog control. This is reasonable if the process dynamics are long compared to the scan interval. However there will be occasions where this is not the case. Compressors, for example, can show vary fast dynamics particularly with surge avoidance controls. Indeed compressor manufacturers will often specify the use of analog or fast scanning digital controllers. However, we will show later that it is possible to tune DCS-based controllers for this application, provided the tuning method takes account of the controller scan interval. Similarly discontinuous instruments, such as on-stream analysers, generate measurements relatively infrequently. If a controller using this measurement only takes action when there is a new value then the scan interval will again be large compared to the process dynamics. To determine whether a proposed tuning method is designed for discrete control we need only check whether it uses the scan interval in its calculations.

So, to summarise, we should be looking for the following features in a tuning method.

- It is designed for the control algorithm. Our preferred algorithm is the noninteractive, proportional-on-PV, derivative-on-PV version. The method must also be suited to any DCS-specific features in the algorithm, particularly if these cannot be disabled. For instance the derivative filter term (a) is not adjustable in many DCS and should therefore be taken account of by the tuning technique.
- It permits the engineer to limit the MV overshoot explicitly when required. Methods that do not take this into account will suggest very aggressive tuning as the θ/τ ratio falls. By checking what tuning results for a zero deadtime process we can determine whether MV movement is taken into account. The engineer-defined performance criterion must be a consistent measure of MV movement, no matter what the process dynamics.
- It is designed for digital rather than analog control. If the scan interval is used in the tuning calculations then this is likely to be the case. But this only becomes an issue if the process dynamics are very fast and approach the scan interval.

It is unlikely that a formula-based approach to tuning is practical. The different versions of the PID controller are not mutually exclusive. We have described three options for the error term included in the integral action. There are the noninteractive and interactive versions. Derivative action can be based on error or PV. Proportional action can be based on error or PV. This already gives 24 possible combinations without all the DCS-specific enhancements. We have two types of process, self-regulating and integrating, and we have not yet considered higher order processes such as those with inverse response. We have to accommodate derivative filtering, the choice of tuning criteria (ITAE etc., MV overshoot), the controller scan interval and whether we want tuning for SP or load changes.

We would thus need several hundred sets of tuning formulae to cover all the options. It is not surprising therefore that no published technique has yet proved effective for all situations.

Instead we should learn from the engineers that have largely adopted trial-and-error as their preferred technique. But instead of the time-consuming exercise of performing this on the real process, we can simulate it. This is provided we have, from plant testing, an understanding of the process dynamics. We could, of course, use one of the many commercial computer based tuning products available. However care should be taken in selecting one of these. Few would meet the criteria listed above. Alternatively we can develop a simulation of our process and the controller either in code or in a spreadsheet, define the tuning criteria and have the computer optimise tuning to meet these criteria. It is this approach that has been used to generate the following figures.

Of course these figures can lead the engineer into the same trap. They again offer a set of standard approaches to cover all situations. However the approaches embody all the recommendations developed in this chapter and, where practical, they leave the engineer some flexibility in their application. Used with care, almost any controller can be optimally tuned.

3.17 Suggested Tuning Method for Self-Regulating Processes

Figures 3.31 to 3.33 give the recommended tuning for the preferred algorithm (noninteractive, proportional-on-PV, integral-on-E_n, derivative-on-PV and no derivative filtering). It is assumed that the scan interval is small compared to the process dynamics. The tuning is designed to minimise ITAE subject to a maximum MV overshoot of 15 % on a self-regulating process.

However, the conditions under which these charts were developed may not apply. The following examples show the impact on the charts of:

- changing to tuning for load changes
- relaxing the MV overshoot constraint
- using only PI control
- changing the scan interval.

3.18 Tuning for Load Changes

Figures 3.34 to 3.36 show the difference between tuning for load and SP changes. We have seen that tuning for load changes can be faster than that for SP changes because the error changes more slowly. The problem with deriving tuning by simulation is that we have to make an assumption about the process dynamics of the PV with respect to the source of the process disturbance. In the absence of any better information, we assume that they are the same as those with respect to the MV. In our example heater, this is the same as saying that the dynamic relationship between outlet temperature and feed rate is the same as that between the outlet temperature and fuel rate. This is unlikely to be the case. Further, different sources of process disturbance are likely to have different dynamics. So any load tuning method must be used with caution.

PID Algorithm 67

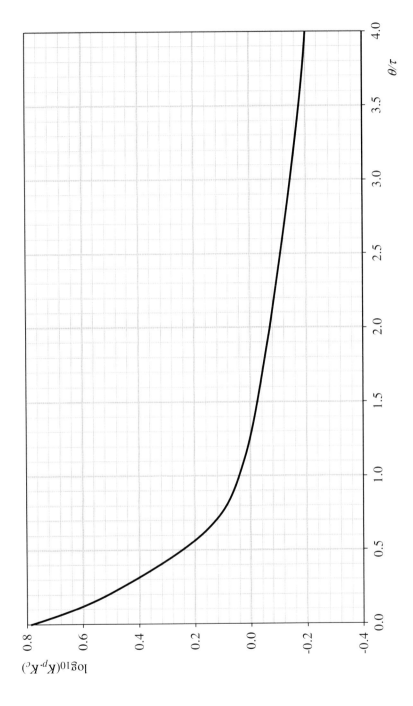

Figure 3.31 Controller gain

68 Process Control

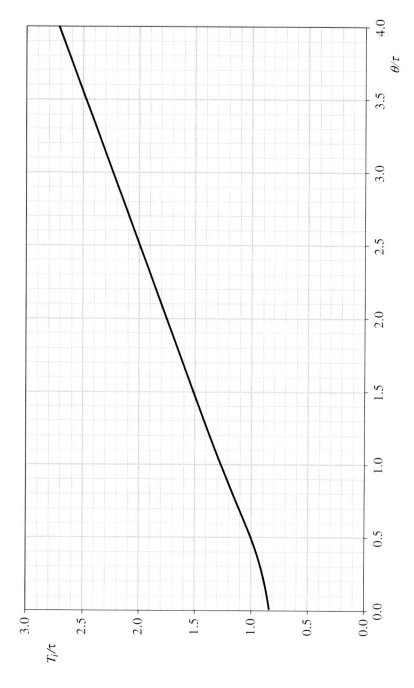

Figure 3.32 Integral time

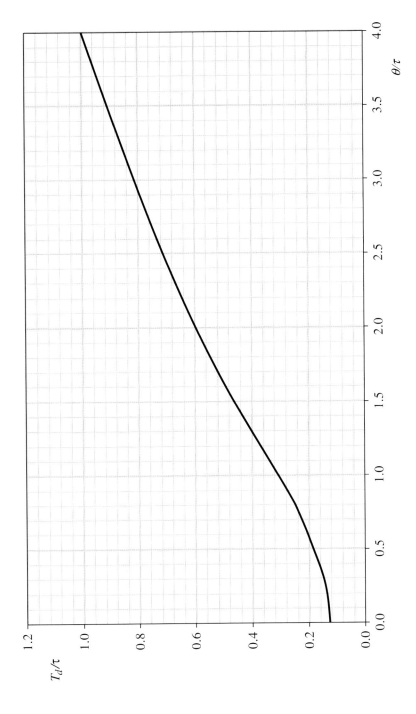

Figure 3.33 Derivative time

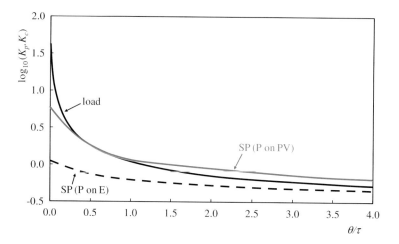

Figure 3.34   SP versus load tuning (controller gain)

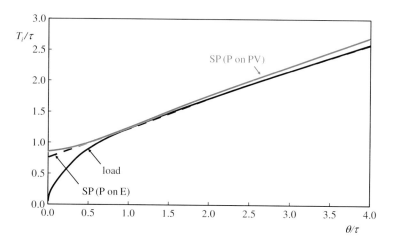

Figure 3.35   SP versus load tuning (integral time)

We have already seen that, if we tune for SP changes, for load changes the proportional-on-PV algorithm outperforms the proportional-on-error algorithm. Figure 3.34 suggests that, for processes with a θ/τ ratio of less than about 0.4, we can tune the controller to be even more aggressive with load changes. However it will then over-react to SP changes. For larger θ/τ ratios the tuning for both disturbances is similar.

Figure 3.35 shows that we should take the same approach to integral action remembering that, as we increase T_i, integral action is reduced. While Figure 3.36 suggests the same argument does not apply to derivative action but there is only a small difference in T_d for the two cases.

Overall then we should always tune for SP changes.

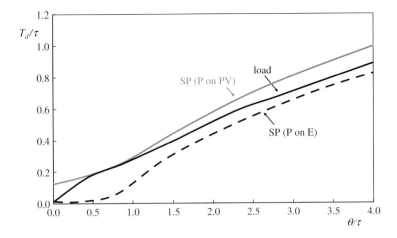

Figure 3.36 SP versus load tuning (derivative time)

3.19 Tuning for Unconstrained MV Overshoot

Figures 3.37 to 3.39 give the tuning for the same controller but this time showing the effect of removing the constraint on MV overshoot. The result is that as the θ/τ ratio approaches zero the tuning is the same as that given by many of the published methods, as shown in Equation (3.84).

While constraining MV movement will clearly slow down the response to disturbances, the impact can be very small. Figure 3.40 shows the impact on ITAE. The chart has been scaled so 100 % ITAE corresponds to the open loop response, i.e. no corrective action is taken over the period in which the process would normally reach steady state ($\theta + 5\tau$). Even when the θ/τ is close to zero, the impact on ITAE is small.

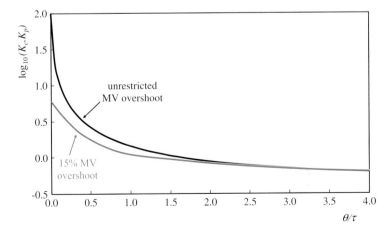

Figure 3.37 Effect of MV overshoot constraint on controller gain

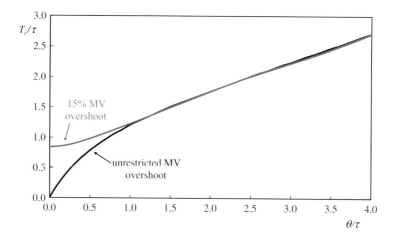

Figure 3.38 Effect of MV overshoot constraint on integral time

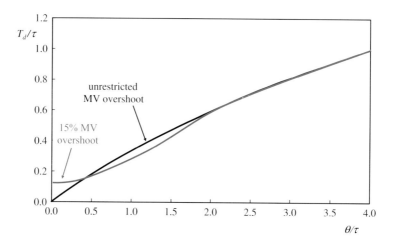

Figure 3.39 Effect of MV overshoot constraint on derivative time

3.20 PI Tuning Compared to PID Tuning

Examination of all the methods described in this chapter shows that different tuning is required for the proportional and integral actions if derivative action is added to an optimally tuned PI controller. Figures 3.41 and 3.42 show the effect of switching from PID to PI. They give the tuning for a SP change using the preferred algorithm (proportional-on-PV, integral on E_n). Since there is no derivative action they apply to both the interactive and noninteractive versions and are not affected by derivative filtering. The tuning is designed to minimise ITAE subject to a maximum MV overshoot of 15 % on a self-regulating process.

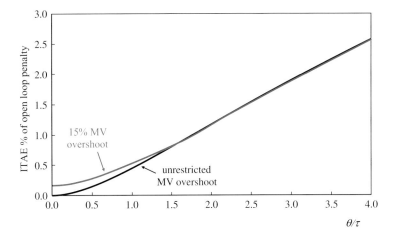

Figure 3.40 Effect of MV overshoot constraint on ITAE

Figure 3.41 shows that the controller gain can be significantly increased if derivative action is included. Derivative action helps reduce MV overshoot and, since we have restricted this to 15 %, it remains advantageous to include it, even if the θ/τ ratio is small. Figure 3.43 shows the impact on ITAE of switching from well-tuned PID to well-tuned PI control. For low θ/τ ratios the impact on SP changes is relatively minor. However the effect on load disturbances is substantial, particularly if θ/τ is less than 0.5. So Figures 3.41 and 3.42 should only be used to derive tuning where derivative **cannot** be included (e.g. because of excessive noise that cannot be removed by filtering).

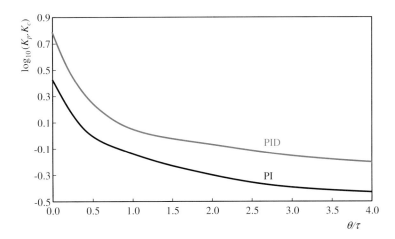

Figure 3.41 PI versus PID tuning (controller gain)

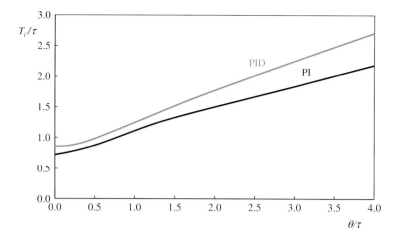

Figure 3.42 PI versus PID tuning (integral time)

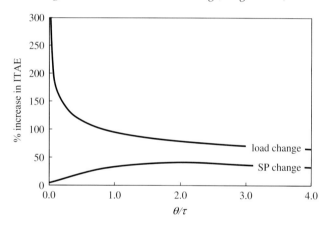

Figure 3.43 Effect of not using derivative action

3.21 Tuning for Large Scan Interval

In most cases, the scan interval (*ts*) will be small compared to the process lag (τ) and the controller can be treated as analog. But Figures 3.44 to 3.46 show that the controller can be tuned successfully even when the scan interval approaches the lag. As a modification to some analog controller tuning methods, the developer has suggested replacing θ in the tuning formulae with $\theta + ts/2$. This is on the basis that digital control will, on average, increase the deadtime by half the scan interval. However, if this were a good approximation, we would expect the curves in Figure 3.44 to be horizontally spaced by a distance of 0.25. Actually the spacing is much larger than this value, showing that the estimate of controller gain is very sensitive to scan interval.

Figure 3.47 shows the impact of increasing the scan interval from zero (analog control) to a value equal to double the process lag (2τ) and retuning the controller to take account of the

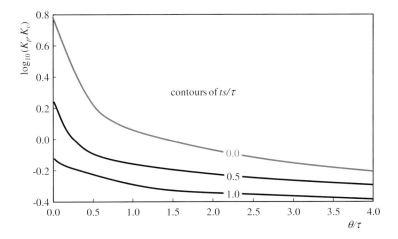

Figure 3.44 Effect of scan interval on controller gain

change. The delay to the PV reaching SP would probably not be noticed on a real process, although perhaps the steps now made by the MV would be. What is important is that the controller still behaves well. It is a misconception that increasing scan interval adversely affects controllability.

We are not suggesting here that scan intervals can be dramatically increased, for example to alleviate the processing load on a DCS. It is still important that the delay in **detecting** a SP or load change is not excessive. While we could readily control a process with a lag of (say) 5 minutes, using a controller with a scan interval of 5 minutes, we would not want the controller to take no action for those 5 minutes if a disturbance happens to occur immediately after a scan. What we are suggesting is that a controller scanning, say, every second or two is capable of controlling processes with dynamics of the same order of magnitude.

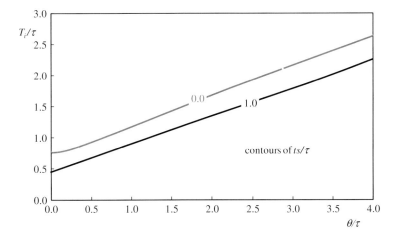

Figure 3.45 Effect of scan interval on integral time

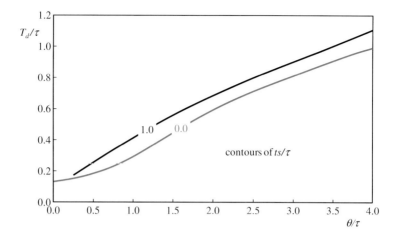

Figure 3.46 Effect of scan interval on derivative time

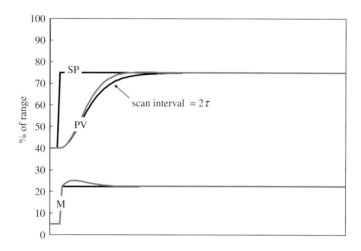

Figure 3.47 Impact of large scan interval

3.22 Suggested Tuning Method for Integrating Processes

We could include a similar set of charts for tuning controllers on integrating processes. However, since they are all straight line relationships we can more easily represent them as formulae. They assume that K_p has units of min^{-1}, θ has units of minutes and ts has units of seconds. K_c will then be dimensionless; T_i and T_d will have the units of minutes.

While Table 3.7 gives tuning for virtually all possible situations the preferred choice of algorithm and disturbance is listed last.

Table 3.7 Tuning for integrating process

Algorithm	Interactive or Non-interactive	Proportional based on	Derivative based on	Disturbance	$K_c = \dfrac{1}{K_p}\left[\dfrac{60A}{\theta + B\frac{ts}{60}}\right]$		$T_i = A\left[\theta + B\frac{ts}{60}\right]$		$T_d = A\left[\theta + B\frac{ts}{60}\right]$	
					A	B	A	B	A	B
P				SP	0.00873	0.697				
PI	I			load	0.00768	0.724	15.5	0.675		
PI	I	E		SP	0.01185	0.854	4.53	0.578		
PI	I	PV		SP	0.01208	0.846	3.38	0.543		
PID	I			load	0.00988	1.729	6.56	0.696	0.796	1.443
PID	I	E	E	SP	0.01503	1.243	2.23	0.867	0.463	0.366
PID	I	E	PV	SP	0.01518	1.392	1.96	0.660	0.516	0.468
PID	I	PV	E	SP	0.01580	1.288	2.06	0.794	0.509	0.339
PID	I	PV	PV	SP	0.01620	1.307	1.95	0.784	0.501	0.347
PID	N			load	0.01470	3.422	7.55	0.761	0.351	5.180
PID	N	E	E	SP	0.01903	1.496	2.51	0.832	0.374	0.274
PID	N	E	PV	SP	0.01947	1.523	2.53	0.784	0.365	0.267
PID	N	PV	E	SP	0.02530	2.078	1.34	0.946	0.305	1.153
PID	N	PV	PV	SP	0.02153	1.587	2.31	0.785	0.387	0.317

3.23 Implementation of Tuning

Caution should be exercised in implementing the tuning constants. If the tuning has been calculated for an **existing** controller then comparison with the current tuning might show large changes in the tuning constants. If the algorithm has been changed from the proportional-on-error to the proportional-on-PV type then a large increase in K_c is to be expected. It is not unusual for this to increase by a factor of two or three. A change larger than this should be implemented stepwise, testing with intermediate values before moving to those calculated. Similarly, if an existing tightly tuned level controller is re-engineered as an averaging controller, as we shall see in Chapter 4 the change in tuning can be one or two orders of magnitude slower. However, if there has been no change in algorithm and no change in tuning objective then changes in K_c should be restricted to around 20 % of the current value. Changes of around 50 % may be made to T_i and T_d. The controller is tested with a SP change following each incremental change to the tuning and, provided it exhibits no problems, the next increment made.

This testing of the controller tuning presents an opportunity to re-identify the process model. Provided a computer-based model identification technique is applied then SP changes made to validate the tuning can also be analysed to determine the process dynamics. Since the data are likely to be collected routinely by the process information system, this re-evaluation takes little additional effort.

Following the tuning method presented here should obviate the need for tuning by trial-and-error. However the method does assume that the model dynamics have been determined accurately and that they are close to first order. This chapter would not be complete therefore without offering some guidance in this area. Controller gain affects all three P, I and D actions and should therefore be adjusted first. Steps of 20 % are reasonable until the optimum value is approached, when smaller changes can be made. Adjustments to integral action can be made initially in much larger steps – either halving or doubling the action. If slightly oscillatory then controller gain may be reduced. Derivative action can be similarly adjusted. All three constants can then be fine-tuned to give optimum performance. In doing so the controller may show *kickback*. This is illustrated in Figure 3.48; the PV turns **before** reaching SP. This indicates that controller gain is too high and integral action is insufficient.

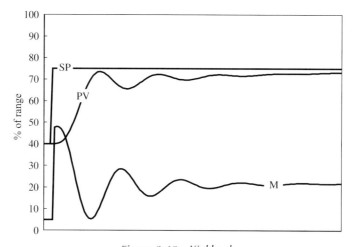

Figure 3.48 Kickback

3.24 Loop Gain

In addition to process gain (K_p) and controller gain (K_c), a term often used is the *loop gain* (K_l). The loop gain is obtained by multiplying all the gain terms in the control loop. In the case of a simple PID controller, the loop gain is given by

$$K_l = K_p.K_c \qquad (3.88)$$

We have seen in all the tuning methods that the product $K_p.K_c$ should be constant. Once we have established what the controller gain should be, we will need to change the value if there is any change which affects the loop gain. For example if the instrument range of either the PV or MV is changed, K_c will need adjustment. From Equations (2.2) to (2.4) we define process gain as

$$K_p = \frac{\left(\frac{\Delta PV}{PV\ range}\right)}{\left(\frac{\Delta MV}{MV\ range}\right)} \qquad (3.89)$$

So if we change the range of the PV or the MV then, to keep the loop gain constant, the controller gain should be recalculated as

$$(K_c)_{new} = (K_c)_{old} \times \frac{(PV\ range)_{new}}{(PV\ range)_{old}} \times \frac{(MV\ range)_{old}}{(MV\ range)_{new}} \qquad (3.90)$$

The same correction would be necessary if we change the MV of the controller, for example changing a primary controller cascaded to a flow controller so that it instead directly manipulates the control valve. This is often a 'quick fix' if the secondary flow transmitter has a problem. We will show in Chapter 6 that adding a ratio-based feedforward can also require recalculation of K_c again because the effective range of the MV may be changed.

3.25 Adaptive Tuning

Adaptive tuning, as the name suggests, automatically changes controller tuning constants as necessary to accommodate changes in process dynamics. One example is *gain scheduling* which changes the gain of the controller as the process gain changes. This may exist as a standard feature within the DCS or may require some custom coding by the engineer. The engineer may define some relationship between controller gain and process conditions. This may be a table of values to be used as circumstances change or it may be some continuous function.

For example, one method of dealing with the highly nonlinear problem of pH control is to split the titration curve shown in Figure 3.49 into several linear sections. As the pH measurement moves between sections the controller would be configured to use a different process gain.

We showed in Chapter 2 that process gain (K_p) for most processes is inversely proportional to feed rate. To keep the loop gain constant the controller gain (K_c) could be scheduled to vary in proportion to feed rate – although we will show in Chapter 6 that there is a more elegant solution to this problem.

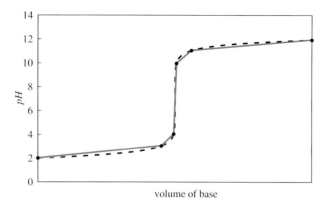

Figure 3.49 Gain scheduling for pH

In some cases it is possible to infer the process gain from process measurements. The controller is tuned for a known process gain. This establishes the value for the loop gain. As the process gain varies the controller gain is automatically adjusted to maintain the loop gain constant.

There are a range of *self-tuners* on the market which attempt to perform online model identification and retune the controller as the process dynamics change. However these can apply tuning methods that do not meet the key criteria that we have identified. Further they should not be seen as a replacement for developing a sound understanding of why and how the process dynamics are changing. With this knowledge it is likely that a more rigorous solution may be engineered.

3.26 Initialisation

Initialisation is the process that takes place when a controller is switched from manual to automatic mode. Its purpose is to ensure that the process is not disturbed by a sudden change in controller output. We first touched on this subject when converting the full position version of the control algorithm to its velocity form. On initialisation the output of a full position controller must be matched to the current value of the MV. On older controllers this exercise was completed manually. However, with the velocity form, we have to ensure only that the incremental change made by the controller is zero. This is achieved by *PV tracking*. When in manual mode the SP is maintained equal to the PV so that, when the mode is changed to automatic, the previous and current errors are zero. Thus the controller output will be zero. Once in automatic mode the SP stops tracking the PV; the controller will respond to any process disturbance and the operator may change the SP as required.

This can occasionally cause problems when switching controllers to automatic. It is advisable when configuring a controller in a DCS to place upper and lower limits on the SP to prevent the operator accidently entering a value that might otherwise cause an operating problem. With PV tracking it is possible for the SP to move outside this acceptable range and some DCS then prevent the controller being switched to automatic. The operator will first need to manually adjust the process until the PV, and hence the SP, moves into range.

Initialisation is not unique to PID controllers. Other algorithms such as ratios and biases also require the equivalent of PV tracking. We cover this in more detail in Chapter 6.

3.27 Anti-Reset Windup

Control valves require *calibration* to convert the signal from the controller into a valve position. This calibration may be required in the valve positioner located in the field or in the DCS. The output range, coinciding to 0–100 % valve position, may be 3–15 psi (in pneumatic systems) or (in electronic systems) 10–50 mA, 4–20 mA or 1–5 V. The actual output is usually permitted to move outside this range. Thus a valve not perfectly calibrated can still be driven fully shut or fully open. Similarly, if the valve is prone to stiction or hysteresis, it also overcomes the mismatch between position and signal that would otherwise prevent the valve from reaching its fully open or fully closed position. This is the main reason why the ranges do not start at zero. Further it creates a distinction between a zero signal and loss of signal. Thus the controller output might vary from −25 % to 125 %, corresponding to a pneumatic signal of 0–18 psi or an electronic signal of 0–60 mA, 0–24 mA or 0–6 V.

There will be occasions when the controller *saturates*. For example a flow controller may encounter a hydraulic limit so that, even with valve fully open, the SP cannot be reached. The integral (or reset) action will respond to this by continually increasing the output but, because the valve is fully open, will have no affect on the flow. This is *reset windup*. Windup should be avoided because, if the process constraint is removed – for example by starting a booster pump, there will be a delay while the controller removes the windup and can begin actually closing the valve. This is resolved by keeping the permitted output range as narrow as possible, typically −5 to 105 %.

However the situation becomes more complex with cascaded controllers. The situation can arise where the secondary is controlling at SP but with its output at minimum or maximum. It is important therefore that the primary makes no changes to the secondary's SP which will cause it to saturate. DCS controllers have *external anti-reset windup protection*, sometimes described just as *external reset feedback*, to prevent this.

A similar technique is required with signal selectors. We cover these in more detail in Chapter 8, but a common use is to have two or more controllers outputting to a low or high signal selector. While one signal will pass through, the other(s) could potentially wind up. There must be logic in the selector that stops the deselected controller(s) from increasing their outputs (if routed to a low signal selector) or decreasing their outputs (if routed to a high signal selector)

3.28 On-Off Control

Before completing this section, it is right that we briefly examine the use of *on-off control* – also called *bang-bang control*. While primarily used for temperature control in domestic systems (such as refrigerators, ovens, home heating etc.) it does have some limited applications in the process industry. The technique is, in a heating application, to switch on the source of energy when the temperature is low and switch it off when the temperature is high.

Although the controller has a SP, there must be a *deadband* around this value within which no control action takes place. Without this the MV would be switched on and off at an unsustainable frequency. In domestic situations this deadband occurs almost accidentally as a result of the mechanism involved. Temperature is generally measured using a bimetallic strip which, as it bends, makes or breaks a contact. The distance it has to move between contacts provides the small deadband necessary. If this were not the case, then a deadband would need to be deliberately designed into the controller.

Figures 3.50 and 3.51 demonstrate the point. As the deadband is reduced, the frequency of MV switching increases. Reducing it to zero would increase the frequency to the maximum the mechanics would allow – almost certainly soon causing damage to the actuator.

In Figure 3.52 a small deadtime has been added to the process. Thus when the temperature reaches the high limit, even though the source of energy is switched off, the temperature continues to rise until the deadtime has elapsed. Similarly the temperature will fall below the low limit. Despite retaining the narrow deadband, the temperature deviates further from SP.

On-off control is thus only applicable on industrial processes where tight control is not necessary and where deadtime is negligible. This restricts its use primarily to some level controllers. Typically it would be implemented using high and low level limit switches that would activate a solenoid valve. It can be emulated with a high gain proportional-only controller with a deadband, but care needs to be taken that excessive control action does not damage the control valve.

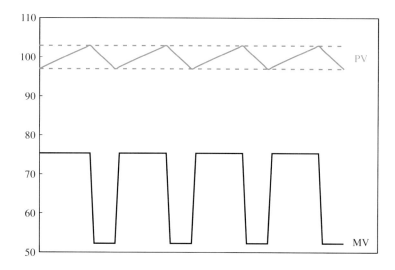

Figure 3.50 On-off control with wide deadband

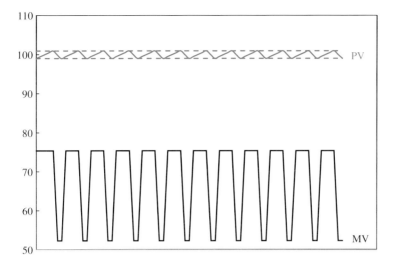

Figure 3.51   On-off control with narrow deadband

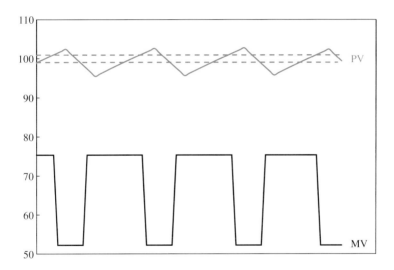

Figure 3.52   On-off control on process with deadtime

3.29   Laplace Transforms for Controllers

While for the most part we have been able to avoid using Laplace transforms to describe controllers, there are many text books that do not. Many of the DCS vendors document their system in this way – which is incorrect since they should only be used for analog controllers. However so that the engineers can recognise them, the transforms for the common types of controller are listed here – along with their time domain equivalents.

a. Noninteractive PID

$$M = K_c\left(E + \frac{1}{T_i}\int E.dt + T_d\frac{dE}{dt}\right) \quad (3.91)$$

$$M = K_c\left[1 + \frac{1}{T_i s} + T_d s\right]E \quad (3.92)$$

b. Noninteractive PID with derivative gain limit

$$M = K_c\left[1 + \frac{1}{T_i s} + \frac{T_d s}{1 + aT_d s}\right]E \quad (3.93)$$

c. Noninteractive PID (with derivative-on-PV)

$$M = K_c\left(E + \frac{1}{T_i}\int E.dt + T_d\frac{dPV}{dt}\right) \quad (3.94)$$

$$M = K_c\left(\left[1 + \frac{1}{T_i s}\right]E + T_d s PV\right) \quad (3.95)$$

$$M = K_c\left(\left[1 + \frac{1}{T_i s} + T_d s\right]PV - \left[1 + \frac{1}{T_i s}\right]SP\right) \quad (3.96)$$

d. Noninteractive PID (with proportional-on-PV and derivative-on-PV)

$$M = K_c\left(PV + \frac{1}{T_i}\int E.dt + T_d\frac{dPV}{dt}\right) \quad (3.97)$$

$$M = K_c\left(\frac{1}{T_i s}E + [1 + T_d s]PV\right) \quad (3.98)$$

$$M = K_c\left(\left[1 + \frac{1}{T_i s} + T_d s\right]PV - \left[\frac{1}{T_i s}\right]SP\right) \quad (3.99)$$

e. Interactive PID

$$M = K_c\left(\left(1 + \frac{T_d}{T_i}\right)E + \frac{1}{T_i}\int E.dt + T_d\frac{dE}{dt}\right) \quad (3.100)$$

$$M = K_c\left[1 + \frac{1}{T_i s}\right][1 + T_d s]E \quad (3.101)$$

f. Interactive PID (with derivative-on-PV)

$$M = K_c\left(E + \frac{T_d}{T_i}PV + \frac{1}{T_i}\int E.dt + T_d\frac{dPV}{dt}\right) \quad (3.102)$$

$$M = K_c \left(\left[1 + \frac{1}{T_i s} \right] E + \left[\frac{T_d}{T_s} + T_d s \right] PV \right) \quad (3.103)$$

$$M = K_c \left(\left[1 + \frac{1}{T_i s} \right] [1 + T_d s] PV - \left[1 + \frac{1}{T_i s} \right] SP \right) \quad (3.104)$$

g. Interactive PID (with proportional-on-PV and derivative-on-PV)

$$M = K_c \left(\left(1 + \frac{T_d}{T_i} \right) PV + \frac{1}{T_i} \int E.dt + T_d \frac{dPV}{dt} \right) \quad (3.105)$$

$$M = K_c \left(\frac{1}{T_i s} E + \left[1 + \frac{T_d}{T_i} + T_d s \right] PV \right) \quad (3.106)$$

$$M = K_c \left(\left[1 + \frac{1}{T_i s} \right] [1 + T_d s] PV - \frac{1}{T_i s} SP \right) \quad (3.107)$$

h. Interactive PID with derivative gain limit

$$M = K_c \left[1 + \frac{1}{T_i s} \right] \left[\frac{1 + T_d s}{1 + a T_d s} \right] E \quad (3.108)$$

i. Interactive PID with derivative gain limit (with derivative-on-PV)

$$M = K_c \left(\left[1 + \frac{1}{T_i s} \right] \left[\frac{1 + T_d s}{1 + a T_d s} \right] PV - \left[1 + \frac{1}{T_i s} \right] SP \right) \quad (3.109)$$

j. Interactive PID with derivative gain limit (with proportional-on-PV and derivative-on-PV)

$$M = K_c \left(\left[1 + \frac{1}{T_i s} \right] \left[\frac{1 + T_d s}{1 + a T_d s} \right] PV - \frac{1}{T_i s} SP \right) \quad (3.110)$$

k. integral only

$$M = \frac{1}{T_i s} E \quad (3.111)$$

3.30 Direct Synthesis

Whether this should be included in a book of this type is debatable – particularly for a tuning method which has a number of limitations. But the reader can easily skip this section. However, others, not daunted by the mathematics, might find it of value in linking IMC tuning techniques to those published elsewhere.

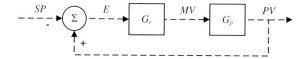

Figure 3.53 Control loop

Figure 3.53 shows the control loop, where G_c is the transfer function of the controller and G_p is the transfer function for the process.

PV is related to E by the combination of the controller and process transfer functions

$$PV = G_c G_p E \tag{3.112}$$

But, by definition

$$E = PV - SP \tag{3.113}$$

Substituting this in Equation (3.112) gives

$$PV = \frac{G_c G_p}{G_c G_p - 1} SP \tag{3.114}$$

This describes how the PV responds to changes in SP. But we require this trajectory to be a first order response with a lag of λ, where the value of λ is selected by the engineer. Therefore

$$\frac{G_c G_p}{G_c G_p - 1} = \frac{1}{1 + \lambda s} \quad \text{or} \quad G_c = \frac{-1}{G_p \lambda s} \tag{3.115}$$

Let us assume that we have a simple first order process with no deadtime, then

$$G_p = \frac{K_p}{1 + \tau s} \tag{3.116}$$

Substituting into Equation (3.115)

$$G_c = \frac{-(1 + \tau s)}{K_p \lambda s} = \frac{-\tau}{K_p \lambda}\left[1 + \frac{1}{\tau s}\right] \tag{3.117}$$

Comparing this to Equation (3.92), this is a PID controller with

$$K_c = \frac{-\tau}{K_p \lambda} \quad T_i = \tau \quad T_d = 0 \tag{3.118}$$

K_c has an opposite sign to K_p because, according to our definitions, a reverse acting controller is required if the process gain is positive.

The result of this method produced a transfer function identical to that of a PID algorithm. Rarely does this occur with other process models. For example, if we introduce deadtime into the process, then the target trajectory becomes

$$\frac{G_c G_p}{G_c G_p - 1} = \frac{e^{-\theta s}}{1 + \lambda s} \quad \text{or} \quad G_c = \frac{-e^{-\theta s}}{G_p(1 + \lambda s - e^{-\theta s})} \qquad (3.119)$$

And the process transfer function becomes

$$G_p = \frac{K_p e^{-\theta s}}{1 + \tau s} \qquad (3.120)$$

Substituting into Equation (3.119))

$$G_c = \frac{-(1 + \tau s)}{K_p(1 + \lambda s - e^{-\theta s})} \qquad (3.121)$$

The presence of $e^{-\theta s}$ means that the result is not a PID controller. We can resolve this by making the first order Taylor approximation

$$e^{-\theta s} = 1 - \theta s \qquad (3.122)$$

Substituting into Equation (3.121)

$$G_c = \frac{-(1 + \tau s)}{K_p(\lambda + \theta)s} = \frac{-\tau}{K_p(\lambda + \theta)}\left[1 + \frac{1}{\tau s}\right] \qquad (3.123)$$

which is a PID controller with

$$K_c = \frac{-\tau}{K_p(\lambda + \theta)} \quad T_i = \tau \quad T_d = 0 \qquad (3.124)$$

An alternative approach is to make the first order Padé approximation

$$e^{-\theta s} = \frac{2 - \theta s}{2 + \theta s} \qquad (3.125)$$

Substituting into Equation (3.121)

$$G_c = \frac{-(1 + \tau s)(2 + \theta s)}{K_p[(1 + \lambda s)(2 + \theta s) - (2 - \theta s)]} = \frac{-[2 + (2\tau + \theta)s + \theta \tau s^2]}{K_p[(2\lambda + 2\theta)s + \theta \lambda s^2]} \qquad (3.126)$$

Since they cannot be part of a PID algorithm, we neglect high order terms of s^2 and above.

$$G_c = \frac{-2 - (2\tau + \theta)s}{K_p(2\lambda + 2\theta)s} = \frac{-(\tau + \theta/2)}{K_p(\lambda + \theta)}\left[1 + \frac{1}{(\tau + \theta/2)s}\right] \qquad (3.127)$$

which is a PID controller with

$$K_c = \frac{-(\tau + \theta/2)}{K_p(\lambda + \theta)} \quad T_i = \tau + \theta/2 \quad T_d = 0 \qquad (3.128)$$

A third approach is to ignore the high order terms only in the denominator of Equation (3.126).

$$G_c = \frac{-[2 + (2\tau + \theta)s + \theta\tau s^2]}{K_p(2\lambda + 2\theta)s} = \frac{-(\tau + \theta/2)}{K_p(\lambda + \theta)}\left[1 + \frac{1}{(\tau + \theta/2)s} + \frac{\theta\tau s}{2\tau + \theta}\right] \quad (3.129)$$

which is a PID controller with

$$K_c = \frac{-(\tau + \theta/2)}{K_p(\lambda + \theta)} \quad T_i = \tau + \theta/2 \quad T_d = \frac{\theta\tau}{2\tau + \theta} \quad (3.130)$$

In summary, if we compare this result with Equations (3.124) and (3.128), we get slightly different tuning formulae – depending on how the approximations are made.

The technique may be applied to any process. For example an integrating process with deadtime is described by

$$G_p = \frac{K_p e^{-\theta s}}{s} \quad (3.131)$$

Substituting into Equation (3.119)

$$G_c = \frac{-s}{K_p(1 + \lambda s - e^{-\theta s})} \quad (3.132)$$

Applying the first order Taylor approximation to $e^{-\theta s}$ gives

$$G_c = \frac{-1}{K_p(\lambda + \theta)} \quad (3.133)$$

This is proportional only controller with

$$K_c = \frac{-1}{K_p(\lambda + \theta)} \quad (3.134)$$

References

1. Process Instrumentation Terminology. The Instrumentation, Systems and Automation Society, ISA-51.1-1979 (R1993).
2. O'Dwyer, A. (2000) A summary of PI and PID controller tuning rules for processes with time delay. IFAC Digital Control: Past, Present and Future of PID Control, Terrassa, Spain.
3. Ziegler, J.G. and Nichols, N.B. (1942) Optimum settings for automatic controllers. *Transactions of the ASME*, **64**, 759–768.
4. Lopez, A.M., Miller, J.A., Smith, C.L. and Murrill, P.W. (1967) Tuning controllers with error-integral criteria. *Instrumentation Technology*, **14**, 57–62.
5. Astrom, K. and Hagglund, T. (1984) Automatic tuning of simple regulators with specifications on phase and amplitude margins. *Automatica*, **20**, 645–651.
6. Cohen, G.H. and Coon, G.A. (1953) Theoretical considerations of retarded control. *Transactions of the ASME*, **75**, 827–834.
7. Smith, C.L. (1972) *Digital Computer Process Control*, Intext Educational Publishers, p. 176.

8. Dahlin, E.B. (1968) Designing and tuning digital controllers. *Instruments and Control Systems*, **2**(6), 77–83.
9. Chien, I.-L. (1988) IMC-PID controller design – an extension. Proceedings of the IFAC Adaptive Control of Chemical Processes Conference, *Denmark*, 147–152.
10. Chien, I.-L. and Fruehauf., P.S. (1990) Consider IMC tuning to improve controller performance. *Chemical Engineering Progress*, **86**, 33–41.
11. Rice, R. and Cooper, D.J. (2002) Design and tuning of PID controllers for integrating (non-self regulating) processes. Procedures of the ISA 2002 Annual Meeting.

4
Level Control

So why do we dedicate a chapter to level control? What makes it so different from controlling other key process parameters such as flow, pressure and temperature? There are several reasons.

- The process behaviour is different. It is the most common example of a non-self-regulating (or integrating) process. It will not, after a change is made to the manipulated flow, reach a new equilibrium. The level will continue moving until either the process operator or a trip system intervenes. This affects the way that we execute plant tests and the way that we analyse the results.
- We may wish to apply very different tuning criteria. It may be more important to minimise disturbances to the manipulated flow than it is to maintain the level close to SP. This type of controller performance is known as *averaging* rather than *tight* level control. Averaging control can dramatically reduce the impact that flow disturbances have on a process.
- Most DCS offer a range of nonlinear algorithms intended to address specifically some of the problems that can arise with level control. While of secondary importance compared to applying the correct tuning, they can be particularly useful in dealing with processes that experience a wide range of flow disturbances.
- Cascade control is usually of benefit but for reasons different from most other situations. Rather than offer the more usual dynamic advantage, it permits more flexibility in tuning and simplifies the calculation of tuning constants.
- While the use of filtering to reduce the effect of measurement noise affects the dynamic behaviour of any process, in the case of level control its impact is usually substantial and ideally should be avoided.

4.1 Use of Cascade Control

Before tuning the level controller we must decide whether it should act directly on the valve or be cascaded to a secondary flow controller, as shown in Figure 4.1. The general

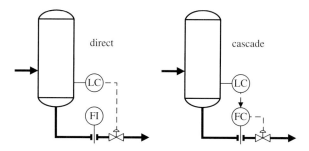

Figure 4.1 Alternative control configurations

rule in applying cascade control is that the secondary should be able to detect and resolve any disturbance before the primary. Failure to adhere to this can result in instabilities caused by the secondary attempting to correct for a disturbance that has already been dealt with by the primary. Since the vessel level will change at almost the same time as the flow there would appear to be no dynamic advantage in applying cascade control. Indeed, this is the case if our objective is tight control. However, for averaging control there is another consideration.

Imagine that a feed surge drum experiences an upstream or downstream fluctuation in pressure. The change in pressure drop across the manipulated flow valve will cause a change in flow. If this valve is under flow control then the disturbance will be dealt with quickly, resulting in little fluctuation to either the drum level or the manipulated flow. However, with no flow controller, the level controller is left to handle the disturbance. Since we want the manipulated flow to be as steady as possible, the level controller will need to be tightly tuned so that the control valve is moved quickly to compensate for the change in pressure drop. This is in conflict with the way we want the level controller to behave if there is a change in the uncontrolled flow. Under these circumstances we would want averaging level controller tuning. Applying a cascade allows us to meet both objectives. The flow controller would respond quickly to pressure changes, while the level controller would respond slowly to flow changes.

There is secondary advantage to using a cascade arrangement when it comes to tuning both tight and averaging controllers. Both calculations require the range of the manipulated flow. This value is a constant if a flow controller is in place; without one the range will vary with operating pressure and stream properties.

Orifice type flow meters require a straight run length equal to 20 pipe diameters upstream and 10 downstream; they can be very costly to retrofit if this does not exist. However the incremental cost of including the measurement in the original process design will be much smaller. If the construction budget is a constraint, the installation can be limited to the orifice flanges and orifice plate. The remainder of the instrumentation can then be added if necessary later without incurring the cost of pipework modification.

The schematic of the process on which most of this chapter is based is included as Figure 4.2. This shows the level controller manipulating the discharge flow from the vessel. In this case the inlet flow is the DV, the outlet flow is the MV and the level is the PV. However there are situations where it is necessary to manipulate inlet flow. This makes no difference to the tuning calculations or controller performance – provided the engineer remembers to reverse the control action!

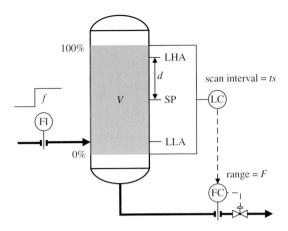

Figure 4.2 Process flow diagram

4.2 Parameters Required for Tuning Calculations

To calculate controller tuning constants we first need to determine the working volume (V) of the vessel. This is the volume between 0 % and 100 % of the range of the level gauge. This can be determined by performing a simple plant test. Starting with the process at steady state we decommission any existing level controller and step either the inlet or outlet flow to cause a flow imbalance (ΔF). We allow this imbalance to exist for a known time (t) and record the change in level indication (ΔL). Of course, because the process is not self-regulating we must end the test by restoring the flow balance before the level violates any alarms. The test result is shown in Figure 4.3.

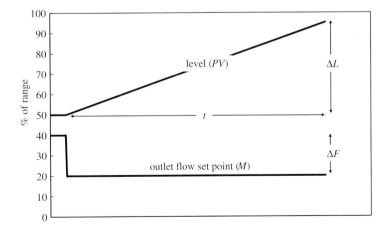

Figure 4.3 Analysis of plant test

We can then calculate the volume using Equation (4.1).

$$V = \frac{100 \Delta F . t}{\Delta L} \qquad (4.1)$$

Care should be taken with the engineering units. ΔL is a percentage (hence the 100 in the expression). The duration of the test (t) should be in units consistent with the flow imbalance (ΔF). So for example, if the flow is measured in m³/hr, t should be in hours. If in USGPM (US gallons per minute) then t should be in minutes and if in BPD (barrels per day), t should be in days.

This calculation assumes a linear relationship between volume and level indication. For vertical drums, assuming no large nozzles or internals, this will be the case. For horizontal drums and spheres the relationship is theoretically nonlinear but, providing the level gauge has been correctly ranged, the effect may generally be ignored. The larger the level change resulting from the test, the more representative will be the estimate of V.

Of course, if the vessel dimensions are known, it is possible to calculate the working volume. For a vertical drum, the calculation is trivial, i.e.

$$V = \pi r^2 (h_{100} - h_0) \qquad (4.2)$$

The radius of the vessel is r, h_0 is the height of the 0 % level indication (measured from the base of the vessel) and h_{100} is the height of the 100 % level indication. Care should be taken in quantifying these values. They will usually **not** correspond to the location of the nozzles to which the level gauge is connected. The difference ($h_{100} - h_0$) is the instrument range, usually found on the instrument datasheet. While a value for h_0 is not required for vertical drums, it is required for other shapes.

Again, care should be taken with units. If flow is measured in m³/hr then r and h should be in m, if flow is in USGPM and r and h are measured in ft then a multiplier of 7.48 is required to convert ft³ to USG. If the flow is in BPD then the multiplier should be 0.178.

Calculation of working volume for a horizontal cylindrical vessel is more complex. Firstly we have to calculate the volume (V_0) between the bottom of the vessel and the 0 % level indication. The length of the vessel (l) is that measured between *tangent lines* – the point where any dished ends are welded to the vessel. The last term in Equation (4.3) determines the volume of liquid held in the dished ends. It assumes a 2:1 ratio between drum radius and depth of each dish. It should be omitted if the vessel has flat ends.

$$V_0 = \left[r^2 \cos^{-1}\left(\frac{r-h_0}{r}\right) - (r-h_0)\sqrt{2rh_0 - h_0^2} \right] l + \frac{\pi h_0^2}{6}(3r - h_0) \qquad (4.3)$$

The working volume (V) may then be derived (again omitting the term for the dished ends if not required).

$$V = \left[r^2 \cos^{-1}\left(\frac{r-h_{100}}{r}\right) - (r-h_{100})\sqrt{2rh_{100} - h_{100}^2} \right] l + \frac{\pi h_{100}^2}{6}(3r - h_{100}) - V_0 \qquad (4.4)$$

The same form of equation can be used to assess the linearity of the volume/height relationship. Equation (4.5) permits the measured volume (V_m) to be calculated as a function of h.

$$V_m = \left[r^2 \cos^{-1}\left(\frac{r-h}{r}\right) - (r-h)\sqrt{2rh - h^2} \right] l + \frac{\pi h^2}{6}(3r - h) - V_0 \qquad (4.5)$$

Similar, somewhat simpler, calculations can be performed if the vessel is spherical.

$$V_m = \frac{\pi}{3}\left[h^2(3r-h)-h_0^2(3r-h_0)\right] \tag{4.6}$$

More usefully, it is better to plot the function in a dimensionless form, i.e. percentage of working volume against percentage level indication.

$$100\frac{V_m}{V} \quad \text{versus} \quad 100\frac{h-h_0}{h_{100}-h_0}$$

Figure 4.4 illustrates the impact of taking the unusual step of mounting the level gauge to operate over the full height of the vessel.

As expected, the horizontal drum and sphere show significant nonlinearity. Omitting here the mathematics involved it can be shown that, for a horizontal drum with the level moving between 1 % and 99 % of the drum height, the process gain varies by ±77 % around the mean. This would raise issues with controller tuning. However, taking the more usual approach of mounting the level gauge so that it does not operate of the full height of the tank, for example locating h_0 and h_{100} at 15 % and 85 % of the drum height, reduces the variation in process gain to ±23 %. This would easily be accommodated by a well-tuned linear controller and, because of the nonlinearity, involves sacrificing only 16 % of the theoretically available capacity.

However, it is equally possible through poor design, to *increase* nonlinearity greatly by poor siting of the level gauge. For example locating h_0 close to the bottom of the vessel and h_{100} at around 25 % of the vessel height would cause significant tuning problems. Figure 4.5 illustrates this.

The problem of nonlinearity is therefore best avoided at the vessel engineering stage. If the vessel is either intended to provide surge capacity, or will provide useful capacity – even if this is not its main purpose, then there are two main design criteria. The first is to position h_0 and h_{100} as far apart as possible without encroaching into any serious nonlinearity. This is

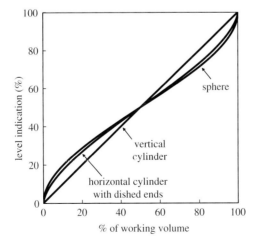

Figure 4.4 Checking linearity of level indicator

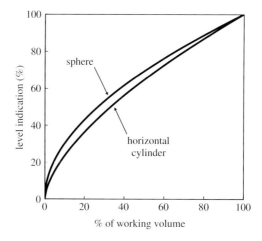

Figure 4.5   Poorly placed level indicator

to make maximum use of the vessel capacity. This seems obvious but it is very common for gauges of a very narrow range to be mounted on very tall vessels. The second, on horizontal cylinders and on spheres, is to position h_0 and h_{100} symmetrically either side of the centre line. The aim is to ensure there is equal capacity either side of the controller SP of 50 %. The controller can thus handle equally both increases and decreases in flow. For example, if there is less capacity above the SP then the controller will need to be tuned for flow increases and will not fully utilise vessel capacity when there is a decrease in flow.

It is common for the control engineer to have to deal with problems inherent to poor design. Should nonlinearity present a problem then this can be resolved with suitable signal conditioning. By definition, the level measurement (%) is given by:

$$L = 100 \frac{h-h_0}{h_{100}-h_0} \qquad (4.7)$$

Rearranging:

$$h = \frac{L}{100}(h_{100}-h_0) + h_0 \qquad (4.8)$$

Substituting for h in Equation (4.5) (or Equation (4.6) for spherical vessels) and building the resulting equation (or look-up table) in the DCS will allow V_m to be continuously determined from L. V_m may then be also determined as a percentage of the working volume (V). Using this value as the measurement of the controller will present to the process operator a more reliable measure of vessel inventory and changes in its value will be repeatable with respect to flow imbalances – no matter what the current inventory.

Other parameters required to permit controller tuning to be calculated are included in Figure 4.2. They include the normally expected flow disturbance (f). Some judgement should be used in selecting this value. If the plant is not yet commissioned, or historical data is not available for any other reason, then a value of 10 % of the maximum flow is a good starting point. If process data do exist then a visual inspection of flow trends (either inlet or outlet) should permit a sensible value to be selected. The important consideration is choosing a **normal** disturbance, i.e. the sort of disturbance we would expect not to generate any process alarms. We are not designing for a catastrophic reduction in feed due, for example, to equipment failure.

It is often the case that a process will largely experience minor flow disturbances but with the occasional larger upset. This might be caused by routine switches in process conditions such as drier swings, reactor regeneration, feed type change, change in operating mode etc. If this is the situation then two values for f should be chosen – f_1 for the small frequent disturbances and f_2 for the larger occasional upset.

Once the controller is commissioned its performance should be closely monitored to confirm that the value(s) chosen are realistic. For example if surge capacity is not being fully utilised by an averaging level control, then a smaller value of f should be chosen and the controller tuning recalculated. The simplest method of doing this is to assess what fraction of the surge capacity is being used and then multiplying the controller gain by this fraction.

The tuning method needs us to define how much of the vessel capacity may be used. This is set by the parameter d which is defined as the maximum deviation (as a percentage) permitted from the level SP. Ideally, to make maximum use of surge capacity, this should be the distance between the SP and the nearest alarm. Placing high and low alarms symmetrically either side of SP will permit maximum use to be made of surge capacity. For tight control a much smaller value of d, for example 1 %, would be selected.

Because controllers generally operate with their input and output in dimensionless form (e.g. percentage of range) we need the factor (F) to convert controller output into engineering units. If the level controller is cascaded to a flow controller then F is simply the range of the flow instrument. However, if the level controller acts directly on a valve, F is the flow with this valve fully open. If there is a flow measurement then this may be estimated by using historical data to correlate flow against valve position. If this is not possible then F may be approximated by multiplying the design flow by a factor of 1.3 – since this is typically the factor used in sizing the valve.

Finally we need the level controller scan interval (ts).

4.3 Tight Level Control

Tight level control is required in situations where holding the level close to its SP is of greater importance than maintaining a steady manipulated flow. This would be applied, for example, to a steam drum level where we want to avoid the risk of routing liquid into the steam header and potentially damaging turbine blades. Similarly, on a compressor suction drum, we want to avoid routing to the compressor any of the liquid collected in the drum. As we shall see in Chapter 12, certain types of level controllers on distillation columns similarly require tight tuning. If reflux drum level is controlled by manipulating reflux flow then we must manipulate the overhead product flow to control product composition. This only has an impact because the drum level controller then takes corrective action and changes the reflux. In order for our composition control to act as fast as possible, the drum level controller must be tightly tuned. This would similarly apply to the level controller on the column base if it is set up to manipulate reboiler duty.

Controller tuning is derived by first assuming that we apply a proportional-only controller.

$$\Delta M = K_c(E_n - E_{n-1}) \qquad (4.9)$$

Let us assume that before the flow disturbance, the level is at steady state and at SP, i.e. E_{n-1} will be zero. Since the flow imbalance (f) will have existed for one controller scan interval (ts), the current error (in dimensionless form) is given by

$$E_n = \frac{f.ts}{V} \qquad (4.10)$$

In order to bring the level back to steady state we need to restore the flow balance and so the controller must change the manipulated flow by the flow disturbance (f). In dimensionless form this means

$$\Delta M = \frac{f}{F} \qquad (4.11)$$

The tightest possible control would be to take this corrective action in the shortest possible time, i.e. the scan interval (ts). By combining Equations (4.9) to (4.11) we can derive the largest possible controller gain (K_{max}).

$$K_{max} = \frac{V}{F.ts} \qquad (4.12)$$

Care should again be taken with the choice of engineering units. Controller scan interval (ts) in most DCS is measured in seconds. So, if the flow range (F) is measured in m^3/hr, the result of this calculation should be multiplied by 3600 to ensure K_{max} is dimensionless. If the flow is in USGPM then a factor of 60 is required. If the flow is in BPD then a factor of 86 400 should be used.

Examination of Equation (4.12) shows K_{max} is independent of f. This means that, no matter what size the flow disturbance, the controller will set the SP of the manipulated flow equal to the variable flow within one scan interval. Of course control valve dynamics and the tuning of the secondary flow controller (if present) will mean the change in actual flow will lag a little, but nevertheless the controller should be effective.

Similar examination of the result shows that K_{max} is dependent on ts. Unlike most controllers a small change in scan interval (e.g. from 1 to 2 seconds) will have a dramatic effect on the required tuning.

Because the controller is proportional-only it cannot return the level to its SP. However the offset, given by Equation (4.10), will be extremely small and would probably not be noticeable – even if there are successive disturbances in the same direction as the first. But integral action may be added. To estimate how much we first determine a vessel time constant (T) – measured with no controls in place. This is defined as the time taken for the level to change by the permitted deviation (d) following the flow disturbance (f). It is given by

$$T = \frac{Vd}{100f} \qquad (4.13)$$

Since we require tight level control, we would select a very small value for d, for example 1 %. Experience shows that, within a sensible range, the level of integral action is not critical to controller performance. Empirically, setting T_i to $8T$ will give good control performance. Again care should be taken with engineering units. With f measured in m³/hr the result for T will be in hours. Although it is system-specific the value of T_i is usually required in minutes and so a factor of 60 must be included. No factor would be needed if the flow is in USGPM. A factor of 1440 should be used if the flow is in BPD.

The additional control action introduced will mean that the controller will now overcorrect. Compensation for the addition of integral action should be made by reducing proportional action. Again empirically, applying a factor of 0.8 to K_{max} works well. Derivative action is not normally beneficial to level control – indeed in the absence of any significant deadtime even a small amount of action will cause instability.

Full controller tuning is therefore:

$$K_c = \frac{0.8 V}{F.ts} \quad T_i = \frac{V}{12.5f} \quad T_d = 0 \tag{4.14}$$

The performance of a typical controller is shown as Figure 4.6. In this case the inlet flow was increased by 20 % at the 8 minute point. The discharge flow was increased by the same amount in less than half a minute. Only the dynamics of the control valve prevented the correction being made more quickly. As a result the disturbance to the level would unlikely to be noticed on a real process. Because the controller includes integral action, the discharge flow briefly exceeds the inlet flow in order to return the level to SP.

While this approach will normally provide effective control the presence of measurement noise may present a problem. The value of K_c derived is likely to be considerably greater than unity and will therefore amplify noise and may ultimately cause damage to the control valve. Controller gain may need to be reduced and larger deviations from SP accepted. The use of filtering can be counter-productive. The filter will add lag to a process which is likely to have almost none. The controller is likely then to be unstable and a large reduction in

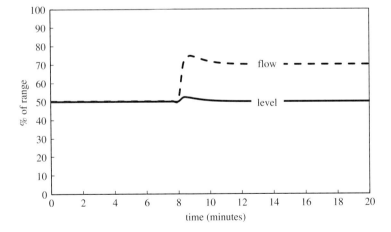

Figure 4.6 Tight level control

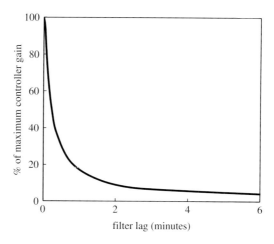

Figure 4.7 Effect of filter

controller gain will be necessary to avoid this. This is illustrated in Figure 4.7. Noise is best dealt with at the vessel design stage. Turbulence in the vessel may be caused by the velocity of liquid entering the vessel through the inlet nozzle or, in the case of flash drums and steam drums, by boiling. The appropriate use of baffles and *stilling wells* will reduce the effect this turbulence has on the level measurement.

4.4 Averaging Level Control

Averaging level control is required in situations where keeping the manipulated flow as steady as possible is more important than keeping the level at its SP. Its aim therefore would be to make full use of the vessel capacity without violating any level alarms. Failure to appreciate the benefit of averaging level control, and how to design it, is one of the most common oversights in the process industry. There are many processes that would benefit from it greatly in terms of *disturbance rejection*.

Its most obvious application is to feed surge drums. These are included in the process design specifically to reduce the effect of upstream flow disturbances on the downstream process. Installing tight level control in this situation makes the drum ineffective.

However, there are many situations where surge capacity is a spin-off benefit from a vessel that is in place for an entirely different purpose. For example, as we shall in Chapter 12, it is common in a sequence of distillation columns for one column to be fed from the reflux drum of the preceding column. Provided the drum level controller manipulates the feed to the downstream column then averaging level control may be applied to minimise feed flow disturbances. Even if the overhead product is routed to storage, if the product is cooled by exchanging heat with another process stream, then disturbances to the energy balance can be reduced. The level controller at the base of the column may similarly be exploited if it is manipulating the flow of the bottom product. However the available surge capacity may be small and therefore offer little opportunity. Also care must be taken if reboiler performance is affected by variations in level.

The main issue with averaging level control is its acceptance by the process operator. To achieve its objective the vessel level will often approach alarm limits and may take several hours to return to SP. The operator may, not unreasonably, be quite concerned by this and not entirely persuaded that the benefit to the downstream unit is worth the apparent risk. A more cautious approach can allay such concerns. Initially tuning the controller to use only part of the available capacity and demonstrating over time that it does not violate this limit will help persuade the operator to accept use of all the available capacity – particularly if the benefit is demonstrable.

There are likely to be other similar issues. Some sites permit the operators to configure process alarms; these will often then be set conservatively and the operator will need to be persuaded to relax them as far as possible. The operator may introduce asymmetry. He may be concerned about potential pump cavitation and therefore more worried by a reduction in level rather than a rise. He will increase the level SP above 50 % and may also increase the position of the low level alarm. This will mean full use is not made of the surge capacity when there is a flow increase. The converse may also apply, for example if the operator is more concerned about overfilling the vessel.

The method used to tune the controller is very similar to that applied to tight level control. We start as before with a proportional-only controller. However, rather than eliminate the flow imbalance as quickly as possible we do so as slowly as possible. In this case the controller will take considerably more than one scan to make the correction, i.e.

$$\Delta M = K_c[(E_n - E_{n-1}) + (E_{n-1} - E_{n-2}) \ldots + (E_1 - E_0)] = K_c(E_n - E_0) \qquad (4.15)$$

To make full use of the capacity we will allow the level to approach the alarm before steady state is reached. In other words we design for an offset of d, i.e.

$$E_n = \frac{d}{100} \qquad (4.16)$$

By combining Equations (4.11), (4.15) and (4.16) we calculate the smallest possible controller gain (K_{min}).

$$K_{min} = \frac{100f}{Fd} \qquad (4.17)$$

This, however, is just a first step in the controller design. Unlike tight level control we cannot retain such a proportional only controller. As we can see in Figure 4.8 the level, as designed, remains at the alarm limit set at 90 %.

We will need integral action to return the level to its SP in preparation for the next disturbance. We determine this using the same method as for tight controller. The full tuning then becomes

$$K_c = \frac{80f}{Fd} \qquad T_i = \frac{Vd}{12.5f} \qquad T_d = 0 \qquad (4.18)$$

Figure 4.9 shows how this controller would respond to the flow disturbance f with a SP of 50 % and high level alarm at 90 %, i.e. d is set at 40 %. The uncontrolled flow was increased as a step change.

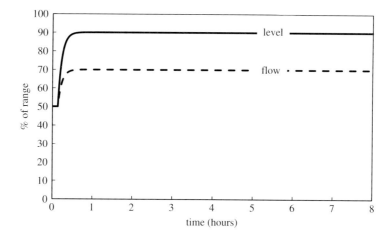

Figure 4.8 Interim proportional only control

As can be seen the manipulated flow was increased as slowly as possible without violating the alarm. This increase took about 30 minutes, compared to the almost instantaneous increase that was made by the tight controller – substantially stabilising the downstream process.

Figure 4.10 illustrates how tuning constants vary, on a typical surge drum, as the maximum deviation (d) is changed from 1 % to the maximum of 50 %. Remembering that integral action is governed by the ratio K_c/T_i, the change in tuning moving from tight to averaging is more than three orders of magnitude.

It is common for this approach to determine a value for T_i which is larger than the maximum supported by the DCS. Under these circumstances one of two approaches may be taken. The first is simply to set T_i to the maximum that the system will support and

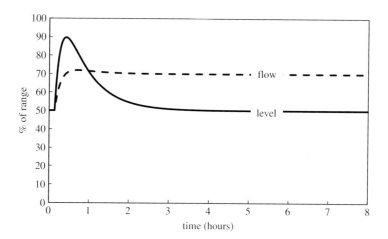

Figure 4.9 Performance of averaging level control

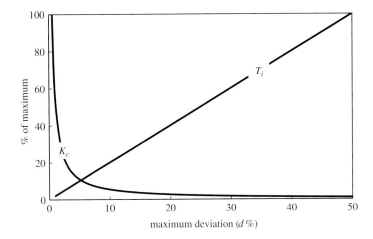

Figure 4.10 Effect of permitted deviation on tuning constants

accept that full use will not be made of the available surge capacity. Clearly, whether this is effective will depend on how much greater the ideal value is compared to the maximum.

The alternative approach is to apply a proportional only controller. Because this will cause an offset, we need to ensure that the offset never violates an alarm. Rather than use the normal disturbance to determine the controller gain we must instead use the minimum and maximum flow. The controller is designed so that the level will be at SP when the flow is midway between these values. The level will be at the low alarm at minimum flow and at the high alarm at maximum flow. The most conservative design basis is to assume the minimum flow is zero and the maximum is F. The maximum deviation from the mean flow is thus $F/2$. Replacing the normal disturbance f in Equation (4.17) with this value gives

$$K_c = \frac{50}{d} \qquad (4.19)$$

Whether this is a more effective solution than using the maximum value of T_i will depend on the pattern of flow disturbances. If the minimum and maximum flows are only approached rarely then the full surge capacity will not be used. This is particularly true if f is small compared to the range of flow variation. Figure 4.11 compares the performance of the proportional controller to the PI controller, in terms of the change made to the manipulated flow.

The proportional controller, since it must have a larger gain, initially changes the flow more rapidly. The PI controller must increase the flow above the steady-state value in order to bring the level back down to SP, but the overshoot is small and can be reduced further if necessary by increasing T_i.

Remember that if a proportional-only controller is configured as proportional-on-PV, it will not respond to changes in SP. This might be considered advantageous since it prevents the operator changing the SP to a value where the offset violates an alarm. However it might create problems with operator acceptance, in which case the proportional-on-error algorithm can be used.

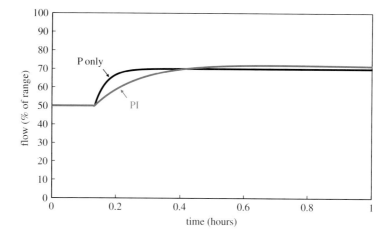

Figure 4.11   Comparison between P only and PI control

One notable difference between the calculation for averaging control and those for tight tuning is the omission of ts from the calculations. Changing controller scan interval has no affect on controller tuning. However K_c, unlike tight control, is now strongly dependent on f. This begs the question as to how the controller will handle disturbances different from design. Figure 4.12 shows that a disturbance 25 % larger than design causes an alarm violation – almost exceeding the instrument range. Similarly a disturbance 25% smaller results in underutilisation of surge capacity.

The simplest approach is to tune the controller based on the largest normally expected disturbance. While this will avoid alarms it will underutilise surge capacity. This will be a significant disadvantage if the larger disturbances are relatively rare. Under these circumstances a better approach would be to use a nonlinear control algorithm. Several

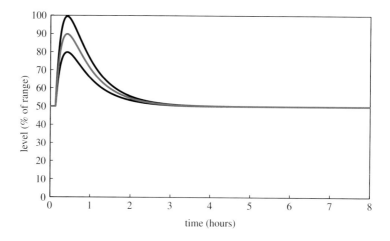

Figure 4.12   Impact of size of flow disturbance on linear controller

different types of algorithm are included in many DCS. They are specifically designed for averaging level control. While they can be tuned to give tight control, they offer no advantage under these circumstances over the normal linear version.

4.5 Error-Squared Controller

The most well-known nonlinear algorithm is *error-squared*. Since the controller works with a dimensionless error scaled between 0 and 1 (or 0 % and 100 %) the square of the error will have the same range. Strictly the error is not squared but multiplied by its absolute value, because we need to retain the sign if the error is negative. The effect is illustrated in Figure 4.13.

As is occasionally stated in some texts, error-squared does **not** compensate for the nonlinearity between level indication and liquid volume in horizontal cylindrical drums (or spheres).

It is not usual to square each error term in the controller individually. The most common approach is to multiply the controller gain by the absolute value of the error. Omitting the derivative term (since we usually do not require this for averaging level control) the control equation becomes:

$$\Delta M = K_c |E_n| \left[(E_n - E_{n-1}) + \frac{ts}{T_i} E_n \right] \quad (4.20)$$

The effect of the additional $|E_n|$ term is to increase the effective controller gain as the error increases. This means the controller will respond more quickly to large disturbances and largely ignore small ones.

Tuning is calculated using the same approach as for the linear algorithm. We first determine K_{min} for a proportional-only controller based on restoring the flow balance when the offset has reached the alarm. In its continuous form we can write the control algorithm as

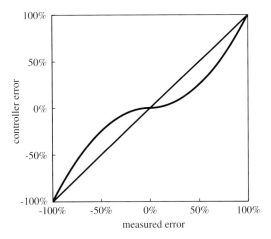

Figure 4.13 Effect of error-squaring

$$\Delta M = K_{min} \int_0^\infty E.dE = K_{min} \left[\frac{1}{2}E^2\right]_0^\infty = K_{min}\left(\frac{1}{2}\left(\frac{d}{100}\right)^2 - 0\right) \quad (4.21)$$

Combining with Equation (4.11) gives

$$K_{min} = \frac{100f}{Fd}\left[\frac{200}{d}\right] \quad (4.22)$$

Following the same approach as the linear algorithm, the full tuning becomes

$$K_c = \frac{80f}{Fd}\left[\frac{200}{d}\right] \quad (4.23)$$

T_i and T_d are determined as in Equation (4.18). Figure 4.14 compares the performance of this controller compared to that of the linear version. It meets the design criterion of fully using the surge capacity without violating the alarm. However, it appears to show some oscillatory behaviour as the level returns to SP. The effect of error-squaring is to reduce the controller gain to zero when the error is zero. As the level returns to SP the small effective controller gain means that very little corrective action is taken and the level overshoots the SP. It is not until sufficient error accumulates that the controller gain increases enough for the flow imbalance to be reversed and the cycle then repeats itself.

In theory this oscillation will also be reflected in the flow. However, these changes will be almost imperceptible, having no effect on the downstream process. The changing level is a minor inconvenience. However, if noticed by an already reluctant process operator, it may cause difficulty in acceptance. And, if a real-time optimiser is installed, its steady-state detection logic may reduce the frequency of executions.

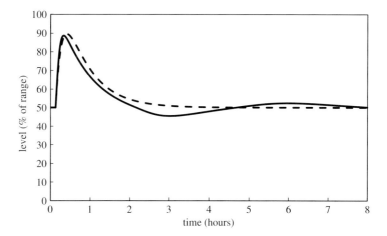

Figure 4.14 Performance of error-squared versus linear

Fortunately a simple solution exists within most DCS. Rather than provide separate linear and error-squared algorithms the DCS will usually include a dual purpose algorithm. A typical example is:

$$\Delta M = K_c(C|E_n| + 1 - C)\left[(E_n - E_{n-1}) + \frac{ts}{T_i}E_n\right] \quad (4.24)$$

The additional term (C) gives the engineer the option of switching between algorithms. Setting C to 1 will give error-squared, while setting it to 0 gives linear performance. But the engineer is free to choose any value between these limits. By choosing a value close to 1, the controller will largely retain the nonlinear performance but the effective controller gain will no longer be zero as the error falls to zero. Controller tuning then becomes:

$$K_c = \frac{80f}{Fd}\left[\frac{200}{200(1-C) + Cd}\right] \quad (4.25)$$

T_i and T_d are determined as in Equation (4.18). Figure 4.15 shows the performance of this controller (with C set at 0.9) for the design disturbance and for disturbances 25 % larger and smaller than design. The addition of the small amount of linear action has removed the oscillatory behaviour and, for the design case, given performance virtually identical to that of the linear controller. This algorithm however outperforms the linear controller for the nondesign cases. Comparing the responses to those in Figure 4.12, for disturbances larger than design the level violates the alarm by less and for a shorter period. For disturbances smaller than design greater use is made of the surge capacity. While it does not completely solve the problem of varying flow disturbances it does offer a substantial improvement in performance.

It should be noted that the tuning calculation presented as Equations (4.23) and (4.25) are for the control algorithms **exactly** as described. DCS contain many variations of the error-squared algorithm. Even relatively minor changes to the algorithm can have significant

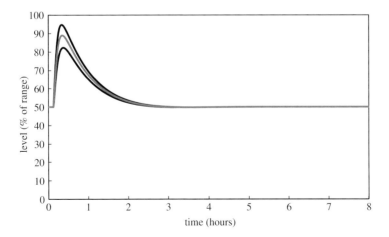

Figure 4.15 Impact of size of flow disturbance on error-squared controller

effects on the required tuning. For example, squaring each error term individually appears to make a minor change to the integral action, i.e.

$$\Delta M = K_c \left[(|E_n|E_n - |E_{n-1}|E_{n-1}) + \frac{ts}{T_i} |E_n|E_n \right] \quad (4.26)$$

Comparing this to the controller described by Equation (4.20), the previous value of the error (E_{n-1}) is now multiple by $|E_{n-1}|$ rather than $|E_n|$. Since the two values are measured only one scan interval apart, they will be almost identical and one would think this would have little impact on controller tuning.

Taking the same approach as Equation (4.15))

$$\Delta M = K_c \left[(E_n^2 - E_{n-1}^2) + (E_{n-1}^2 - E_{n-2}^2) \ldots + (E_1^2 - E_0^2) \right] = K_c (E_n^2 - E_0^2) \quad (4.27)$$

Combining this with Equations (4.11) and (4.16), and applying the 0.8 factor, gives

$$K_c = \frac{80f}{Fd} \left[\frac{100}{d} \right] \quad (4.28)$$

Comparing this result to that in Equation (4.23) shows that a very minor change to the algorithm requires that the controller gain be **halved** to give the same performance. Other changes offered within some DCS include the option to apply error-squaring selectively to each of the proportional, integral and derivative actions. There are also forms of the control algorithm that include other parameters to allow the engineer to specify the type of nonlinearity, for example:

$$\Delta M = K_c (C|E_n| + K_n) \left[(E_n - E_{n-1}) + \frac{ts}{T_i} E_n \right] \quad (4.29)$$

C may be set between 0 and 1. By setting it to 1 and the nonlinear gain term (K_n) to zero gives the same form as error-squared algorithm as described by Equation (4.20). Similarly setting C to 0 and K_n to 1 reduces the controller to the linear form. The controller described in Equation (4.24) can be emulated by setting K_n to $(1 - C)$. But other values of K_n may be used – although probably with little benefit. Controller tuning is determined from:

$$K_c = \frac{80f}{Fd} \left[\frac{200}{200 K_n + Cd} \right] \quad (4.30)$$

T_i and T_d are determined as in Equation (4.18).

4.6 Gap Controller

An alternative approach to introducing nonlinearity into the controller is by introducing a gap. In its simplest form this introduces a deadband around the SP within which no control action takes place. Outside the deadband the controller behaves as a conventional linear controller. The gap is configured by the engineer as a deviation from SP (G %).

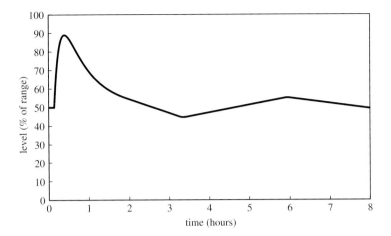

Figure 4.16 Performance of gap controller with deadband

Tuning is given by

$$K_c = \frac{80f}{F(d-G)} \qquad (4.31)$$

T_i and T_d are determined as in Equation (4.18). Figure 4.16 shows the performance of this control with G set at 5 % either side of SP. In this form it exhibits to behaviour similar to that of the error-squared controller in that it will never settle at SP. Within the deadband no control action is taken and so any flow imbalance will be maintained until the level reaches the edge of the band. At which point corrective action is taken to reverse the direction. While again this has little impact on the downstream process, it is undesirable for the same reasons as described for the error-squared controller, i.e. operator acceptance and steady state detection.

The solution is to apply a nonzero gain within the gap. To preserve the required nonlinear behaviour the value chosen should be substantially less than that used outside the deadband. Most DCS permit the engineer to define the value as a ratio (K_r) where

$$K_r = \frac{(K_c)_{gap}}{K_c} \qquad (4.32)$$

in which case the tuning is derived from

$$K_c = \frac{80f}{F(d-(1-K_r)G)} \qquad (4.33)$$

Typically K_r is chosen to be about 0.1, which will give performance much the same as the error-squared controller – including its ability to better handle nondesign disturbances.
Alternatively a value for $(K_c)_{gap}$ may be chosen and used in the following tuning method.

$$K_c = \frac{80f - FG(K_c)_{gap}}{d-G} \qquad (4.34)$$

But the gap algorithm is better used in situations where flow disturbances can be classified into two types – relatively small changes (f_1) which take place frequently and much larger more intermittent changes (f_2). The controller is designed to deal with the smaller disturbance within the gap, thus

$$(K_c)_{gap} = \frac{80 f_1}{FG} \tag{4.35}$$

The balance of the disturbance is then dealt with using the remaining vessel capacity, hence

$$K_c = \frac{80(f_2 - f_1)}{F(d - G)} \tag{4.36}$$

Substituting Equations (4.35) and (4.36) into Equation (4.32))

$$K_r = \frac{f_1}{f_2 - f_1} \left[\frac{d - G}{G} \right] \tag{4.37}$$

Key to the performance of this controller is the choice of G. Firstly the same value should be used for both positive and negative variations from SP. This symmetry, combined with symmetrically placed high and low level limits, ensures that we do not have to tune the controller for disturbances in the more demanding direction and thus underutilise surge capacity for disturbances in the opposite direction.

For the gap to be beneficial K_r must be less than 1. Applying this constraint to Equation (4.37) results in

$$G \geq d \frac{f_1}{f_2} \tag{4.38}$$

Applying a more realistic limit on K_r (e.g. 0.1) results in

$$G \geq d \frac{10 f_1}{9 f_1 + f_2} \tag{4.39}$$

The wider we make G, the smaller we make the drum capacity which the controller can use to deal with the larger disturbance. A larger controller gain will therefore be required. This gain given by Equation (4.36) should not exceed that required for tight control as determined by Equation (4.14).

$$G \leq d - \frac{100(f_2 - f_1) ts}{V} \tag{4.40}$$

Again care should be taken with the choice of engineering units. Between these constraints the choice of G is a compromise. Larger values will make better use of surge capacity during small disturbances but will leave little capacity to smooth larger flow changes. T_i and T_d are determined as in Equation (4.18).

Figure 4.17 shows the performance of a well-tuned gap controller. In this case G is set at 30 % and the flow changed by f_1. The coloured line shows that the surge capacity is used as specified. The black line shows the result of a flow change of f_2, which is 4 times larger than f_1. The level deviation peaks at 40%, the value in this case for d. From Equation (4.37), we can see that K_r is set at 0.11.

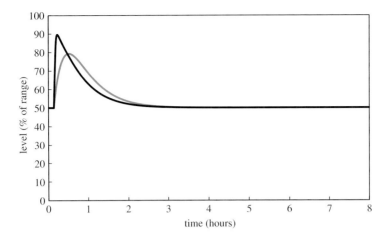

Figure 4.17 Performance of gap controller with very different disturbances

4.7 Impact of Noise on Averaging Control

The effect of measurement noise on averaging level control is somewhat different to its effect on tight control. Transmission of noise to the control valve is less likely to be a problem because the controller gain is substantially smaller. Nor is it likely that introducing a filter and its associated lag will give stability problems. However, when using the full surge capacity, as the level approaches alarm limits, the noise will cause nuisance alarms. This can be avoided by reducing the value chosen for d to take account of the noise amplitude. This will increase K_c and reduce the use of surge capacity. Filtering will not eliminate the need to increase K_c. It will have no affect on alarms unless the filtered value is used by both the controller and the alarm. The lag introduced will delay the response of the controller. As the size of the flow disturbance approaches the design value, the true level will violate the alarm before the controller can complete its correction.

It might be thought that nonlinear controllers deal better with noise and might therefore be considered for tight level control, where the high controller gain would otherwise amplify the noise. In theory, for small disturbances, the effective controller gain is small and hence little noise will be passed to the manipulated flow. However such controllers, to compensate for the little action taken at the beginning of a disturbance, require a gain higher than that for a linear controller. This means that, as the level moves away from SP, noise amplification will become worse than that from a linear controller.

With controllers that are nonlinear over the full range of error, such as error-squared, noise can cause oscillatory behaviour. Different gains will be applied to negative and positive spikes of noise – so the **average** output from the controller will be different from that if there was no noise. This is illustrated in Table 4.1. It shows the situation where, at time $= t$, the SP is changed from 60 % to 50 %. A spike of noise, of ± 1 % around the PV, then occurs over the next four controller scan intervals. Since the algorithm is proportional-on-PV, the proportional action in response to the change in SP should be zero. But the noise causes proportional action of $0.02K_c$. And the integral action causes a change $0.02K_c.ts/T_i$

112 Process Control

Table 4.1 Effect of error squared control on noise

Time	SP	PV	$|E_n|(PV_n-PV_{n-1})$	$|E_n|E_n$
t	50	60	0.00	1.00
t + ts	50	59	−0.09	0.81
t + 2.ts	50	60	+0.10	1.00
t + 3.ts	50	61	+0.11	1.21
t + 4.ts	50	60	−0.10	1.00
total			+0.02	5.02

larger than it would be without noise. While small, these changes will be repeated for every noise spike. They speed up the return to SP and can trigger cyclic behaviour.

This is illustrated in Figure 4.18, where noise of ±1 % of measurement range has been added to the example of the error-squared controller shown as Figure 4.14. The controller still responds well to flow changes but as it returns to SP the nonlinearity appears to amplify the noise to something in excess of ±10 % – despite the controller gain approaching zero. In fact the combination of noise and nonlinearity is triggering an oscillation with a period of about two hours.

The effect can be reduced by the use of the dual purpose algorithm described in Equation (4.24) but to eliminate it C would have to be set close to 1, almost removing the nonlinearity completely. The frequency of oscillation is too low for there to be any noticeable impact on the manipulated flow, so the controller is still meeting the objective of maintaining this as steady as possible. However, it will appear to the operator as if it is not working well. While filtering the PV can eliminate the cause of the problem we have shown that this results in less of the available surge capacity being utilised.

The better solution is to use a gap controller set up as described in Equations (4.35) to (4.37). Since the nonlinearity only exists when the level crosses in or out of the gap then for most of the time the same gain is applied to both positive and negative spikes of noise.

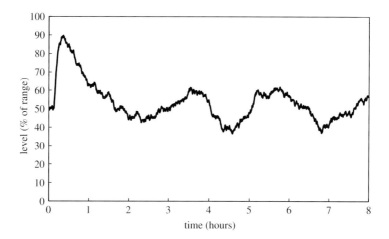

Figure 4.18 Performance of error-squared controller with noise

Indeed, if the gap controller is not required to handle very different flow disturbances, the deadband can be set at slightly larger than the noise amplitude so that noise is completely ignored when the level is close to SP. Some systems support more than one band to be defined. For example gain scheduling, as described in Chapter 3, could be configured to use different values of controller gain for different deviations of the level from SP. The gains would be determined from Equations (4.35) and (4.36).

4.8 General Approach to Tuning

So far we have adopted an approach to controller tuning which is specific to level control. It cannot be applied to other integrating processes such as some applications of pressure and temperature control. We have done this because conventional tuning methods do not readily lend themselves to averaging control or to nonlinear control algorithms. This does not mean that we cannot apply conventional methods to tight level control using the linear algorithm. Indeed we can predict the process gain that we would otherwise need to obtain from plant testing. Consider the general equation for an integrating process:

$$PV = K_p \int MV.dt + bias \quad \text{or} \quad \frac{dPV}{dt} = K_p.MV + bias \tag{4.41}$$

We can write this (in dimensionless form) for our vessel:

$$\frac{dL}{dt} = K_p \frac{\Delta F}{F} \tag{4.42}$$

But we can predict the rate of change of level from the working volume of the vessel:

$$\frac{dL}{dt} = \frac{\Delta F}{V} \tag{4.43}$$

Combining Equations (4.42) and (4.43) enables us to predict the process gain. If F is measured in m^3/hr and V in m^3 then

$$K_p = \frac{F}{V}\text{hr}^{-1} = \frac{F}{60\,V}\text{min}^{-1} = \frac{F}{3600\,V}\text{sec}^{-1} \tag{4.44}$$

To calculate tuning constants we also need the process deadtime (θ). For most level controllers this will be small. Choosing a value of a few seconds will result in controller tuning that will give a performance similar to that given in Equation (4.14).

There are level controllers that have substantial deadtimes. Consider the process in Figure 4.19. Level in the base of the distillation column is controlled by manipulating the reboiler duty. Unlike most level controllers it would be difficult (and probably unreliable) to predict the relationship between PV and MV. Further the reboiler introduces a large lag. The only practical way of identifying the process dynamics would be a plant test, as described in Chapter 2. The controller would then be tuned by applying one of the methods described in Chapter 3. This, unlike most level controllers, is likely to benefit from the use of derivative action.

However, whether this level control strategy should be selected requires careful consideration. The process dynamics will restrict how tightly the level can be controlled

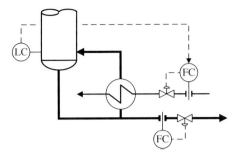

Figure 4.19 Control of column level by manipulation of reboiler duty

without becoming unstable. While in some cases there may be no practical alternatives, its slow response to disturbances may restrict the performance of other controllers. For example product composition would be controlled by adjusting the SP of the bottoms flow controller shown. It is unlikely that the LC could cope with rapid changes to this flow, or to the reflux flow, and correction of off-grade composition could only take place slowly. Full details of alternative approaches are given in Chapter 12.

4.9 Three-Element Level Control

Three-element level control is most commonly applied to the control of water level in steam drums on boilers. However it is applicable to many other situations where tight level control is required and is made difficult by unusual dynamics.

The first most commonly encountered problem is *swell*. The water in the steam drum contains vapour bubbles which expand if the pressure in the drum is reduced, thus increasing the liquid level. So, if there is an increase in steam demand which causes a transient drop in drum pressure, the level controller will reduce the flow of water in order to correct for the apparent increase in level. Of course, on increasing steam demand we need an **increased** water flow. The pressure in the drum will ultimately be restored, for example by a pressure controller on the steam header increasing the boiler duty, and the level controller will ultimately increase the water flow. However, for the controller to be stable, its initial behaviour means that it will have to act far more slowly than the tight controller defined in Equation (4.14).

This problem may be solved by using a dp type level instrument that effectively measures the mass of liquid in the drum rather than its volume. Since the effect is caused by a reduction in the fluid density, rather than an increase in its inventory, an instrument measuring the head of liquid will respond correctly. However some of the increase in level may be due to bubbles expanding in the tubes supplying the drum which causes additional water to enter the drum. Further local legislation may dictate, for safety reasons, that actual liquid level must be measured and used for control. Under these circumstances the problem can be alleviated by applying a correction term to the level measurement.

$$L_{corrected} = L_{measured} + K(P_{measured} - P_{normal}) \tag{4.45}$$

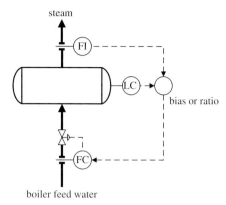

Figure 4.20 Three-element level control

The term K is determined empirically from process data and has the effect of increasing the level measurement transmitted to the controller when the measured pressure increases above the normal operating pressure.

The level controller may also have to cope with inverse response. The boiler feed water ideally is heated in the economiser to the boiling point of water at drum pressure. However, this is often not achieved so that when the cooler water enters the drum it will cause a drop in temperature, thus causing bubbles to collapse and the water level to drop. While controllers can generally be tuned to handle inverse response, they have to be tuned to act more slowly to avoid instability.

Three-element level control (Figure 4.20) is a technique which introduces a feedforward element into the controller. It includes a measurement of the steam flow leaving the drum. Any change in this flow is immediately passed to the water flow so as to maintain the mass balance. This largely meets the objective of tight level control. The level controller is retained as a feedback controller to compensate for any flow measurement errors and to allow the operator to change the SP if required. But it may now be tuned to act relatively slowly.

The feedforward and feedback signals are traditionally combined by using a *bias* algorithm which simply adds the signals. However this requires that the two flow measurements are in the same engineering units. Where water flow is measured in m³/hr and steam flow in te/hr this is already taken care of. If inconsistent units are used then a scaling factor will be required. Alternatively a *ratio* algorithm may be used. This multiplies the two signals, effectively keeping the water flow as an adjustable proportion of the steam flow – where this proportion can be in any units. Details of these algorithms are included in Chapter 6.

Care should be taken if this feedforward ratio scheme is implemented as an addition to an existing level controller. Since the level controller will now be manipulating the ratio target rather than flow controller SP its controller gain may need adjusting. The controller, working in dimensionless form, will generate a change in output (ΔM) which is converted to flow change in engineering units (ΔF) using the ranges of the flow and level controllers, i.e.

$$\Delta F = \Delta M \times \frac{FC_{range}}{LC_{range}} \tag{4.46}$$

With the ratio in place the range of the ratio algorithm replaces that of the flow controller and the change in ratio target is converted to change in water flow by multiplying it by the measured steam flow (F_{steam}), i.e.

$$\Delta F = \Delta M \times \frac{R_{range}}{LC_{range}} \times F_{steam} \qquad (4.47)$$

This will change the effective controller gain. To compensate for this, the existing controller gain should be multiplied by

$$\frac{FC_{range}}{R_{range} \times F_{steam}} \qquad (4.48)$$

Since F_{steam} is not a constant, this would suggest we need to retune the controller as the steam flow changes. However, since the level controller is less critical with the feedforward scheme in place, the use of the range of the steam meter instead of F_{steam} will result in conservative tuning that will be stable over the whole operating range. Further, if the instrumentation is well engineered then the range of the steam meter will be similar to that of the water. The range of the ratio is chosen on configuration. Since the actual ratio will change little then it should be possible to choose a range so that Equation (4.48) generates a correction factor close to unity – thus avoiding any adjustment of tuning. This is of particular benefit if the operator is permitted to selectively disable the feedforward part of the scheme, for example because of a problem with the steam flow instrument, since it would avoid the need to switch between two values for controller gain.

5
Signal Conditioning

Signal conditioning is manipulation of the input measurement to (or output signal from) a controller. A mathematical function is applied in order to improve controller performance. It may be required to compensate for nonlinear behaviour. Alternatively other process parameters may be incorporated into the PV to improve the accuracy of control.

5.1 Instrument Linearisation

The most frequent application of signal conditioning is linearisation. Many of the common functions may not be obvious to the control engineer since they are often built into the DCS or transmitter as standard features. For example, where c_d is the discharge coefficient, d the orifice diameter, dp the pressure drop across the orifice and ρ the fluid density, the flow (F) through an orifice flow meter is given by

$$F = c_d \frac{\pi d^2}{4} \sqrt{\frac{dp}{\rho}} \qquad (5.1)$$

The flow can therefore be measured by measuring dp but, to ensure that there is a linear relationship between this and the flow, the square root of dp is used. This is known as *square root extraction* and is usually an option within the DCS, or it might be performed by the field transmitter. Its effect is illustrated in Figure 5.1.

There is a similar need for linearisation of temperature measurements by thermocouples. Although over much of the range the relationship between the temperature and the voltage they produce is linear, this is not the case for temperatures below zero. Standard conversion tables are published for each thermocouple type and these are usually incorporated into the DCS or transmitter. Some examples of the more common types are shown in Figure 5.2.

Resistance Temperature Detectors (*RTD*) use a different linearisation function. The *Callendar-van Dusen Equation* relates resistance (R) to temperature (T) according to

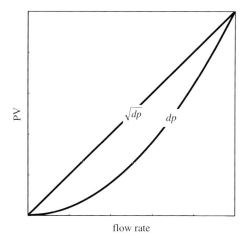

Figure 5.1 Square root extraction for flow meter

$$R = R_0\left[1 + AT + BT^2 + C(T-100)T^3\right] \quad \text{for} \quad -200\,°C \leq T \leq 0\,°C \tag{5.2}$$

$$R = R_0\left[1 + AT + BT^2\right] \quad \text{for} \quad 0\,°C \leq T \leq 850\,°C \tag{5.3}$$

The coefficients A, B and C depend on the metal used. The most common is commercial grade platinum with a nominal resistance of $0.385\,\Omega/°C$, in which case

$$A = 3.9083 \times 10^{-3} \quad B = -5.775 \times 10^{-7} \quad C = -4.183 \times 10^{-12} \tag{5.4}$$

The resistance at $0\,°C$ (R_0) is determined by the thickness of the wire. For the most common type, described as Pt100, R_0 is $100\,\Omega$. Thicker wire, Pt10, is often used for very high temperatures. Figure 5.3 shows some typical calibrations.

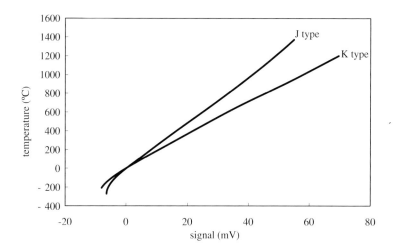

Figure 5.2 Thermocouple calibration

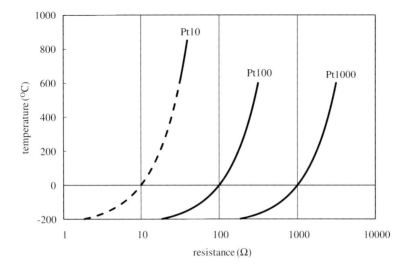

Figure 5.3   RTD calibration

5.2   Process Linearisation

Signal conditioning can also be applied to compensate for nonlinear process behaviour. For example in Chapter 4 we covered the linearisation of a poorly engineered level gauge so that it truly represented the percentage utilisation of the vessel's working volume and would thus be linearly related to the manipulated flow.

Perhaps the most challenging control problem is that of pH. Figure 5.4 shows the curves for a strong base, of pH 13, being titrated against a strong acid, of pH 2, and against another of pH 1.5. This illustrates two problems. For a constant acid pH, over the whole operating

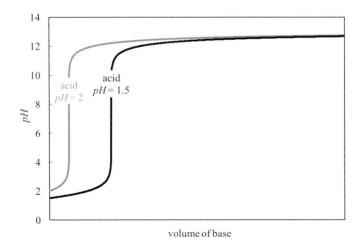

Figure 5.4   Nonlinearity of pH

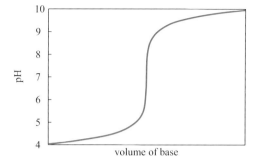

Figure 5.5 Nonlinearity of pH when close to neutrality

range the process gain (in engineering units) varies from a value of about 0.00013 to around 200, i.e. a factor about 10^6 times larger than can handled by a linear controller! Secondly, if the flow of base is correct for neutrality (i.e. the pH is 7) a change in the pH of the acid would cause a large change in the process gain.

It might appear that the nonlinearity might be characterised by dividing the titration curve into several sections that can be treated as linear. For example the section between a pH of 4 and 10 would appear to be a straight line. However, as Figure 5.5 shows, zooming in on this section of the curve, it is actually very nonlinear. The process gain changes by a factor of around 50.

And again it is tempting to assume that between a pH of 6 and 8, the line is straight. But, as Figure 5.6 shows, this is not the case; the process gain varies by a factor of around 6.

To derive a linearising function we first need to understand the process in more detail. By definition pH is the negative logarithm of the concentration (in kg-ions/m^3) of hydrogen ions, i.e.

$$pH = -\log_{10}[H^+] \quad \text{or} \quad [H^+] = 10^{-pH} \tag{5.5}$$

Pure water ionises as

$$H_2O \leftrightarrow H^+ + OH^- \tag{5.6}$$

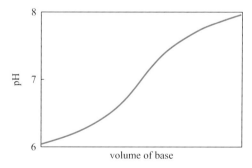

Figure 5.6 Nonlinearity of pH when even closer to neutrality

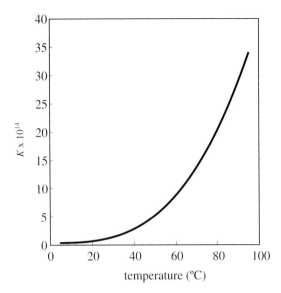

Figure 5.7 Effect of temperature on dissociation of water

where the equilibrium constant (K) is defined as

$$K = \frac{[H^+][OH^-]}{[H_2O]} \tag{5.7}$$

Water is only weakly ionised and so $[H_2O]$ is effectively 1. Figure 5.7 shows how K varies with temperature.

At 25 °C, the ionisation is such that the equilibrium constant for water (K_w) is 10^{-14} and so, from Equations (5.5) and (5.7)

$$[OH^-] = 10^{pH-14} \tag{5.8}$$

If K_a is the equilibrium constant for the ionisation of an acid HA then

$$[HA] = \frac{[H^+][A^-]}{K_a} = 10^{-pH}\frac{[A^-]}{K_a} \tag{5.9}$$

Similarly, if K_b is the equilibrium constant for the ionisation of a base BOH then

$$[BOH] = \frac{[B^+][OH^-]}{K_b} = 10^{pH-14}\frac{[B^+]}{K_b} \tag{5.10}$$

Total acid concentration is given by

$$[HA] + [H^+] = 10^{-pH}\left(\frac{[A^-]}{K_a} + 1\right) \tag{5.11}$$

Total base concentration is given by

$$[BOH] + [OH^-] = 10^{pH-14}\left(\frac{[B^+]}{K_b} + 1\right) \tag{5.12}$$

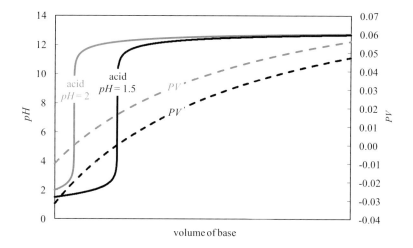

Figure 5.8 Linearisation of pH

We can define a PV as the difference between the base and acid concentrations, i.e.

$$PV = 10^{pH-14}\left(\frac{[B^+]}{K_b}+1\right)-10^{-pH}\left(\frac{[A^-]}{K_a}+1\right) \quad (5.13)$$

This would have a value of zero at neutrality. For a mixture of strong acid and strong base

$$K_a \rightarrow \infty \quad \text{and} \quad K_b \rightarrow \infty \quad (5.14)$$

and so

$$PV = 10^{pH-14} - 10^{-pH} \quad (5.15)$$

The dashed lines in Figure 5.8 show the result of applying this formula. While this does not give a perfectly linear relationship, it is a considerable improvement. The much more modest change in process gain should not present a tuning problem. Further the lines for each of the cases are approximately parallel and so the process gain will change little as acid strength changes.

5.3 Constraint Conditioning

Signal conditioning can be used to extend the apparent range of a measurement. It is common in constraint control applications to use the output (M) of a PID controller as an indication of valve position. This is a measure of how close the process is to a hydraulic limit. The problem is that, if the constraint is being violated, the controller output will be 100% – no matter how bad the violation.

If the controller is saturated then the PV will not usually be at SP. The size of the error (E) gives an indication of the severity of the problem. We can incorporate this into the measurement of the constraint (PV).

$$PV = M + K.E \quad (5.16)$$

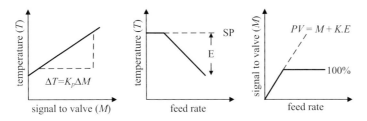

Figure 5.9 Constraint conditioning (hard constraint)

This is illustrated as an example in Figure 5.9. Imagine that we wish to maximise the feed rate to our case study heater (Figure 2.9) and that the constraint in doing so is a hydraulic limit on the fuel. As feed rate is increased the temperature controller will take corrective action and increase the signal (M) to open the fuel valve – usually via a cascade to a flow controller. This is a *hard constraint*, i.e. it can only be approached from one side – it is mechanically impossible for the valve to exceed an opening of more than 100%. If the heater was operating close to this limit and there was a process disturbance, for example a drop in the heater inlet temperature, the temperature controller would increase its signal to the valve, potentially taking it to the 100% limit.

The problem is that, if we wish to alleviate the constraint, the 100% indication does not tell us how far the constraint has been violated. The heater could be operating exactly at the true maximum feed rate, or could be well beyond it. If well beyond it, the heater outlet temperature will be below its SP. We can incorporate the temperature controller error (E) as a measure of the severity of the violation. In this example K would be set to the reciprocal of K_p – the process gain between the outlet temperature and signal to the valve. In doing so the PV, as defined in Equation (5.16) can now exceed 100% and its relationship to feed rate will have the same process gain as it does so.

A similar approach can be applied to the measurement of flue gas oxygen. If the air-to-fuel ratio falls below the stoichiometric requirement then the oxygen analyser will indicate zero – no matter how bad the problem. In Chapter 10 we show how incorporating a measurement of carbon monoxide (CO) can apparently extend the range into negative values of oxygen content.

There are occasions where a nonlinear response is preferred. We may want a controller to respond more quickly if the PV moves away from SP in a particular direction. For example, we can use a larger value of K in Equation (5.16) so that violation of a constraint is dealt with more quickly than it is approached. Similarly, even if the measurement stays within range, we may be more concerned about a high PV and a low one. We could again use the error to condition the measurement.

$$PV = PV_{measured} + K(PV_{measured} - SP) \qquad (5.17)$$

In this example, K is set to zero if the PV is less than the SP : otherwise it is set to 0.3 to increase K_c by 30%. Care should be taken in introducing such nonlinearities so that control remains stable when operating in the region where the conditioning is active.

Again, as an example, let us imagine that the constraint on increasing feed rate to our case study heater is now a limit on maximum burner pressure. Unlike the fuel valve position, this is a *soft constraint*. Although violation is undesirable it is physically possible. Burner

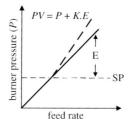

Figure 5.10 Constraint conditioning (soft constraint)

pressure will continue to rise as feed rate is increased. However high burner pressure can extinguish the flame and would be considered hazardous. So, any violation should be dealt with more urgently than exploiting spare capacity. Figure (5.10) illustrates how Equation (5.17) would be applied, increasing the apparent severity of the violation.

5.4 Pressure Compensation of Distillation Tray Temperature

Many process measurements are sensitive to pressure changes. By incorporating the pressure measurement into the PV we can ensure that the controller takes the correct action. We have already covered one example of this as Equation (4.45) – a means of reducing the effect of swell in steam drums by conditioning the level measurement to reduce its sensitivity to pressure.

We can adopt a similar approach to tray temperature controllers on distillation columns. They provide some control of product composition because this correlates with the bubble point of the liquid. However, changing pressure changes this relationship. Figure 5.11 shows the effect pressure has on bubble point, in this case water, but all liquids show a similar behaviour.

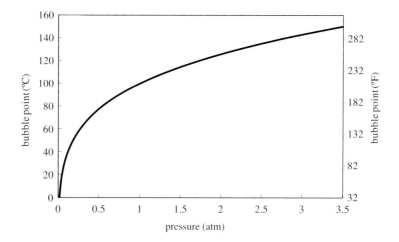

Figure 5.11 Effect of pressure on bubble point

If a distillation tray temperature controller keeps the temperature constant as the pressure changes, the composition will move away from target. We can resolve this by using the pressure to condition the temperature measurement. The subject of *pressure compensated temperatures* is covered in full in Chapter 12.

5.5 Pressure Compensation of Gas Flow Measurement

Gas flow measurements, using an orifice type of flow meter, are also sensitive to pressure. The instrument range, configured in the DCS, was determined assuming a calibration molecular weight (MW_{cal}), pressure (P_{cal}) and temperature (T_{cal}). If the current conditions (MW, P and T) are different from these then we must apply correction to the measured flow ($F_{measured}$) to obtain the true flow (F_{true}). The form of correction depends on the units of measure. The equation given below should only be applied to orifice type meters, with the flow recorded in **volumetric** units at **standard conditions**, for example nm³/hr or SCFM (standard cubic feet per minute). Pressure and temperature should be on an absolute basis:

$$F_{true} = F_{measured} \sqrt{\frac{MW_{cal}}{MW} \times \frac{P}{P_{cal}} \times \frac{T_{cal}}{T}} \qquad (5.18)$$

If the flow measurement is in **actual** volumetric units, i.e. reported at actual (rather than standard) pressure and temperature, then the formula becomes

$$F_{true} = F_{measured} \sqrt{\frac{MW_{cal}}{MW} \times \frac{P_{cal}}{P} \times \frac{T}{T_{cal}}} \qquad (5.19)$$

And if the flow measurement is on a **weight** basis then

$$F_{true} = F_{measured} \sqrt{\frac{MW}{MW_{cal}} \times \frac{P}{P_{cal}} \times \frac{T_{cal}}{T}} \qquad (5.20)$$

However these formulae should be applied with care. In Chapter 10 we show how their application to gaseous **fuels** can worsen problems with combustion control.

Similarly applying them to gas mixtures, where the aim is to maintain the flow of a **single component** as composition changes, also requires special consideration. For example, if we wished to control the flow of hydrogen supplied as a mixture with other gases then we can infer the mole fraction of hydrogen (x). Knowing the molecular weight of the gas mixed with the hydrogen (MW_{other}) and that of hydrogen itself ($MW_{hydrogen}$) gives

$$x = \frac{MW_{other} - MW}{MW_{other} - MW_{hydrogen}} \qquad (5.21)$$

So the flow of hydrogen, in standard volumetric units, is given by

$$F_{true} = F_{measured} \sqrt{\frac{MW_{cal}}{MW} \times \frac{P}{P_{cal}} \times \frac{T_{cal}}{T}} \times \frac{MW_{other} - MW}{MW_{other} - MW_{hydrogen}} \qquad (5.22)$$

Whether the correction term for temperature should be included should also be given consideration. If, for example, the calibration temperature is 50 °C (around 120 °F) and the actual temperature varies by 10% then, because we convert temperature to an absolute basis and then take square root of the ratio, the error introduced is less than 1%. This is probably within the measurement repeatability. Involving another measurement introduces an additional source of potential instrument problems. Temperature correction is only worthwhile therefore if the operating temperature is very high, or the change can be very large.

A similar argument applies to pressure compensation if the operating **gauge** pressure is close to zero. For example a 10% change in a pressure of 0.3 barg (around 4 psig), when converted to absolute pressure, causes a flow measurement error of about 1%.

5.6 Filtering

Another common form of signal conditioning is filtering – used to reduce measurement noise. Noise may be genuine in that the instrument is faithfully reproducing rapid fluctuations in the measurement. Examples include measuring the level of a turbulent liquid or the flow of a mixed phase fluid. Noise may be introduced mechanically by vibration or electrically though interference. While filtering may reduce the problem, it is unlikely to remove it completely and it will distort the base signal.

Whatever the cause, efforts should be made to eliminate the noise at source. The use of baffles or stilling wells around the level sensor can prevent turbulence affecting the measurement. Ensuring that flows are measured where liquid is below its bubble point will avoid flashing across the orifice plate. Placing transmitters away from vibrating equipment and having signal cables properly screened and not routed close to large electrical equipment will avoid induced noise.

Filtering will change the apparent process dynamics, usually in a way detrimental to controller performance. This is often explained in text books, using Figure 5.12, as a phase lag.

If a sinusoidal signal is injected into a conventional DCS filter then the output will be reduced in amplitude and shifted in time. This is not particularly helpful in the process industry, where the engineer rarely comes across sinusoidal signals. Perhaps a more pragmatic approach is to consider the noisy measurement trended in Figure 5.13.

The challenge is to remove the noise from the underlying base signal, without distorting it. This would appear straightforward enough; most could add the base signal to the trend in

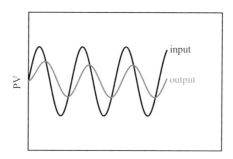

Figure 5.12   Filter phase lag

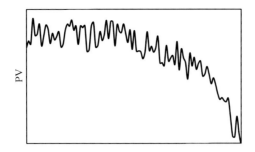

Figure 5.13 Typical noisy measurement

Figure 5.13. But doing so involves looking back in time. If the second half of the trend had yet to be drawn it would be quite difficult to decide whether the downward movement is a genuine reduction or just another noise spike. It is not until more information is provided that the distinction is clear. Filters have the same problem; it is not until well after the base signal has changed that the filter can recognise it. There will therefore be a delay of some sort, before the change is passed to the controller.

While a noisy measurement may not look good when trended, this no reason to add a filter. The criterion on which the decision should be made is what noise is passed though the controller to the final actuator, for example the control valve. If there is a danger of mechanical damage then a filter may offer the only practical solution. In particular, filtering should be used if derivative action is justified since this would otherwise greatly amplify the noise.

The problem is that no filter is perfect. While all filters can be tuned to suit the process, there will always be a compromise between noise reduction and base signal distortion. Whether a filter is effective depends on the relative impact these two problems have on the controller. Filter lag may be of little concern if the process already has a very large lag. Noise reduction may not be critical if the controller gain is small and there is no need for derivative action.

5.7 Exponential Filter

DCS have generally standardised on the *first order exponential filter*. This introduces an engineer-configurable lag (with time constant τ_f) on the PV. It is implemented as

$$Y_n = P \cdot Y_{n-1} + (1-P)X_n \tag{5.23}$$

Y_n is the current output from the filter, Y_{n-1} the previous output and X_n the current input. P is a tuning parameter set by the engineer in the range 0 to 1. If set to 0 the current output will be equal to the current input and no filtering takes place. If set to 1 the current output will be equal to the previous output and any change in measurement is ignored. Some systems permit any value within this range; others limit P to predefined values such as 0, 0.5, 0.75 or 0.85. Other systems accept the time constant (τ_f) where this is related to P by

$$P = e^{-ts/\tau_f} \quad \text{or} \quad \tau_f = \frac{ts}{-\ln(P)} \tag{5.24}$$

Some texts will define P as

$$P = 1 - \frac{ts}{\tau_f} \quad \text{or} \quad \tau_f = \frac{ts}{1-P} \tag{5.25}$$

This is based on the first order approximation to the Taylor expansion

$$e^{-ts/\tau_f} = 1 - \frac{ts}{\tau_f} \tag{5.26}$$

Others will define P as

$$P = \frac{2\tau_f - ts}{2\tau_f + ts} \quad \text{or} \quad \tau_f = \frac{ts(1+P)}{2(1-P)} \tag{5.27}$$

This is based on the first order Padé approximation

$$e^{-ts/\tau_f} = \frac{2 - \frac{ts}{\tau_f}}{2 + \frac{ts}{\tau_f}} \tag{5.28}$$

Remembering that ts is likely to be measured in seconds and τ_f in minutes, ts will be very much smaller than τ_f. Higher order terms in the Taylor and Padé approximations will then rapidly approach zero. The performance of a filter based on either approximation will be indistinguishable from the exact version unless τ_f is very small – in which case one would question whether the filter is necessary.

Care needs to be taken when calculating P from τ_f, or vice-versa, to work in consistent units of time. The relationship between P and τ_f depends on the controller scan interval (ts seconds), as shown in Figure 5.14. While it is unusual to change the scan interval of the DCS it is common for controllers to be moved from one system to another that may have a different scanning frequency. The filter will then perform differently both in terms of noise reduction and the effect it has on the apparent process dynamics. Hence the performance of the controller may degrade.

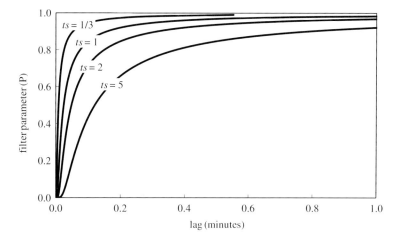

Figure 5.14 Impact of controller scan interval

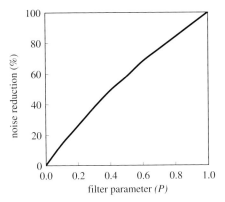

Figure 5.15 Performance of the exponential filter

The effectiveness of the filter is shown in Figure 5.15. While the actual level of noise reduction is dependent on the type and frequency of the noise, it shows that it is approximately linear with P. However the impact on process dynamics is highly nonlinear. Figure 5.14 show the lag introduced by the filter increases sharply as the value of P exceeds 0.9. Should this level of filtering be required then the additional lag may be large compared to the process lag. If it exceeds around 20% of the process lag then, even if the controller is retuned to accommodate it, the degradation in performance will be noticeable. If filter lag is excessive then it would be better to adopt a different technique, such as that covered later in this chapter.

5.8 Higher Order Filters

Higher orders of filter are possible. For example a second order exponential filter (with both lags set to τ_f) can be developed by connecting two first order filters in series. The first would be as Equation (5.23) generating an intermediate output Y^* that becomes the input to a second filter, i.e.

$$Y_n^* = P \cdot Y_{n-1}^* + (1-P)X_n \tag{5.29}$$

$$Y_n = P \cdot Y_{n-1} + (1-P)Y_n^* \tag{5.30}$$

Substituting Equation (5.29) into Equation (5.30)

$$Y_n = P \cdot Y_{n-1} + (1-P)(P \cdot Y_{n-1}^* + (1-P)X_n) \tag{5.31}$$

Writing Equation (5.30) for the preceding scan

$$Y_{n-1} = P \cdot Y_{n-2} + (1-P)Y_{n-1}^* \quad \text{or} \quad Y_{n-1}^* = \frac{Y_{n-1} - P \cdot Y_{n-2}}{1-P} \tag{5.32}$$

Substituting into Equation (5.31)

$$Y_n = 2P \cdot Y_{n-1} - P^2 Y_{n-2} + (1-P)^2 X_n \tag{5.33}$$

In theory high order filters of this sort will distort the base signal less for the same amount of noise reduction. However, the difference in performance would go unnoticed on a real process and the effort required in installing and tuning it is not justified. Higher order techniques using a combination of different lags, such as *Butterworth*, are available but their design requires detailed knowledge of the frequency distribution of the noise and real changes in the base signal. The approach is too sophisticated for the process industry and certainly beyond the scope of this book.

Equation 5.33 can be written in the form

$$Y_n = a_1 \cdot Y_{n-1} + a_2 Y_{n-2} + b_1 X_n + b_2 X_{n-1} \quad (5.34)$$

$$a_1 = 2e^{-ts/\tau_f} \qquad a_2 = -e^{-2ts/\tau_f} \quad (5.35)$$

$$b_1 = 1 - 2e^{-ts/\tau_f} + e^{-2ts/\tau_f} \qquad b_2 = 0 \quad (5.36)$$

The definitions of b_1 and b_2 are somewhat different from those obtained by setting $PV = Y$, $MV = X$, $\tau = \tau_f$ and $\tau_3 = 0$ in the second order process model shown as Equations (2.35) to (2.37). These were developed by applying z-transforms. It demonstrates the sensitivity of some definitions to the method by which they are obtained.

5.9 Nonlinear Exponential Filter

It is possible to modify the first order exponential filter, described by Equation (5.23), to make it nonlinear. Instead of the parameter P being set by the engineer it is changed automatically according to the formula

$$|X_n - Y_{n-1}| > R \quad \text{then} \quad P = 0 \quad (5.37)$$

$$|X_n - Y_{n-1}| \leq R \quad \text{then} \quad P = 1 - \frac{|X_n - Y_{n-1}|}{R} \quad (5.38)$$

The engineer selects the value for R. If the difference between the current input and the last output is greater than this value then the change passes through unfiltered. Changes less than R are filtered depending on their size. Very small changes are heavily filtered while larger changes are filtered less. The objective of this enhancement is to reduce the lag caused by filtering. Figure 5.16 shows its effectiveness at noise reduction for different types of noise.

The filter works well if the noise amplitude is predictable. By setting R to a value slightly higher than this amplitude, real changes in the base signal will be little affected by the filter. This would be the situation, for example, if the noise in a flow measurement is caused by flashing across an orifice plate. However the filter offers little advantage if the noise is less predictable or 'spiky' – such as level measurement of a boiling liquid. Such noise is often described as *Gaussian*, reflecting its statistical distribution. To prevent the spikes passing through the filter the value of R has to be set so large that the filter behaves much like the unmodified exponential filter. Its performance on our example process is shown in Figure 5.17.

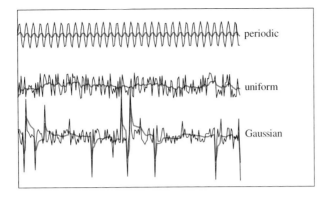

Figure 5.16 Impact of noise type on performance of nonlinear exponential filter

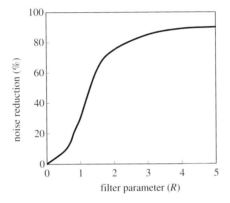

Figure 5.17 Performance of the nonlinear exponential filter

5.10 Averaging Filter

Simple averaging can be used as a filter. The filter can be represented as

$$Y_n = \frac{\sum_{r=1}^{N} X_{N-r+1}}{N} \tag{5.39}$$

N is the number of historical values included in the average and is the tuning constant defined by the engineer. The filter can also be written as

$$Y_n = B_1 X_N + B_2 X_{N-1} \ldots + B_r X_{N-r+1} \ldots + B_N X_1 \tag{5.40}$$

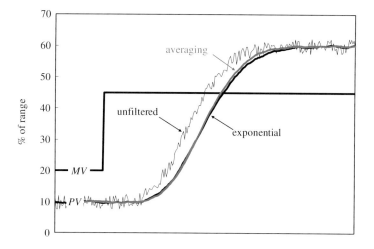

Figure 5.18 Comparison of averaging versus exponential filter

The filter should have a gain of 1 and so

$$\sum_{i=1}^{N} B_i = 1 \qquad (5.41)$$

and because it is a simple linear average

$$B_1 = B_2 \ldots = B_{N-r+1} \ldots = B_N = \frac{1}{N} \qquad (5.42)$$

This filter offers little advantage over the standard exponential filter. Rather than a lag, it introduces a ramp function where the duration of the ramp is given by $N.ts$. This would be visible if the input were a step change. But when superimposed on the process lag the result, with N adjusted to give equivalent noise reduction, is virtually indistinguishable from the exponential filter – as shown in Figure 5.18.

Given that the averaging filter is unlikely to be standard feature of the DCS and that its impact on process dynamics is not quite so easy to predict, it offers little advantage.

5.11 Least Squares Filter

We can choose different coefficients for B in Equation (5.40) provided they sum to 1. An approach which does this is known as the *least squares filter*. It gets its name from the least squares regression technique used to fit a line to a set of points plotted on an XY (scatter) chart. Its principle, as shown in Figure 5.19 is to fit a straight line to the last N points. The end of this line is Y_n.

We can show that the filter is of the form of Equation (5.40). The development of the formula to estimate the coefficients $B_1, B_2, \ldots, B_N$ is quite complex. However the end result is simple to apply. Indeed the reader could skip to Equation (5.56) if happy to just to accept the result.

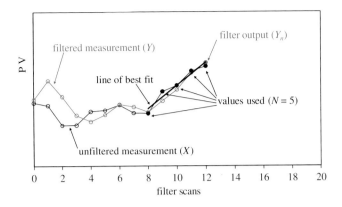

Figure 5.19 Principle of least squares filter

Figure 5.19 shows the last N values of a process measurement (Y). We define the time axis as r, where $r=1$ for the most recent measurement and $r=N$ for the oldest. The filter is based on predicting the value of Y based on the equation of the line of best fit, where m is the gradient of the line and c the intercept on the Y axis, i.e.

$$\hat{Y} = mr + c \qquad (5.43)$$

The equation of the line is developed to minimise the sum of the squares between the predicted value of Y and the actual value, i.e.

$$\sum_{i=1}^{N} E_i^2 = \sum_{i=1}^{N} (\hat{Y}_i - Y_i)^2 = \sum_{i=1}^{N} (mr_i + c - Y_i)^2 \qquad (5.44)$$

Partially differentiating with respect to each of m and c, and setting the derivative to 0 will identify the best choice of these values, i.e.

$$\frac{\partial \sum_{i=1}^{N} E_i^2}{\partial m} = \sum_{i=1}^{N} (2r_i^2 m + 2rc - 2rY_i) = 0 \qquad (5.45)$$

$$\therefore m \sum_{i=1}^{N} r_i^2 + c \sum_{i=1}^{N} r_i - \sum_{i=1}^{N} r_i Y_i = 0 \qquad (5.46)$$

$$\frac{\partial \sum_{i=1}^{N} E_i^2}{\partial c} = \sum_{i=1}^{N} (2r_i m + 2c - 2Y_i) = 0 \qquad (5.47)$$

$$\therefore m \sum_{i=1}^{N} r_i + Nc - \sum_{i=1}^{N} Y_i = 0 \qquad (5.48)$$

Solving Equations (5.46) and (5.48) gives

$$m = \frac{\sum_{i=1}^{N} r_i \sum_{i=1}^{N} Y_i - N \sum_{i=1}^{N} r_i Y_i}{\left(\sum_{i=1}^{N} r_i\right)^2 - N \sum_{i=1}^{N} r_i^2} \tag{5.49}$$

$$c = \frac{\sum_{i=1}^{N} r_i \sum_{i=1}^{N} r_i Y_i - \sum_{i=1}^{N} r_i^2 \sum_{i=1}^{N} Y_i}{\left(\sum_{i=1}^{N} r_i\right)^2 - N \sum_{i=1}^{N} r_i^2} \tag{5.50}$$

We wish to predict the current value of Y, i.e. when $r = 1$ and so

$$\hat{Y} = m + c \tag{5.51}$$

and so

$$\hat{Y} = \frac{\sum_{i=1}^{N} r_i \sum_{i=1}^{N} Y_i - N \sum_{i=1}^{N} r_i Y_i + \sum_{i=1}^{N} r_i \sum_{i=1}^{N} r_i Y_i - \sum_{i=1}^{N} r_i^2 \sum_{i=1}^{N} Y_i}{\left(\sum_{i=1}^{N} r_i\right)^2 - N \sum_{i=1}^{N} r_i^2} \tag{5.52}$$

This is a linear function of previous values of Y and so can be written in form of Equation (5.40). To determine the coefficients (B) we use the formula for the sum of a series of consecutive integers:

$$\sum_{i=1}^{N} r_i = \frac{N(N+1)}{2} \tag{5.53}$$

and the sum of a series of squares of consecutive integers

$$\sum_{i=1}^{N} r_i^2 = \frac{N(2N+1)(N+1)}{6} \tag{5.54}$$

Substituting these into Equation (5.52) we can determine the value of each coefficient; so

$$B_r = \frac{\frac{N(N+1)}{N} - Nr - \frac{N(N+1)}{N} r + \frac{N(2N+1)(N+1)}{6}}{\left(\frac{N(N+1)}{2}\right)^2 - N \frac{N(2N+1)(N+1)}{6}} \tag{5.55}$$

This simplifies to

$$B_r = \frac{4N - 6r + 4}{N(N+1)} \tag{5.56}$$

Figure 5.20 shows its performance in terms of noise reduction on our example process. A larger value of N is required to achieve the same level of noise reduction as the averaging filter.

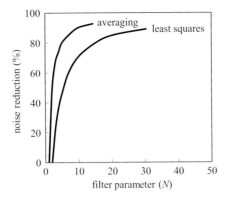

Figure 5.20 Performance of averaging and least squares filters

The advantage of this filter is that, because it uses the trend of the unfiltered measurement to predict the filtered value, it has no lag. Indeed its predictive nature introduces lead which partially counteracts the process lag. This effect is shown in Figure 5.21. Both filters have been tuned to give the same level of noise reduction. The least squares filter not only outperforms the exponential filter in that it tracks the base signal more closely but actually overtakes the base signal.

The disadvantage of applying any filter other than the standard first order exponential type is that it will require custom coding in the DCS. However in the case of the least squares type the calculation is quite simple and the coefficients (B) can be determined outside of the DCS and stored as constants. While it is likely that the value required for N will be larger, as an illustration, Equation (5.56) has been used to generate the coefficients in Table 5.1 for values of N up to 8.

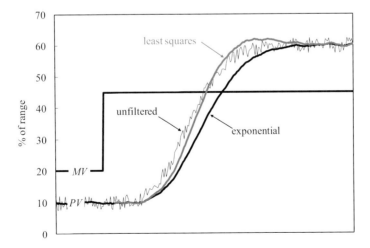

Figure 5.21 Comparison of least squares versus exponential filter

Table 5.1 Coefficients for least squares filter

N	B_1	B_2	B_3	B_4	B_5	B_6	B_7	B_8
1	1.000							
2	1.000	0.000						
3	0.833	0.333	−0.167					
4	0.700	0.400	0.100	−0.200				
5	0.600	0.400	0.200	0.000	−0.200			
6	0.524	0.381	0.238	0.095	−0.048	−0.190		
7	0.464	0.357	0.250	0.143	0.036	−0.071	−0.179	
8	0.417	0.333	0.250	0 167	0.083	0.000	−0.083	−0.167

5.12 Control Valve Characterisation

Nonlinear valves are an example of *output conditioning*. It is often the case that we do not want the valve position to move linearly with controller output. The most common type of nonlinear control valves are of the *equal percentage* type, often used when controlling the flow of fluid from a centrifugal pump or compressor as shown in Figure 5.22.

Figure 5.23 shows how pressure varies with flow. Assuming that the pressure at the process exit is constant, for example because the product is routed to storage, then the process inlet pressure will increase with the square of the flow. Also shown is the pump curve. With no control valve in place, the pump discharge pressure must be the same as the process inlet pressure and the flow is set by where the pump curve and the process curve intersect. But we want to control the flow at a desired value. To do so the drop in pressure (Δp) across the control valve must be equal to the pump discharge pressure less the process inlet pressure.

Figure 5.24 shows the effect that varying the flow has on the pressure drop across the valve. Ideally we want a linear relationship between flow and valve position, as would be the result if the pressure drop across the valve were constant. The constant process gain would allow a gain to be implemented in the flow controller that would be valid over the whole operating range.

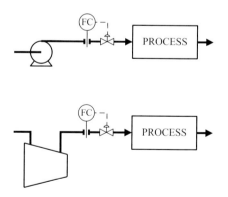

Figure 5.22 Control of pump discharge flow

Signal Conditioning 137

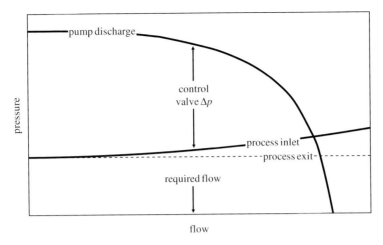

Figure 5.23 System pressures

Figure 5.25 shows how the controller output should be conditioned to compensate for the nonlinearities introduced by the process and pump curves. Some DCS permit the definition of a look-up table. For example, the DCS would interpolate between the five selected points – effectively joining them with straight lines to closely match the required curve.

5.13 Equal Percentage Valve

An alternative approach, also shown in Figure 5.25, is the use of an equal percentage valve. This may be manufactured to behave in a specified nonlinear behaviour or it may be a conventional linear valve fitted with a positioner in which the engineer can define the valve

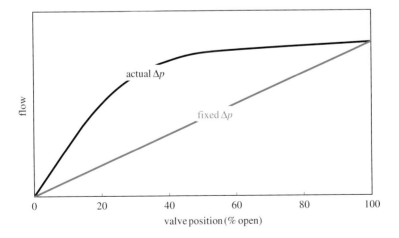

Figure 5.24 Relationship between flow and valve position

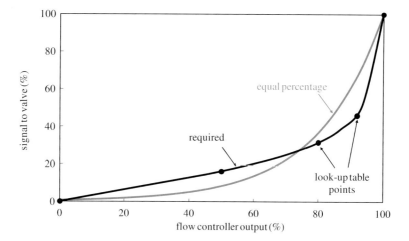

Figure 5.25 Valve characterisation

characterisation. The definition of 'equal percentage' is that the same change in valve position will give the same percentage change in flow. This illustrated in Figure 5.26; the change in valve position required to increase the flow from 40 to 60 (i.e. a 50% increase) is the same as that required to increase the flow from 10 to 15 (i.e. also a 50% increase).

A small compromise has to be made to equal percentage to ensure the flow through the valve is zero when the valve is fully shut but, in general, if F is the flow and V the percentage valve position then, by definition

$$\frac{\Delta F}{F} \propto \Delta V \qquad (5.57)$$

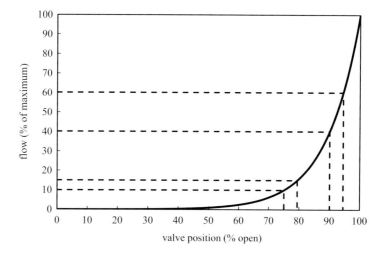

Figure 5.26 Equal percentage valve

and so, by introducing the *valve constant* (k)

$$\frac{dF}{F} = k\,dV \tag{5.58}$$

integrating gives

$$\ln(F) = kV + A \quad \text{or} \quad F = e^{(kV+A)} \tag{5.59}$$

The constant A can be eliminated by defining F_{max} as the flow with the valve 100% open, i.e.

$$F_{max} = e^{(100k+A)} \tag{5.60}$$

and so

$$\frac{F}{F_{max}} = \frac{e^{(kV+A)}}{e^{(100k+A)}} = e^{k(V-100)} \tag{5.61}$$

If ρ is the liquid density then control valves are characterised by the coefficient (C_v) defined as

$$C_v = F\sqrt{\frac{\rho}{\Delta p}} \tag{5.62}$$

For a linear valve C_v is constant, but for an equal percentage valve

$$C_{v0} = F_{max}\sqrt{\frac{\rho}{\Delta p}} \quad \text{and so} \quad \frac{C_v}{C_{v0}} = e^{k(V-100)} \tag{5.63}$$

Figure 5.27 shows the effect of the valve constant (k). By choosing the correct value a reasonably close match can be achieved to the output conditioning required (see Figure 5.25).

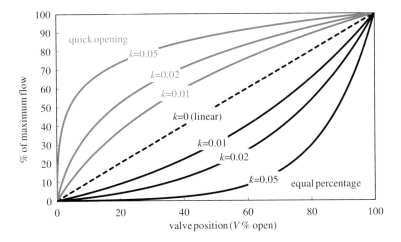

Figure 5.27 Effect of valve constant

If the pressure drop across the valve exceeds the pressure drop across the rest of the process then it will change relatively little as the flow changes. A linear valve will then provide better linearity. If the pressure drop across the valve is relatively small then an equal percentage will perform better. But oversizing a linear valve has a more detrimental effect than oversizing an equal percentage valve; so the latter is often chosen to provide more robust control as process conditions vary.

It is also possible to apply the inverse of this transformation to give the valve a *quick opening* characteristic. Such valves would be selected where the speed of opening is more important than linearity. A common example is their use in antisurge recycles on compressors.

5.14 Split-Range Valves

While valves are generally calibrated to move over their full range as controller output varies from 0 to 100%, other options are possible. For example we may wish the valve to fail open on loss of signal – in which case we calibrate it to operate over the range 100 to 0%. We can also calibrate the valve to move over its full range as the controller output changes over only part of its range, for example 0 to 50%. We could then calibrate a second valve to move over its full range as the controller output varies from 50 to 100%. This *split-range* approach would then cause the valves to open and close in sequence.

Before describing possible applications it is important to distinguish between split-ranging and *dual-acting* valves. Rather than act in sequence dual acting valves move simultaneously. Figure 5.28 show two ways in which pressure might be controlled in a distillation column.

The first scheme controls pressure by changing the condenser duty through manipulation of the bypass. On increasing pressure the controller simultaneously begins to close the bypass and open the valve on the condenser, as shown in Figure 5.29. The valves have been paired to act like a three-way valve – but are a lower cost approach.

The second scheme in Figure 5.28 would first open the condenser valve and, if the pressure remains high, begin opening the valve venting vapour from the process. This has been achieved by calibrating the first valve to operate over 0 to 50% of controller output and the second to operate over 50 to 100%, as shown in Figure 5.30

While split-ranging is common in industry it does have some limitations. Figure 5.31 shows another method of controlling pressure in a distillation column, often used when vapour production is intermittent. In the absence of sufficient vapour the scheme is designed to allow a noncondensible gas into the column. So, on increasing pressure, the controller will first begin to close valve A until it is fully closed. If the pressure does not fall sufficiently, it will then begin to open valve B.

Figure 5.32 shows how the valves have been calibrated. As is common the range has been split equally between the two valves.

One of the problems with split-ranging is that there can be a large change in process gain as control switches from one valve to the other. To avoid the controller becoming oscillatory it has to be tuned for the range where the process gain is higher. It will thus respond sluggishly when operating in the other part of the range. It is possible to alleviate this problem by redefining the split. The method for doing so is shown in Figure 5.33. The solid

Signal Conditioning 141

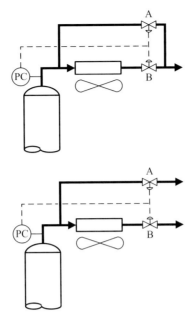

Figure 5.28 Dual-acting versus split-range

line represents the process behaviour. It is obtained from historical data or plant tests as described in Chapter 2. In this case it is likely that the pressure will be an integrating process and so, in this example, rate of change of pressure is plotted against controller output. The coloured line is the required relationship. It would be provided by moving the split to around 30%.

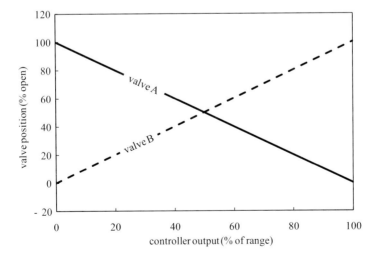

Figure 5.29 Dual-acting valves

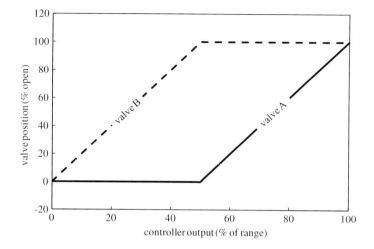

Figure 5.30 Split-range valves

While this technique deals with differences in process gain (K_p), it does not account for any difference there might be in the θ/τ ratio. Even with the split optimised it is possible that the pressure controller will require different tuning as its output moves from the 0 to 30% part of the range to the 30 to 100% part. In extreme cases it might not be possible to choose a compromise set of tuning that gives acceptable performance over the whole range.

A further potential source of problems is the accuracy of the valve calibration. For example if valve A actually travelled its full range as the controller output moved from 0 to 25%, because of some inaccuracy, then there would be a deadband between 25% and 30% where the process gain will be zero – as shown in Figure 5.34.

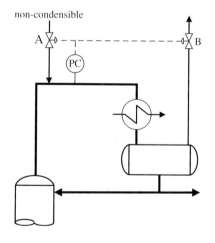

Figure 5.31 Split-range pressure controller

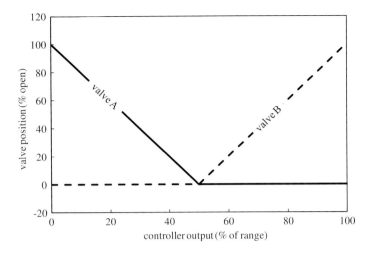

Figure 5.32 Split-range valve calibration

Conversely if valve A were inaccurately calibrated to operate over the controller output range of 0 to 35% then there would be an *overlap* between 30 and 35% where the process gain is doubled – as shown in Figure 5.35.

Both poor calibration situations will severely impact controller performance. Further, if the gas, being used to pressurise the column through valve A, is valuable and depressurisation through valve B is to some waste gas system, then the overlap would mean that both valves are open simultaneously and incurring an unnecessary cost.

Operators generally interpret a controller output of 0% as a signal to close the valve and 100% to open a valve. With split-ranging it may not be possible to display the signal in this way; so requiring special attention to operator training. This can be complicated by some

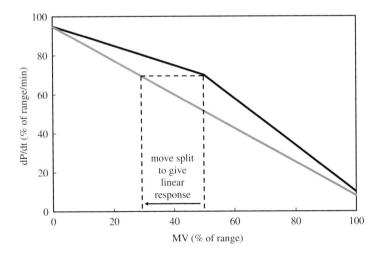

Figure 5.33 Redefining the split to give a linear response

144 Process Control

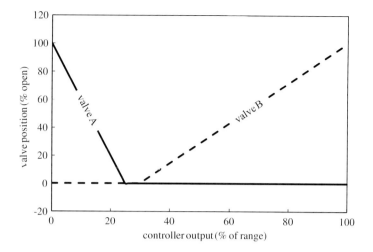

Figure 5.34 Inaccurate calibration causing a deadband

control valves being configured as air fail close (or air to open) and others as air fail open (or air to close) depending on what behaviour is required in the event of loss of signal.

In many cases split-ranging is a perfectly satisfactory way of meeting the control objectives. However, poor performance may go unnoticed if the controller spends the majority of the time in one part of the range. It is also possible to split the controller output into more than two parts if there are more MVs available to extend the operating range. With more splits, accurate valve calibration becomes more important and it is more likely that the there will a greater variation in process dynamics.

With pneumatic control systems one advantage of split-ranging is that both valves can be connected to the same pneumatic signal line, thus saving the cost of the second line. With

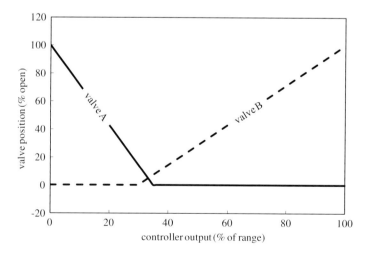

Figure 5.35 Inaccurate calibration causing an overlap

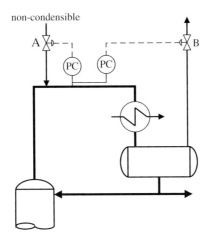

Figure 5.36 Alternative to split-range pressure controller

multicore cabling or networked 'smart' systems, the incremental cost of the second connection is small. A cost-effective alternative approach is then to design separate controllers, as shown in Figure 5.36. In our example we would have one pressure controller manipulating valve A and another manipulating valve B. Both valves are calibrated normally. This allows us to tune the controllers independently, allowing for any difference in dynamics. It is important that the controllers share the same PV; using independent transmitters with even the slightest measurement error will cause the controllers to fight each other. To achieve the required sequential operation, the controller manipulating valve B would have a slightly higher SP.

6
Feedforward Control

Figure 6.1 shows a simple *feedback* scheme. The objective is control the temperature (T) of mixture of two streams of temperatures T_a and T_b by manipulating the flow of one of them. We have no control over the flow of stream A – indeed changes in its flow are the main disturbance to the process.

The feedback scheme is limited in that it can only take corrective action once it has detected a deviation from temperature SP. This is particularly important if there is significant deadtime between the PV and the MV, for example if the temperature measurement was a long way downstream of the mixer. No matter how well-tuned the feedback controller, it cannot have any impact on the PV until the deadtime has elapsed. During this time the error (E) will increase to

$$E = (K_p)_a \Delta F_a \left[1 - e^{-\theta_b/\tau_a}\right] \tag{6.1}$$

$(K_p)_a$ is the process gain and τ_a the lag of the temperature in response to a change in the flow of stream A (ΔF_a). θ_b is the deadtime of the temperature in response to its MV, the flow of stream B.

6.1 Ratio Algorithm

Figure 6.2 shows a possible *feedforward* scheme. We introduce an additional measurement – known as the disturbance variable (DV). In this case we have incorporated the measurement of the flow of stream A. By monitoring this flow we can predict that it will disturb the temperature and so can take action **before** the temperature changes. To do so we include a *ratio algorithm*. This algorithm generates an output by multiplying the two inputs. One input is the measured flow of stream A; the other is the operator-entered target for the ratio of the flow of stream B to stream A. As the flow of stream A changes the ratio algorithm maintains the flow of stream B in proportion, thus keeping the temperature

---

Process Control: A Practical Approach Myke King
© 2011 John Wiley & Sons, Ltd

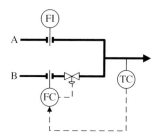

Figure 6.1 Feedback control

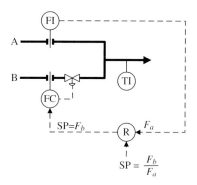

Figure 6.2 Feedforward control

constant. If the liquids have the specific heats of $(c_P)_a$ and $(c_P)_b$ then, in order to meet the target temperature (T), this ratio (R) will be set as

$$R = \frac{(c_P)_a}{(c_P)_b}\left[\frac{T - T_a}{T_b - T}\right] \tag{6.2}$$

The ratio algorithm is found in most DCS as a standard feature. Strictly, since it does not have feedback, it is not a controller – although it is often described as such. It is however more than just a simple multiplier. It incorporates the equivalent of PV tracking. When in manual mode the ratio SP tracks the actual ratio. If R is the ratio SP, I the input measurement and M the algorithm's output then the algorithm performs the calculation

$$R = \frac{M}{I} \tag{6.3}$$

When switched to automatic the ratio SP is fixed at the current value and the calculation changes to

$$M = R \times I \tag{6.4}$$

Once initialised in this way the operator may change the ratio SP, or may cascade a controller to adjust it as necessary. This means that we do not need to measure the stream temperatures and specific heats, as used in Equation (6.2). Provided that, when the ratio is switched to automatic, the target temperature is being met then the ratio SP will be automatically initialised to the correct value. Further, to make it appear like a controller, the DCS will permit both the ratio SP and the actual ratio to be displayed. In some DCS, the ratio algorithm is incorporated into the PID controller and can be configured as an option.

In this particular case, which is an example of blending, a better approach is to manipulate the ratio between the controlled flow and the **total** flow. If F_a and F_b are the flows of the two streams then, if the specific heats of the two streams are the same,

$$F_a T_a + F_b T_b = (F_a + F_b)T \quad (6.5)$$

If we define the ratio, as shown in Figure 6.2, as F_b/F_a then

$$T = \frac{T_a + RT_b}{R+1} \quad \text{and} \quad K_p = \frac{dT}{dR} = \frac{T_b - T_a}{(R+1)^2} \quad (6.6)$$

This shows that K_p is not constant. For example, if R is typically 0.5 but varies outside the range 0.35 to 0.65, K_p will vary by more than $\pm 20\%$ and cause problems with tuning the temperature controller. However, if we define R as $F_b/(F_b + F_a)$, then

$$T = (1-R)T_a + R.T_b \quad \text{and} \quad K_p = \frac{dT}{dR} = T_b - T_a \quad (6.7)$$

The process gain no longer varies with R. So, if the ratio is likely to be highly variable, the scheme shown as Figure 6.3 would be preferable.

An alternative approach is shown in Figure 6.4. The ratio is calculated from the two flow measurements and used as the PV of a true ratio controller. In addition to the nonlinearity described above between temperature and ratio, the process gain of the ratio with respect to

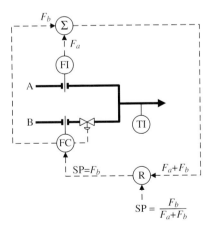

Figure 6.3 Preferred blend ratio control

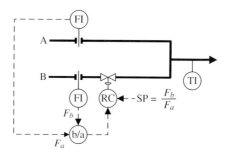

Figure 6.4 Alternative configuration for feedforward control

changes in valve position is now inversely proportional to the uncontrolled flow. In theory therefore the ratio controller gain should be kept in proportion to the uncontrolled flow. But, unless the variation is extremely large, it is unlikely that this is required in practice. The scheme can also be modified to use the preferred blend ratio.

Feedforward is not however a **replacement** for the feedback scheme. It does not incorporate all possible disturbances. For example it would not compensate for a change in the temperature of either stream. While in theory we could include these measurements as additional DVs, it is unlikely that we could include all possible sources of disturbance. For instance, measuring changes in the specific heat of either liquid is unlikely to be practical. Further, feedforward relies on the instrumentation being accurate. If there was a bias error in the measurement of either flow, maintaining a fixed proportion would not meet the target temperature. Finally we need to provide the process operator with a practical way of changing the target temperature. It should not be necessary for him to have to calculate the revised target ratio.

We therefore treat feedforward as an **enhancement** to feedback control. We combine them to give a *feedforward-feedback* scheme where, in this case, the temperature controller now manipulates the ratio target – as shown in Figure 6.5.

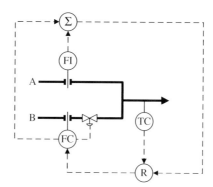

Figure 6.5 Feedforward-feedback control

6.2 Bias Algorithm

The ratio algorithm provides a means by which two strategies can manipulate the same variable. In our example both the feedforward and feedback parts of the scheme can change the flow of stream B as necessary without fighting each other.

Rather than multiply them together, an alternative means of combining two signals is to add them. In addition to the ratio algorithm, the DCS is likely to include a *bias algorithm*. This performs in a similar way to the ratio. When in manual mode the bias (B) is determined by

$$B = M - I \tag{6.8}$$

When switched to automatic this calculation is replaced by

$$M = I + B \tag{6.9}$$

Using a bias algorithm would be incorrect in our mixing example. The temperature would not be kept constant by fixing the total (or the difference) of the two flows. However there are occasions where the use of the bias is correct and the ratio not so. Figure 6.6 shows one such situation. The heater uses a combination of two fuels. The first (fuel A), perhaps a by-product from another part of the process, can vary. The second (fuel B) can be manipulated to control the heater outlet temperature. Without feedforward in place, any change in the flow of fuel A would disturb the temperature. Given that the process will have a significant deadtime the disturbance will not be detected immediately by the temperature. And, for the same reason, the controller will not support very fast tuning. The deviation from SP is therefore likely to be large and sustained.

With the flow of fuel A incorporated as a DV, we can apply the bias algorithm to maintain the total fuel constant. The bias algorithm supports the addition of a scaling factor on its input. In this case we set this factor to -1. If we think as the output from the temperature controller being the total fuel required, the bias now subtracts the flow of fuel A from this

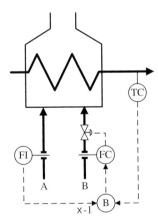

Figure 6.6 Bias feedforward example

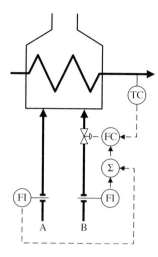

Figure 6.7 Alternative configuration for bias feedforward

and the result is the SP of the fuel B flow controller. Thus any change in the flow of fuel A causes an immediate compensation in fuel B and the temperature will remain constant. In Chapter 10 we show how this technique can be applied even if the fuels are quite different – for example one a gas and the other a liquid.

As with ratio feedforward, some DCS include the bias as an option within the PID algorithm. Some also permit configuration of the operator display to make the algorithm appear like a true controller with both the actual bias and the bias SP.

An alternative approach is shown in Figure 6.7. The two fuel flow measurements are summed, with suitable scaling factors to ensure the result is in units consistent with the total energy supplied. The result is then used as the PV of a total energy flow controller.

In this example it would be incorrect to apply a ratio algorithm. We do not wish to keep the two fuel flows in proportion. However there are situations were either algorithm may be used. For example, in Chapter 4, we described how the three-element level control scheme for a steam drum may be adapted to either approach.

6.3 Deadtime and Lead-Lag Algorithms

So far we have ignored process dynamics; we have assumed that the controller should change the MV **at the same time** as the DV changes. This is only correct if the dynamics of the PV with respect to changes in the MV are the same as its dynamics with respect to the DV. In our examples this is the case, but this is not always so. Consider the modification made to the blending process shown in Figure 6.8. A surge drum has been added, fitted with an averaging level control, so that fluctuations in the flow of stream A are reduced. However the temperature control scheme is unchanged.

An increase in the flow of stream A will cause an immediate increase in the flow of stream B. But, because the additional stream A is largely accumulated in the drum, this will cause the combined temperature to change. The feedback controller will detect this and will

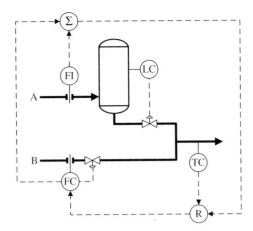

Figure 6.8 Need for dynamic compensation

bring the temperature back to SP by reducing the ratio target and hence the flow of stream B. However, the drum level controller will ultimately increase the flow of stream A to the mixer, causing a second disturbance to the temperature. The temperature controller will compensate for this by bringing the ratio SP back to its starting point.

The feedforward controller therefore made the correct change; it just did so too soon. The dynamics of the PV with respect to the DV are now much slower that its dynamics with respect to the MV. We therefore need to include some dynamic compensation that, in this case, delays the feedforward correction. Failure to properly include such compensation can result in the addition of feedforward causing the scheme to perform less well than the standalone feedback controller.

Of course, this is a contrived example. Had the process design department consulted the plant's control engineer, the drum would have been installed upstream of the measurement of the flow of stream A! Further the averaging level controller could then properly be cascaded to a flow controller.

The dynamic compensation required uses algorithms that are provided by the DCS. These are the *deadtime algorithm* and the *lead-lag algorithm*. The deadtime algorithm generates a pure delay, just like process deadtime, except the delay is a tuning constant (θ) configurable by the engineer.

The lead-lag algorithm has three tuning constants – gain (K), lead ($T1$) and lag ($T2$). It should not be necessary for the engineer to know how the DCS vendor has defined the algorithm, but if Y is the output and X the input, then strictly it should be of the form

$$Y_n = e^{-ts/T2} Y_{n-1} + K \frac{T1}{T2} X_{n-\theta/ts} - K \left(\frac{T1-T2}{T2} + e^{-ts/T2} \right) X_{n-\theta/ts-1} \qquad (6.10)$$

While one might think that the formula for the algorithm could be derived from Equations (2.23) to (2.26) by setting $Y=PV$, $X=MV$, $\tau_1=T2$, $\tau_2=0$ and $\tau_3=T1$. However the performance of the result only partially matches that of the analog version. Equation (6.10) was derived by making the same changes to the Laplace form **before** converting it to the z-transform. This illustrates the sensitivity of some definitions to the method by which they are obtained.

Generally θ will not be an exact multiple of ts and so the values of $X_{n-\theta/ts}$ and $X_{n-\theta/ts-1}$ are linearly interpolated between adjacent values of X.

$$X_{n-\theta/ts} = X_{n-\text{int}(\theta/ts)} - \left(\frac{\theta}{ts} - \text{int}\left(\frac{\theta}{ts}\right)\right)\left(X_{n-\text{int}(\theta/ts)} - X_{n-(\text{int}(\theta/ts)-1)}\right) \quad (6.11)$$

$$X_{n-\theta/ts-1} = X_{n-\text{int}(\theta/ts)-1} - \left(\frac{\theta}{ts} - \text{int}\left(\frac{\theta}{ts}\right)\right)\left(X_{n-\text{int}(\theta/ts)-1} - X_{n-\text{int}(\theta/ts)-2}\right) \quad (6.12)$$

The algorithm is coded in different ways in different DCS. For example, we can assume that $ts \ll T2$ and so make the first order Taylor approximation

$$e^{-ts/T2} = 1 - \frac{ts}{T2} \quad (6.13)$$

and so we get

$$Y_n = \frac{T2-ts}{T2} Y_{n-1} + K \frac{T1}{T2} X_{n-\theta/ts} - K \left(\frac{T1-ts}{T2}\right) X_{n-\theta/ts-1} \quad (6.14)$$

Alternatively we can make the more accurate first order Padé approximation

$$e^{-ts/T2} = \frac{2 - \frac{ts}{T2}}{2 + \frac{ts}{T2}} \quad (6.15)$$

and so we get

$$Y_n = \frac{2T2-ts}{2T2+ts} Y_{n-1} + K \frac{T1}{T2} X_{n-\theta/ts} - K \left(\frac{2T1}{2T2+ts}\right) X_{n-\theta/ts-1} \quad (6.16)$$

Like the PID algorithm, there are a variety of versions of the lead-lag algorithm depending on the method used to convert the analog version into its closest discrete equivalent. Few DCS vendors disclose exactly how the algorithm is coded. The approach to tuning must therefore be to assume that it is theoretically correct and, if it does not behave as exactly as expected, to adjust the tuning constants by trial and error.

Figure 6.9 shows the effect of the algorithm. The dashed line shows a step change in input (ΔDV).

The gain term (K) determines the steady state change in output. The impact of adjusting K is shown in Figure 6.10.

The output is delayed by θ. Figure 6.11 shows the impact of adjusting θ.

The output then changes as a step with the height of the step determined by the ratio of $T1$ to $T2$. Figure 6.12 shows the effect of varying the lead term ($T1$). With $T1$ set to zero the step is eliminated. If $T1$ is less than $T2$ then the step is less than the steady state change, if greater than $T2$ then the output overshoots the steady state change.

After the step the output approaches the steady state condition with a lag of $T2$. Figure 6.13 shows the effect of adjusting $T2$ – simultaneously adjusting $T1$ to maintain the ratio.

Feedforward Control 155

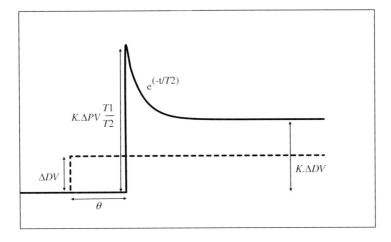

Figure 6.9 Effect of deadtime and lead-lag algorithms

If $T1$ is equal to $T2$ the step in the input passes through the algorithm as a step. By also setting K to 1 and θ to zero the output of the algorithm will be the same as the input.

6.4 Tuning

The inclusion of this algorithm has added four tuning constants. When feedforward is added to an existing feedback scheme we will show later that it may be necessary to retune the PID controller. Tuning seven constants by trial and error would be extremely time-consuming. While a little **fine tuning** may be necessary we should use the process dynamics to obtain the best possible estimate. Ideally we should be able to obtain an estimate which works first time.

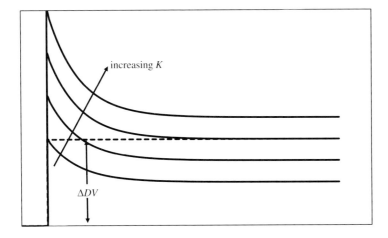

Figure 6.10 Effect of K

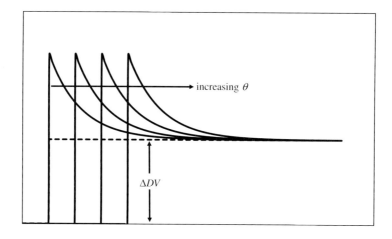

Figure 6.11   Effect of θ

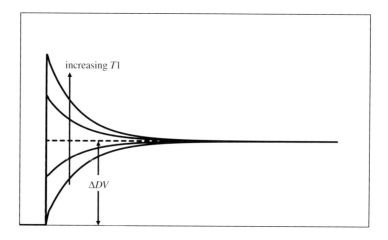

Figure 6.12   Effect of T1

As an example we will add a feedforward ratio scheme to our case study heater. Its schematic is shown in Figure 6.14.

The principle behind the scheme is that it will maintain a constant fuel-to-feed ratio. This is not a blending problem, as described in Figure 6.3, and so this definition of the ratio will not give tuning problems. Indeed, we shall see later that it resolves one. From a steady state point of view it is a good approximation. It is unlikely to be perfect because the heater efficiency will change a little as feed rate is changed. However the feedback controller will correct for this by trimming the ratio SP.

Without dynamic compensation, the scheme will immediately change the fuel rate as the feed rate changes. If the temperature responds more quickly to fuel changes than it does to feed changes, then the correction will have been made too soon and the temperature will

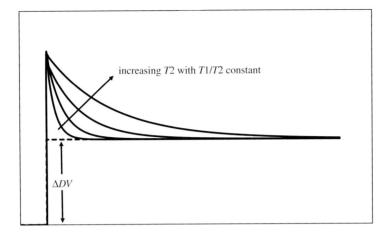

Figure 6.13   Effect of T2

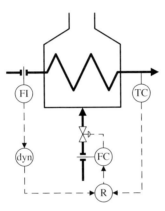

Figure 6.14   Feedforward-feedback schematic

show a transient deviation from SP. The feedback controller does not 'know' that the temperature will eventually return to its starting point and will take corrective action. This unnecessary action will later result in another temperature disturbance. Overall the impact on the process may be greater than it would have been with only feedback control.

We need first to check whether dynamic compensation is necessary and, if so, obtain estimates for the tuning constants. The approach, as usual, is to first develop a full understanding of the process dynamics. By steptesting the fuel flow SP (MV) we obtain the dynamic behaviour of the temperature (PV). We may already have these dynamics from steps conducted to tune the PID controller. These dynamics we define as $(K_p)_m$, θ_m and τ_m. In addition we need the dynamics for the DV. With the temperature controller in **manual** we step the feed rate (DV) and obtain the dynamic behaviour of the temperature – giving us $(K_p)_d$, θ_d and τ_d.

In general we can draw this scheme as the block diagram shown in Figure 6.15.

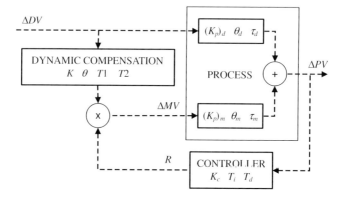

Figure 6.15 Feedforward-feedback block diagram

The objective of the scheme is to ensure there is no change in heater outlet temperature ($\Delta PV = 0$) when the feed rate is changed. For this to be the case, the temperature change caused by ΔDV must be exactly balanced by the change caused by ΔMV, i.e.

$$\Delta DV.(K_p)_d = \Delta DV.K.R.(K_p)_m \tag{6.17}$$

By definition

$$(K_p)_m = \frac{\Delta PV}{\Delta MV} \quad (K_p)_d = \frac{\Delta DV}{\Delta MV} \quad R = \frac{\Delta MV}{\Delta DV} \tag{6.18}$$

Substituting these into Equation (6.17) shows that K should be set to 1. This is always the case for **ratio** feedforward. For **bias** feedforward

$$K = -\frac{(K_p)_d}{(K_p)_m} \tag{6.19}$$

However, in addition to making a correction of the right magnitude, we must ensure it arrives at the same time as the disturbance. It must have the same deadtime and so

$$\theta_d = \theta + \theta_m \quad \text{or} \quad \theta = \theta_d - \theta_m \tag{6.20}$$

We also have to ensure the lag is the same. The lag between temperature and fuel is τ_m while that between temperature and feed is τ_d. We first cancel out τ_m this by setting the lead term ($T1$) equal to this value and then replace it with τ_d by setting the lag term ($T2$) equal to this value.

In summary, for ratio feedforward, we have tuning as follows:

$$K = 1 \quad \theta = \theta_d - \theta_m \quad T1 = \tau_m \quad T2 = \tau_d \tag{6.21}$$

An alternative derivation of the tuning method, using Laplace transforms, is presented at the end of this chapter.

Before implementation there are a number of checks to make. Firstly there is no guarantee that θ_d is greater than θ_m. Thus θ can be negative; this means that we would have to change the fuel before the feed rate changes. Such a requirement is described as not *realisable*. As a compromise, θ is set to zero and $(\theta_m - \theta_d)$ is added to $T1$. We effectively make up for delaying the fuel change by increasing the 'spike'.

Of course, if θ is close to zero and $T1$ is close to $T2$, the dynamic compensation may be omitted from the configuration because the dynamic response of the PV to the DV is the same as that to the MV. However, this may not always be the case. It may be worthwhile including the algorithm, but setting θ to zero and $T1$ equal to $T2$. This would permit compensation to be added easily if required.

We need to consider any noise that may be present in the DV. In our case this was previously only an indication and so any noise was not passed through to a control valve. This will no longer be the case and so filtering may be necessary. Ideally the filter should be put in place before steptesting but if this has been overlooked then we can compensate for its addition by increasing $T1$ by the filter time constant (τ_f).

No matter how fast the response to the DV, $T2$ should not be set to zero. This would cause a full-scale kick to the MV when the DV changes! We should check the $T1/T2$ ratio in any case. For example, if this is greater than 1.15 then the MV overshoot will be greater than 15 %. We may need to reduce $T1$ and make the same compromise that we do for PID control, i.e. accept a slower return to SP in order to avoid harming the process with excessive changes to the MV. Or, if θ is nonzero, it is possible to partially compensate for the reduction in $T1$ by reducing θ.

We should also remember that the tuning has been based on the assumption that the process is first order plus deadtime. It is theoretically possible to implement a second order equivalent of the lead-lag algorithm but this would require the identification of second order models for the DV and MV, and the calculation of additional tuning constants. It is unlikely therefore to be practical. It would be easier to fine tune the dynamic compensation. This also takes account of any abnormalities in the way in which the DCS vendor may have coded the lead-lag algorithm.

Tuning needs to be approached systematically before the feedback controller is commissioned. Otherwise we could be simultaneously adjusting up to seven tuning constants! If bias feedforward has been configured then the value of K can be adjusted by examining the steady state behaviour. If, after a change to the DV, the PV returns to its starting value, then the value for K is correct. If not then, by determining whether feedforward has under or over compensated, K should be increased or decreased. If ratio feedforward has been selected then, as we have seen, K should be fixed at 1. If the PV does not return to its starting point, this will be due to nonlinearities. Adjusting K will improve the performance of the controller for changes in one direction but worsen it in the other.

Once K is correct, θ is next adjusted as required. This is determined by considering how the PV would have changed with no feedforward in place. If, when the DV changes, the PV initially moves in the same direction as it would without feedforward, then θ should be reduced. If the PV moves in the opposite direction then θ should be increased. The next stage is to adjust $T1$, following the same method – but increasing it if the PV moves in the same direction as the open loop response, decreasing it otherwise. $T2$ should be reduced if the PV appears to only slowly to return to steady state; $T1$ should be also changed to maintain a constant $T1/T2$ ratio.

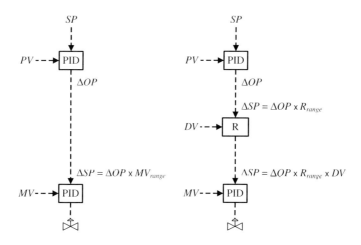

Figure 6.16 Change of control configuration

It is common that ratio feedforward is added to an **existing** feedback controller. If this is the case then we should check whether any change is necessary to its tuning. Since it is now manipulating the ratio SP, rather than the fuel SP, we may have to compensate for the fact that we have changed the range of its MV. Figure 6.16 shows the effect of the change.

While internally controllers operate over a dimensionless range, ratio and bias algorithms normally work in engineering units. Prior to the implementation of feedforward the PID controller output was multiplied by the range of the MV (MV_{range}) to convert it to engineering units. After implementation it is multiplied by the range of the ratio (R_{range}) and by the DV. This will change the loop gain; to compensate we therefore have to adjust the controller gain.

$$(K_c)_{new} = K_c \times \frac{MV_{range}}{R_{range} \times DV^*} \quad (6.22)$$

Of course, DV is not a constant; however it is correct to use DV^* – the value of DV at which $(K_p)_m$ was determined.

We have some control over this correction factor since we choose the range of the ratio when configuring it in the DCS. Ideally we would like the factor to be 1 and so

$$R_{range} = \frac{MV_{range}}{DV^*} \quad (6.23)$$

This not only means that we do not have to retune any existing feedback controller but it also means that we can permit the operator to switch off the feedforward, if for example there is an instrument problem, and retain the feedback scheme without retuning. However we do have to set the range to accommodate the lowest and highest ratios that might be expected during normal operation. This is given by

$$R_{range} = \frac{\text{maximum value of MV}}{\text{minimum value of DV}} - \frac{\text{minimum value of MV}}{\text{maximum value of DV}} \quad (6.24)$$

If the result is larger than the result of Equation (6.23) then this must take precedence and K_c recalculated using Equation (6.22).

Of course, an existing feedback controller may benefit from retuning in any case. It may be configured with the wrong algorithm, such as proportional-on-error rather than proportional-on-PV, or it may be that attention has never been paid to optimising its tuning. New tuning can be derived from the charts presented in Chapter 3, using $(K_p)_m$, θ_m and τ_m and modifying the resulting K_c according to Equation (6.22). Alternatively new dynamics could be obtained by commissioning only the feedforward scheme and stepping the ratio SP. This would then take account of any change of instrument range.

In addition to more quickly responding to process disturbances, **ratio** feedforward on feed rate offers another, less immediately obvious, advantage. We showed in Chapter 2 that process gain is usually inversely proportional to feed rate. This means, that as feed rate changes, we should adjust controller gain in proportion to feed rate in order to keep the loop gain constant. Examination of Figures 6.14 and 6.15 shows that, with feedforward in place, the PID controller output is multiplied by feed rate – effectively increasing the controller gain in proportion, as required. The performance of the feedback controller will therefore be the same at any feed rate. In Chapter 2 we showed that controllers on a process with a turndown ratio greater than 1.5 are likely to need retuning as feed rate changes. Under these circumstances ratio feedforward on feed rate should be considered, even if feed rate disturbances are relatively rare, since this would avoid the need to retune.

It should also be noted that nowhere in our tuning calculations for ratio feedforward is the value of $(K_p)_d$ used. There is thus no need to take account of any changes that might occur in its value because, for example, the process is highly nonlinear. In fact its value can even change sign without causing any control problem. $(K_p)_m$, although not used in calculating the feedforward tuning, does influence feedback tuning. If it changes for reasons other than changes in feed rate then we have to deal with this as normal.

Process gains on integrating processes do not change with feed rate. The use of ratio feedforward would therefore bring a disadvantage in that the feedback controller would require retuning if feed rate changes by more than 20 %. If the use of either bias or ratio feedforward makes good process sense, then bias feedforward would have the advantage.

6.5 Laplace Derivation of Dynamic Compensation

One of the key aims of this book is to present the subject using the minimum of mathematics. Laplace transforms are a very convenient way of representing dynamic behaviour but can be daunting to control engineers unfamiliar with this branch of mathematics. We therefore have largely avoided using them, but include here an example of how they can be used effectively.

The disturbance made to the PV as a result of the change in DV is given by

$$PV_d = \frac{(K_p)_d e^{-\theta_d s}}{\tau_d s + 1} DV \qquad (6.25)$$

The transform for the dynamic compensation comprises a gain (K), deadtime (θ) and lead-lag ($T1$ as lead, $T2$ as lag). It has the form

$$MV = Ke^{-\theta s}\frac{T1s+1}{T2s+1}DV \tag{6.26}$$

The change in PV caused by the feedforward action is

$$PV_m = \frac{(K_p)_m e^{-\theta_m s}}{\tau_m s + 1} MV \tag{6.27}$$

For there to be no change in PV then

$$PV_d + PV_m = 0 \tag{6.28}$$

or

$$\frac{(K_p)_d e^{-\theta_d s}}{\tau_d s + 1} = -\frac{(K_p)_m e^{-\theta_m s}}{\tau_m s + 1} Ke^{-\theta s}\frac{T1 \cdot s + 1}{T2 \cdot s + 1} \tag{6.29}$$

Equating coefficients gives

$$K = -\frac{(K_p)_d}{(K_p)_m} \quad \theta = \theta_d - \theta_m \quad T1 = \tau_m \quad T2 = \tau_d \tag{6.30}$$

By definition

$$(K_p)_d = \frac{\Delta PV}{\Delta DV} \quad (K_p)_m = \frac{\Delta PV}{\Delta MV} \tag{6.31}$$

Therefore

$$K = -\frac{\Delta MV}{\Delta DV} \tag{6.32}$$

For ratio feedforward, $(K_p)_d$ and $(K_p)_m$ must have opposite signs and so K is positive. By definition the SP (R) of the algorithm is the ratio of the MV to the DV. So the algorithm already provides the necessary feedforward gain term and K should be set to 1.

7
Deadtime Compensation

We have seen in Chapter 3 that as the θ/τ ratio increases we have to substantially reduce the controller gain (K_c) to maintain stability. Thus, not only is there a delay in detecting the disturbance, the controller can only respond slowly. We therefore are likely to see large and sustained deviations from SP. There are a number of techniques published which help overcome this problem. However, they rely on estimates of the process dynamics and only offer an advantage if these can be quantified with reasonable accuracy. If only poor estimates are used or if the dynamics can change unpredictably then performance can degrade to less than that that can be achieved with conventional PID control.

7.1 Smith Predictor

Perhaps the earliest, and most well-known technique, is that developed by Smith (Reference 1). It still employs a PID controller but in addition includes a *fast model* that predicts how the process will behave in the future. The controller uses the output of this model, rather than the actual PV, and can therefore be tuned as if there is no deadtime. A *plant model* is also included. Its purpose is to check whether the actual PV eventually matches the prediction and, if not, generate a correction term. Figure 7.1 shows the configuration.

The fast model is simply a standard lag block. Its gain is set to the process gain and its lag set to the process lag. Its input is the output of the PID controller. Thus, when the controller takes corrective action, the output of the fast model will begin changing immediately. As far as the controller is concerned the θ/τ ratio is zero. It can then therefore be tuned very tightly – in theory only restricted by any MV overshoot limit. Of course, it is not sufficient merely to control a model of the process. We have to ensure that the fast model truly represents future process behaviour. This is the purpose of the plant model. It is the same as the fast model but, in addition, includes deadtime set equal to the process deadtime. Rather than predicting the future it now models the **current** process behaviour. Because it is otherwise the same as the fast model, it can be used to compare what the fast model predicted against what is actually happening now. The difference between the actual PV and the predicted ('model') PV is

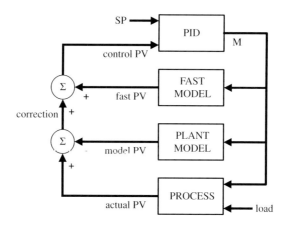

Figure 7.1 Configuration of the Smith predictor

added to the output of the fast model as correction term. The PID controller should be tuned as usual but remembering the θ/τ ratio is now zero so little derivative action will be required.

The Smith is an example of a *predictor-corrector* controller. The output of each part of the model, in response to an **open loop** change in the PID controller output, is shown in Figure 7.2.

The trend for the control PV shows a typical transient disturbance immediately the process deadtime has elapsed. This comes from the corrector and is caused by a mismatch between the actual and the predicted PV. It might arise because of a difference between the process deadtime and the deadtime in the plant model. It might also be caused by the plant model being first order plus deadtime whereas the process probably has a higher order. It is this transient that can limit how large a gain may be used in the PID controller. The controller will respond to it as if it is a load disturbance and the action it takes will later, when the deadtime has elapsed, cause a disturbance in the actual PV. This will then repeat at

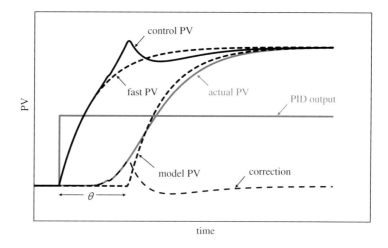

Figure 7.2 Outputs in Smith predictor

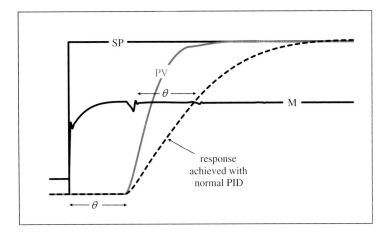

Figure 7.3 Performance of Smith predictor

an interval equal to the process deadtime. If the controller gain is too high each transient will be greater than the one before and the controller will become unstable.

Figure 7.3 shows the closed loop behaviour. Compared to an optimally tuned conventional PID controller, the Smith performs considerably better. In fact, it can be tuned much more aggressively – limited only by the accuracy of the model and the MV overshoot. While the MV shows no overshoot its trend however does show regular disturbances caused by modelling error. In increasing controller gain care would have to be taken in ensuring the PV does not then exhibit similar disturbances, potentially increasing in magnitude after each deadtime.

Figure 7.4 shows the impact the process deadtime falling by about 20 % without a correction made to controller tuning – typically what might occur if there was an increase in feed rate. The conventional PID is considerably more robust than the Smith predictor,

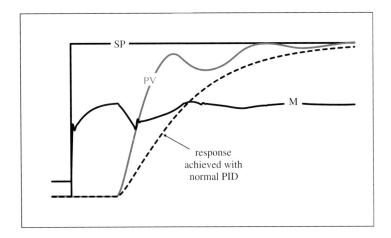

Figure 7.4 Performance of Smith predictor with deadtime error

emphasising the importance of not applying this type of technique unless the process dynamics are accurately known. However, if the dynamics change predictably, adaptive tuning can be applied to the plant model – for example by making its deadtime a function of feed rate. If only deadtime varies then no changes need to be made to the tuning of the fast model or the PID controller.

There are a number of ways in which the Smith predictor may be modified. In addition to deadtime, if the time constant is removed from the fast model, it can also compensate for lag. However removing it completely will make indeterminate the θ/τ ratio as 'seen' by the controller. Most tuning methods require this ratio. Setting τ instead to a small value will assist with tuning the PID. It is also possible to change to second order or higher models, provided the additional time constant(s) can be accurately determined. However this will make configuration in the DCS more complex, potentially undermining the benefit.

7.2 Internal Model Control

IMC (internal model control) replaces the PID controller completely. Its name derives from the fact that it can be shown that it is effectively the same approach as tuning a PID controller using the IMC tuning method. However, this is only exactly true if there is no deadtime. With deadtime IMC will outperform IMC tuning in a PID controller.

Figure 7.5 shows its configuration. Any changes in SP are divided by the process gain (K_p) and pass via a lead-lag algorithm to the process. The lead term ($T1$) is set to the process lag. The lag term ($T2$) is configured to give the desired controller response. If set equal to the process lag then the change is sent to the process as a step. The PV will then approach the SP with a time constant equal to the open loop process lag (τ). If $T2$ is set to a value less than τ then a more aggressive change is made and the MV will overshoot its steady state value. Increasing $T2$ above τ will result in a slower approach to SP with no MV overshoot.

If the gain is accurate then no further control action is required. If not, then there will be a mismatch between the process and the plant model. This model is used in the same way as the Smith controller, to generate a correction term which, in this scheme, is subtracted from the SP.

Performance is equivalent to that of the Smith predictor and it too will perform badly if there is significant model error, particularly with deadtime. As with the Smith, higher order models can be used if required.

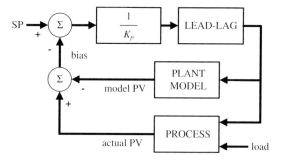

Figure 7.5 Configuration of IMC

7.3 Dahlin Algorithm

The Dahlin algorithm (Reference 2) follows the form

$$MV_n = a_0 E_n + a_1 E_{n-1} + \ldots + b_1 MV_{n-1} + b_2 MV_{n-2} + \ldots \quad (7.1)$$

This is a generalised version of the control algorithm. For example, by setting

$$a_0 = K_c \left[1 + \frac{ts}{T_i} + \frac{T_d}{ts}\right] \quad a_1 = -K_c \left[1 + 2\frac{T_d}{ts}\right] \quad a_2 = K_c \frac{T_d}{ts} \quad b_1 = -1 \quad (7.2)$$

we get

$$\Delta MV = K_c \left[(E_n - E_{n-1}) + \frac{ts}{T_i} E_n + \frac{T_d}{ts}(E_n - 2E_{n-1} + E_{n-2})\right] \quad (7.3)$$

This is the equation for the proportional-on-error, derivative-on-error noninteractive controller. We can however choose coefficients to produce almost any control algorithm. For a first order plus deadtime processes we use

$$MV_n = a_0 E_n + a_1 E_{n-1} + b_1 MV_{n-1} + b_{N+1} MV_{n-(N+1)} \quad (7.4)$$

where

$$a_0 = \frac{1 - e^{-ts/\lambda}}{K_p [1 - e^{-ts/\lambda}]} \cdot \quad a_1 = \frac{-e^{-ts/\tau}[1 - e^{-ts/\lambda}]}{K_p [1 - e^{-ts/\tau}]} \quad (7.5)$$

$$b_1 = e^{-ts/\lambda} \quad b_{N+1} = 1 - e^{-ts/\lambda} \quad N = \frac{\theta}{ts} \quad (7.6)$$

Generally θ will not be an exact multiple of ts and so the value of $MV_{n-(N+1)}$ is linearly interpolated between adjacent values of MV.

$$MV_{n-(N+1)} = MV_{n-\text{int}(N+1)} + (N - \text{int}(N))\left(MV_{n-\text{int}(N+1)} - MV_{n-\text{int}(N+2)}\right) \quad (7.7)$$

The tuning constant (λ) is the required time constant for the trajectory of the approach to SP, as it is in the Lambda tuning method.

The performance of the Dahlin algorithm is similar to that of the Smith predictor and IMC. It is equally sensitive to the accuracy of the deadtime (θ) used in deriving N and hence the value of $MV_{n-(N+1)}$. It too can be extended to higher order models.

The algorithm was originally developed for use when the controller scan interval (ts) is significant compared to the process dynamics. This makes it suitable for use if the PV is discontinuous, such as that from some types of on-stream analysers. Analysers are a major source of deadtime. They may located well downstream of the MV and their sample systems and analytical sequence can introduce a delay. An optimally tuned PID controller would then have a great deal of derivative action. However this will produce the spiking shown in Figure 7.6.

We covered derivative spikes in Chapter 3 and, by switching derivative action to be based on PV rather than error, were able to eliminate them when the **SP** was changed. However

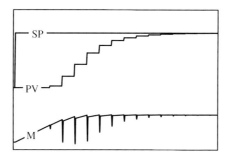

Figure 7.6 Derivatives spikes caused by discontinuous measurement

this change has no effect if the step change in error arises because of a similar change in **PV**. There are however several solutions.

- Simply retuning the controller as PI only would eliminate the spikes but the PV would return to SP more slowly.
- Filtering the measurement would smooth the steps but the level of filtering required would change the overall dynamics, requiring the PID controller to be retuned – probably with a much reduced gain.
- No derivative action is required if the Smith predictor is used. However the discontinuities will cause a mismatch between the process and the plant model, which in turn will cause spurious corrections. IMC will have a similar problem.

The most elegant solution is to take control action only when the analyser generates a new reading – effectively making the controller scan interval the same as the analyser sample interval. Most analysers have a *read-now* contact that can be brought into the DCS and used to initiate a control scan. However the analyser sample interval may not be constant. While some operate with a timed sequence, others will only move to the next step of the sequence once the previous one is complete. Steps like cooling or heating to a required temperature can vary in duration. We should not then apply a technique like PID which assumes a fixed sample interval. Techniques, such as Dahlin, readily permit the coefficients to be based on the actual sample interval.

This also has the advantage that control action is only taken if the analyser produces a new reading. Relying only on the PV from the analyser's sample-and-hold logic will mean that control action will continue even if the measurement is not being refreshed, causing the controller eventually to saturate.

References

1. Smith, O.J.M. (1957) Closer control of loops with deadtime. *Chemical Engineering Progress*, **53** (5), 217–219.
2. Dahlin, E.B. (1968) Designing and tuning digital controllers. *Instruments and Control Systems*, **2** (6), 77–83.

8
Multivariable Control

8.1 Constraint Control

As the name suggests, constraint controllers are designed to drive the process towards operating limits in a direction known to be profitable. Constraints are either hard or soft. Hard constraints are those which can only be approached from one side. This might be because it is mechanically impossible; for example, a control valve cannot open beyond 100 %. Or a hard constraint might be imposed for safety reasons; for example, operating pressure can theoretically be increased above a relief valve setting but doing so would be extremely hazardous. Finally there may be strong operational reasons for not violating a constraint; for example, minimising the flow of recycle around a compressor should not cause the machine to surge. While the machine itself might tolerate this for some period, the loss of gas flow to the process would give severe operational problems.

Soft constraints can be approached from either side. Violation of a soft constraint does not give a major problem, provided corrective action is taken promptly. An example is the quality of liquid products leaving a process. The product will ultimately be routed to storage where small amounts of off-spec production will be mixed with material with *giveaway*. Another example is liquid level; while low and/or high limits may be set, these may be violated briefly with no impact on the process. A maximum limit is applied to skin (tube metal) temperatures on some fired heaters. This may be either to keep coking at a reasonable level or to prevent the tubes from creeping. Both are long-term cumulative factors and are unlikely to be measurably worse if the maximum temperature is occasionally violated.

Some MVC permit the definition of hard and soft constraints. In addition to their meaning as described above, often a soft constraint is used as a more conservative limit on a hard constraint. The controller will violate soft constraints if this is the only way that it can satisfy all the hard constraints.

Constraint controllers fall into three categories.

- Single-input single-output (SISO) controllers, as the name suggests, comprise controllers that can manipulate only one MV to approach only one constraint.

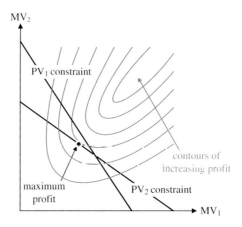

Figure 8.1 Optimisation problem

- Multi-input single-output (MISO) controllers are required when there is more than one constraint that is approached by adjusting just one MV. Such controllers require some form of logic to select the most limiting constraint. Once selected the controller behaves just like a SISO controller.
- Multi-input multi-output controllers adjust multiple MVs in order to satisfy multiple constraints. Generally, although not necessarily, the controller can reach as many constraints as there are MVs. Thus, if there are more constraints than MVs, some selection is necessary. MVC packages generally incorporate a LP for this purpose. This permits some simple economics to be defined by the engineer to drive the controller to select the most profitable set of constraints to approach. This establishes the SPs. More than one constraint will be affected by each MV; otherwise we would only require a set of SISO controllers. Thus the controller must handle the interactions. If there are fewer constraints than MVs then the controller will have *degrees of freedom* – equal to the number of MVs less the number of constraints. There will multiple ways in which the constraints can be approached. Selection of the most profitable becomes an economic optimisation problem, usually outside the scope of MVC – although most support optional add-ons to resolve this.

Care must be taken when determining the number of degrees of freedom. Figure 8.1 shows a 2×2 constraint control problem. Given that it is possible to simultaneously reach both constraints by adjusting the MVs, it would be tempting to assume that there are no degrees of freedom. In fact, the most profitable operating point lies on only one of the constraints. At this point there is then **one** degree of freedom.

8.2 SISO Constraint Control

Very simple constraint control is possible using the conventional PID control algorithm. Figure 8.2 shows a fired heater that has a single constraint – a hydraulic limit on fuel. It is profitable to run the heater at maximum feed rate, subject to this constraint.

Multivariable Control   171

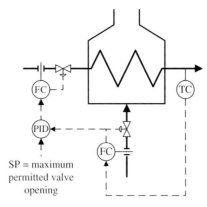

Figure 8.2   Example of SISO constraint controller

The first indication that the process is operating at the maximum feed rate is the fuel valve fully opening. A single-input single-output (SISO) controller, with this valve opening as its PV, manipulates the SP of the feed flow controller until the fuel valve reaches the desired maximum opening. On most processes, valve position transmitters are only installed in special circumstances. What is actually used here is the output of the fuel flow controller. We assume that the valve positioner is properly calibrated and working well.

On our case study heater, the process operator will typically enter a SP of around 90 %. This effectively converts a hard constraint to a soft constraint. In order to maintain control of the outlet temperature during minor process disturbances some leeway is required. This means that the process capacity is not fully utilised. Conditioning the constraint, as described in Chapter 5, offers an alternative method of converting it to a soft constraint and would permit this leeway to be reduced.

Tuning the PID controller is carried out in much the same as normal. First the fuel flow and outlet temperature controllers should be properly tuned since their tuning affects how the fuel control valve moves when feed rate is changed. Model identification is completed by stepping the SP of the feed flow controller and observing how the fuel flow controller output varies. The tuning method outlined in Chapter 3 is followed although it is usual to tune less aggressively than normal by using a PI controller, with the gain reduced to around 25 % of the calculated value.

8.3   Signal Selectors

Signal selectors are used in the most basic form of multivariable control, i.e. multi-input single-output (MISO) applications. The fired heater described also has a maximum limit on burner pressure. This is also approached by increasing feed rate. In order to operate at maximum feed rate the controller must be able to continuously identify which is the more limiting constraint. Figure 8.3 illustrates one possible configuration.

Instead of the error ($PV - SP$) being calculated in the controller it is calculated in each of the two bias algorithms. The two errors are compared in the signal selector and the selected

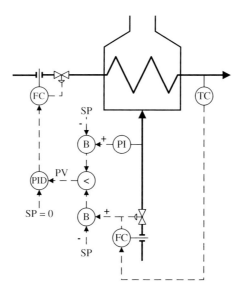

Figure 8.3 Example of MISO constraint controller using single PID

value becomes the PV of the PID controller. The controller SP is fixed at 0 and so it will manipulate the SP of the feed flow controller until the first of the process limitations is reached.

We choose whether to use a low or high signal selector based on what would make sense to the process operator. In this case the feed rate would be determined by the constraint requiring the **lower** value, and so a low selector would seem preferable. However because of our definition of error and the constraints are **upper** limits, the scheme would require a **high** signal selector that would pass through the higher **decrease** in feed and the PID controller would be configured as **reverse**-acting. The biases algorithms therefore include a negative gain; this allows the use of a low signal selector; the controller is then configured as direct-acting.

While simple in principle, controller tuning requires special attention. While it is clear which is the more limiting constraint if one is being violated and the other not, it is not so clear if both show that there is spare capacity. Selection needs to be based on which constraint will be reached first as feed rate is increased. It is tempting the use the process gains to assist this selection. If the process gain between fuel valve position and feed rate is $(K_p)_v$ and that between burner pressure and feed rate is $(K_p)_p$, then the gain term included in the valve bias algorithm is given by $-1/(K_p)_v$ and that in the burner pressure bias algorithm is $-1/(K_p)_p$. The output of each bias algorithm is then the permitted increase in feed rate. Choosing the lower of these two values would certainly result in controlling against the more limiting constraint.

However, we also have to consider the tuning of the PID controller. If the dynamics of each constraint are different then the controller will require different tuning depending on which constraint is selected. A better approach is to move the controller gain from the PID to the biases. Process dynamics are obtained as usual by steptesting the feed flow SP. The dynamics of the fuel valve position are used to develop a set of tuning constants and the resulting controller gain (with its sign changed) entered as the gain in its bias algorithm. The dynamics of the burner pressure are used to develop a second set of tuning constants. The controller gain (with its sign changed) is used as the gain in its bias algorithm. The controller gain in the direct-acting PID algorithm can now be set to 1, although a lower value may be used as in the SISO example.

Derivative action is usually excluded. The integral time used in the controller is the average of the two determined for each constraint. If the values are far apart this might result in compromise tuning that gives poor response no matter which constraint is active. An approach which resolves this problem is shown in Figure 8.4.

In this configuration each constraint has its own PID controller – a pressure controller on burner inlet and a 'valve position' controller on the fuel valve. In this example both constraint controllers are configured as reverse-acting. The DCS must support certain features. The signal selector must include anti-reset windup to prevent the output of the unselected controller from saturating. Secondly the PID controllers must be of the incremental type, otherwise bumpless initialisation is difficult to achieve. These features are present in most DCS, but should be checked.

Measurement noise can create a problem with this configuration. With the full position form of the PID controller any noise superimposed on the output will have little impact on signal selection. But with the incremental form signal selection is based on the much smaller **change** in MV. The change in controller output caused by noise can readily exceed the change made to correct a deviation from SP.

Imagine that the process is operating exactly at the fuel control valve limit and the burner pressure is well away from its limit. Noise **peaks** in the pressure measurement will be treated by the pressure controller as increases in pressure and generate a negative change in controller output. This will be less than the zero output from the valve position controller and so will pass through the signal selector and cause a reduction in feed rate. Noise **troughs** will have no effect so, on average, the heater will then operate at a feed rate lower than that necessary to satisfy the constraints. In some situations this effect can be substantial, actually

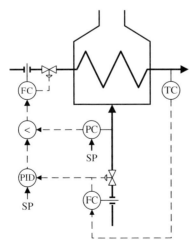

Figure 8.4 *Example of MISO constraint controller using multiple PIDs*

driving the feed rate well below where the operator would have set it. Filtering the pressure measurement would resolve this problem but the additional lag introduced may have a noticeably detrimental effect on controller tuning. If noise is not a problem in terms of excessive valve movement, the filtered measurement can be used to drive the signal selector and the unfiltered measurement used for control.

While it is possible to include additional constraints and signal selection, this type of strategy realistically is restricted to single output controllers. If, for example, it was permitted to adjust the outlet temperature controller SP in order to approach the capacity limits, the resulting MIMO controller would be extremely complex. Since both constraints would be affected by both MVs, simple selection logic cannot be applied. Decoupling would also be required to prevent the controllers from fighting each other. While theoretically a DCS block-based scheme could be designed, it is not an approach recommended.

8.4 Relative Gain Analysis

Relative gain analysis was developed for multivariable control (Reference 1) to assist with 'pairing' each PV with a MV and to assess the level of interaction. Relative gain is defined as the ratio of the process gain with all other controllers on manual to the same process gain with all other controllers in automatic mode. Ideally, placing other controllers on automatic should not affect the process gain of the first and so the relative gain would be 1.

For simplicity, consider first a 2 × 2 system. Plant testing, with all controllers on manual, has determined process gains as follows

	MV_1	MV_2
PV_1	$(K_p)_{11}$	$(K_p)_{12}$
PV_2	$(K_p)_{21}$	$(K_p)_{22}$

The relative gain array (RGA) is

	MV_1	MV_2
PV_1	λ_{11}	λ_{12}
PV_2	λ_{21}	λ_{22}

Where λ is defined by

$$\lambda_{11} = \frac{\left(\frac{\Delta PV_1}{\Delta MV_1}\right)_{\Delta MV_2=0}}{\left(\frac{\Delta PV_1}{\Delta MV_1}\right)_{\Delta PV_2=0}} \qquad (8.1)$$

While we have, from steptesting, the numerator (the process gain with all other controllers on manual) we cannot use steptesting to determine the denominator (the process gain with all other controller on auto) since we have yet to design the controllers. However the process can be described by

$$\Delta PV_1 = (K_p)_{11}\Delta MV_1 + (K_p)_{12}\Delta MV_2 \tag{8.2}$$

$$\Delta PV_2 = (K_p)_{21}\Delta MV_1 + (K_p)_{22}\Delta MV_2 \tag{8.3}$$

From Equation (8.2), if $\Delta MV_2 = 0$, then

$$\left(\frac{\Delta PV_1}{\Delta MV_1}\right)_{\Delta MV_2=0} = (K_p)_{11} \tag{8.4}$$

From Equation (8.3), if $\Delta PV_2 = 0$, then

$$\Delta MV_2 = -\frac{(K_p)_{21}}{(K_p)_{22}}\Delta MV_1 \tag{8.5}$$

Substituting this into Equation (8.2)

$$\left(\frac{\Delta PV_1}{\Delta MV_1}\right)_{\Delta PV_2=0} = (K_p)_{11} - \frac{(K_p)_{12}(K_p)_{21}}{(K_p)_{22}} \tag{8.6}$$

Substituting Equations (8.4) and (8.6) into Equation (8.1)

$$\lambda_{11} = \frac{1}{1 - \dfrac{(K_p)_{12}(K_p)_{21}}{(K_p)_{11}(K_p)_{22}}} \tag{8.7}$$

Each row and each column in the RGA sum to 1; so

$$\lambda_{12} = \lambda_{21} = 1 - \lambda_{11} \quad \text{and} \quad \lambda_{22} = \lambda_{11} \tag{8.8}$$

No matter what units are used for process gains, provided they are consistent, the relative gains are dimensionless.

If there are no interactions then the RGA will be

$$\Lambda = \begin{pmatrix} 1 & 0 \\ 0 & 1 \end{pmatrix} \tag{8.9}$$

This is clearly the ideal situation since it tells us that PV_1 can be controlled by MV_1 and PV_2 by MV_2 with no interaction between the controllers.

If the RGA is given by

$$\Lambda = \begin{pmatrix} 0 & 1 \\ 1 & 0 \end{pmatrix}, \tag{8.10}$$

this again tells us that there is no interaction but that PV_1 must be controlled by MV_2 and PV_2 by MV_1 – i.e., the pairing should be reversed.

For λ_{11} to be 1 or 0, at least one element of the process gain matrix must be zero. If only one process gain is zero the controllers will display a one-way interaction. Two diagonal elements must be zero for there to be no interaction in either direction.

The worst possible case is

$$\Lambda = \begin{pmatrix} 0.5 & 0.5 \\ 0.5 & 0.5 \end{pmatrix} \quad (8.11)$$

Control of both and PV_1 and PV_2 may be impossible.

A similar problem arises if $\lambda >> 1$. For example if λ_{11} was 100 then

$$\Lambda = \begin{pmatrix} 100 & -99 \\ -99 & 100 \end{pmatrix} \quad (8.12)$$

Since it is the absolute value of λ that determines the level of interaction, control of PV_1 and PV_2 would again be very difficult. Control of both is **impossible** if a column in the process gain matrix is an exact multiple of another. Known as *parallel* both PV_1 and PV_2 will respond equally to changes in MV_1 and MV_2. Sometimes described at the *controllability*, the determinant of the matrix will be zero and the relative gain indeterminate.

While the RGA for a 2 × 2 system gives some information about interactions, it usually helps little with pairing other than confirm the decision already made by the engineer. However, for larger problems pairing is not always so obvious. For such problems we apply matrix techniques. Consider the process gain matrix for a $n \times n$ problem:

$$K = \begin{pmatrix} (K_p)_{11} & (K_p)_{21} & \cdot & (K_p)_{n1} \\ (K_p)_{21} & (K_p)_{22} & \cdot & (K_p)_{n2} \\ \cdot & \cdot & \cdot & \cdot \\ (K_p)_{1n} & (K_p)_{2n} & \cdot & (K_p)_{nn} \end{pmatrix} \quad (8.13)$$

It can be shown that

$$\Lambda = K * (K^{-1})^T = \begin{pmatrix} \lambda_{11} & \lambda_{12} & \cdot & \lambda_{1n} \\ \lambda_{21} & \lambda_{22} & \cdot & \lambda_{2n} \\ \cdot & \cdot & \cdot & \cdot \\ \lambda_{n1} & \lambda_{n2} & \cdot & \lambda_{nn} \end{pmatrix} \quad (8.14)$$

Note that the operator * is not the conventional row-by-column matrix multiplication but an element-by-element multiplication – resulting in what is known as the *Hadamard product*.

While this calculation looks complex it is readily configurable in a spreadsheet package – most of which support matrix functions. Alternatively there are a number of commercially available packages which will also analyse the RGA and suggest control strategies.

When using the RGA to decide pairing, any PV/MV combination where λ is negative should be avoided. This means that the process gain for the proposed controller reverses sign as other controllers are switched between auto and manual – leading immediately to the controller saturating. The procedure is to first check if any row or column has only one positive element. If so, then this decides the first pairing. If not the case then the λ in the matrix that is closest to 1 is chosen to provide the first pairing. All other elements in the same column or row as this element are the ignored and the process repeated until all pairings are complete. It is quite possible that not all variables may be paired.

When checking for interactions, each element in a column is divided by the relative gain of the pairing chosen in that column, taking the reciprocal if the result less than 1. As a guide, if the absolute value of a result is between 1 and 1.25 then this indicates an interaction strong enough to require decoupling to avoid instability. If the value is between 1.25 and 1.5 the interaction is significant and some form of decoupling would be beneficial. For values between 1.5 and 2, the interaction is mild and decoupling would provide no noticeable improvement.

There are mixed views about the value of relative gain analysis. It can be useful in checking the design of MVC applications. Large MVC are difficult for the process operator to understand and difficult to maintain. The RGA can be used to identify for consideration any controllers that can be moved outside the MVC as SISO controllers. It can also be used to see if the MVC can be broken into separate smaller controllers. However, while all the process gains are known, the work involved can still be cumbersome. Relative gain analysis can only be applied to square matrices. It is usually the case in MVC applications that there are more CVs than MVs. This would entail selecting a number of CVs, equal to the number of MVs, and varying the selection depending on which variables are likely to constrain the controller.

Relative gain analysis takes no account of process dynamics. Variables can show no interaction at steady state but during disturbances their transient behaviour can cause control problems. Further, the analysis assumes linear behaviour. As operating conditions change it is likely that some of the process gains will vary. Recalculating the RGA under these conditions may alter the conclusions in terms of both pairing and decoupling. For these reasons the RGA should only be considered as technique for eliminating some of the pairing options and suggesting configurations that need to be evaluated further.

8.5 Steady State Decoupling

Figure 8.5 shows an example of a highly interactive control problem. Two streams of different temperature are blended to meet a required temperature but also must be

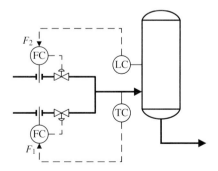

Figure 8.5 Highly interactive controllers

manipulated to control the drum level. Changing either flow will change both the temperature and the level. If both controllers were independently tuned for tight control the likelihood is that, when both on auto, they would be unstable.

One possibility is that one of the controllers could be de-tuned so that it reacted very slowly to disturbances. For example, it might be acceptable for the level to deviate from SP for long periods and changes in its SP are likely to be rare. Priority would then be given to the temperature controller. While the interaction still exists, this would avoid instability.

We can apply relative gain analysis to determine the level of interaction. If F_1 is the flow and F_2 the other then, if the cross-sectional area of the drum is A, then the rate of change of level is given by

$$\frac{dL}{dt} = \frac{(F_1 + F_2)}{A} \tag{8.15}$$

Differentiating to obtain the process gains

$$(K_p)_{11} = \frac{d\left(\frac{dL}{dt}\right)}{dF_1} = \frac{1}{A} \tag{8.16}$$

$$(K_p)_{12} = \frac{d\left(\frac{dL}{dt}\right)}{dF_2} = \frac{1}{A} \tag{8.17}$$

If T_1 is the temperature of one stream and T_2 the other then, if the specific heat is the same for both streams, the combined temperature is given by

$$T = \frac{F_1 T_1 + F_2 T_2}{F_1 + F_2} \tag{8.18}$$

Again differentiating to obtain the process gains

$$(K_p)_{21} = \frac{dT}{dF_1} = \frac{F_2(T_1 - T_2)}{(F_1 + F_2)^2} \tag{8.19}$$

$$(K_p)_{22} = \frac{dT}{dF_2} = \frac{F_1(T_2 - T_1)}{(F_1 + F_2)^2} \tag{8.20}$$

and, from Equations (8.7) and (8.8),

$$\Lambda = \begin{pmatrix} \frac{F_1}{F_1 + F_2} & \frac{F_2}{F_1 + F_2} \\ \frac{F_2}{F_1 + F_2} & \frac{F_1}{F_1 + F_2} \end{pmatrix} \tag{8.21}$$

We can see that as F_1 approaches F_2, i.e. because the average of the two stream temperatures is close to the target temperature, the relative gains all approach 0.5 and the process becomes uncontrollable using the chosen MVs.

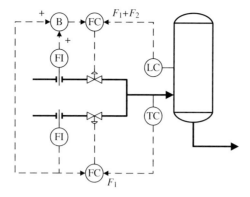

Figure 8.6 Partially decoupled controller (using bias)

Figure 8.6 shows a partial solution. As in Figure 8.5, the TC manipulates F_1, but as the flow changes this is fed forward via a bias algorithm so that F_2 is changed by the same amount in the opposite direction. Thus the drum level will remain unchanged. Interaction has been broken in one direction; $(K_p)_{11}$ is now zero, and so the RGA becomes the identity matrix. If the LC takes corrective action this will change the temperature but the partial decoupling would be sufficient to stop the interaction causing instability.

Figure 8.7 shows an alternative partial solution. In this case the LC shown in Figure 8.5 has been retained, manipulating F_2. For the reasons given in Chapter 6 (see Figure 6.3) the temperature controller now manipulates the ratio between F_1 and the total flow. This ensures that when the LC takes corrective action, the temperature remains unchanged. Thus $(K_p)_{22}$ is now zero. Corrections made by the TC will affect the vessel level and so the interaction has been broken in only one direction.

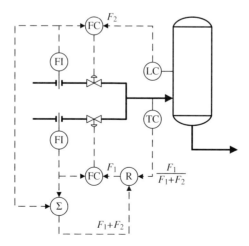

Figure 8.7 Partially decoupled controller (using ratio)

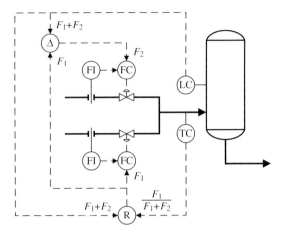

Figure 8.8 Fully decoupled controller

Figure 8.8 shows a fully decoupled controller. Rather than use the flow measurements in the decoupling calculations, the controller outputs have been used. This is an option. It gives a slight dynamic advantage because the changes are fed forward sooner. However, it does assume that the flow controllers can achieve their SPs. If either controller saturates, or is switched to manual, then the decoupling is likely to cause windup problems.

8.6 Dynamic Decoupling

Dynamic decoupling, first installed using DCS function blocks, is now largely provided by proprietary MVC packages. We here describe the DCS approach primarily to help understand the principles of decoupling using nonproprietary techniques. For reasons that will become apparent the MVC package approach, although initially more costly, is usually the better solution economically.

We will consider first a 2×2 system. Decoupling effectively feeds forward corrective action made by each controller to the MV of the other so that the other PV is undisturbed. In principle this results in two noninteracting controllers which can then be tuned conventionally. Figure 8.9 illustrates how both PVs are affected by both MVs. We assume we wish to operate where the contours of constant PV intersect. We start with PV_2 at SP_2, but PV_1 away from SP_1. The challenge is to move PV_1 to its SP without disturbing PV_2. To do so we have to make a compensating change to MV_2. Decoupling is tuned to determine the size of this compensating move. If made **at the same time** as the adjustment to MV_1 then we will have achieved steady-state decoupling. Both PVs will ultimately reach their SPs but PV_2 is likely to show a transient deviation, i.e. it will start and finish on the PV_2 contour but not necessarily follow it. To ensure it does so we also require dynamic compensation.

Figure 8.10 shows the first of the decouplers. When PID_1 takes corrective action, the decoupler applies dynamic compensation to the change in output (ΔOP_1) and makes a change to MV_2 that counteracts the disturbance that the change in MV_1 would otherwise cause to PV_2. Dynamic compensation is provided by a deadtime/lead-lag algorithm.

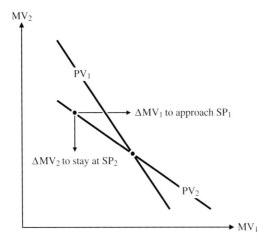

Figure 8.9   Decoupling principle

The tuning method for the decoupler is exactly that described for **bias** feedforward in Chapter 6, i.e.

$$K_1 = -\frac{(K_p)_{21}}{(K_p)_{22}} \quad \theta_1 = \theta_{21} - \theta_{22} \quad T1_1 = \tau_{22} \quad T2_1 = \tau_{21} \tag{8.22}$$

A second decoupler makes compensatory changes to MV_1 when PID_2 takes corrective action, as shown in Figure 8.11.

Tuning for the second decoupler is given by

$$K_2 = -\frac{(K_p)_{12}}{(K_p)_{11}} \quad \theta_2 = \theta_{12} - \theta_{11} \quad T1_2 = \tau_{11} \quad T2_2 = \tau_{12} \tag{8.23}$$

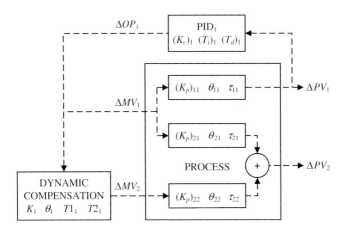

Figure 8.10 MV_1 to MV_2 decoupler

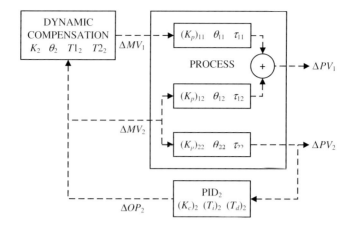

Figure 8.11 MV_2 to MV_1 decoupler

Figure 8.12 shows both decouplers, but without the PID controllers.

The addition of each decoupler has added another feedback path to each controller and so changed the apparent process gain. In the example shown in colour, the output from PID1 (ΔOP_1) passes to the process as before as ΔMV_1 but also passes through the first dynamic compensation algorithm to become part of ΔMV_2. Since both ΔMV_1 and ΔMV_2 both affect ΔPV_1 the process gain changes to

$$\frac{\Delta PV_1}{\Delta OP_1} = (K_p)_{11} + K_1(K_p)_{12} \qquad (8.24)$$

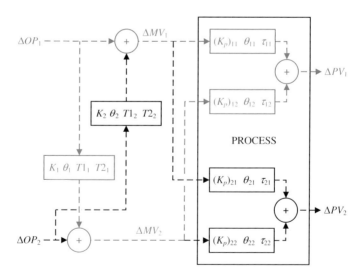

Figure 8.12 2×2 dynamic decoupler

Multivariable Control

Substituting from Equation (8.22) for K_1

$$\frac{\Delta PV_1}{\Delta OP_1} = (K_p)_{11} - \frac{(K_p)_{12}(K_p)_{21}}{(K_p)_{22}} \tag{8.25}$$

By definition, the process gain *without* the decoupler in place is given by

$$\frac{\Delta PV_1}{\Delta OP_1} = (K_p)_{11} \tag{8.26}$$

Therefore, dividing Equation (8.26) by Equation (8.25), to maintain the loop gain constant the controller gain must be multiplied by

$$\frac{1}{1 - \frac{(K_p)_{12}(K_p)_{21}}{(K_p)_{11}(K_p)_{22}}} = \lambda_{11} \tag{8.27}$$

The same applies to the apparent process gain of PID_2. The process gain with the decoupler is

$$\frac{\Delta PV_2}{\Delta OP_2} = (K_p)_{22} + K_2(K_p)_{21} \tag{8.28}$$

Substituting from Equation (8.23) for K_2

$$\frac{\Delta PV_2}{\Delta OP_2} = (K_p)_{22} - \frac{(K_p)_{12}(K_p)_{21}}{(K_p)_{11}} \tag{8.29}$$

By definition, the process gain without the decoupler in place is given by

$$\frac{\Delta PV_2}{\Delta OP_2} = (K_p)_{22} \tag{8.30}$$

Therefore, dividing Equation (8.30) by Equation (8.29), to maintain the loop gain constant the controller gain must be multiplied by

$$\frac{1}{1 - \frac{(K_p)_{12}(K_p)_{21}}{(K_p)_{11}(K_p)_{22}}} = \lambda_{22} \tag{8.31}$$

If it is not the case that the decouplers are added to **existing** PID controllers, or the opportunity to optimally tune existing controllers is to be exploited, then these can be tuned according to the procedures described in Chapter 3. The process model to tune PID_1 is

$$K_p = (K_p)_{11} + K_1(K_p)_{12} \quad \theta = \theta_{11} \quad \tau = \tau_{11} \tag{8.32}$$

And that for PID_2 is

$$K_p = (K_p)_{22} + K_2(K_p)_{21} \quad \theta = \theta_{22} \quad \tau = \tau_{22} \tag{8.33}$$

Alternatively these models could be obtained by steptesting OP_1 and OP_2 with the decouplers in place.

As a partial solution to the problem of interaction, a one-way decoupler might be considered. So if, for example, it was more important to keep PV_1 rather than PV_2 close to SP, then decoupler 2 would be implemented without decoupler 1. This would be sufficient to prevent the interaction causing instability and would be considerably easier to implement and maintain. The disadvantage of course is that the control of PV_2 would be poor.

While the amount of steptesting required for the DCS approach is identical to that for a proprietary MVC package, and the tuning calculations very simple, implementation in the DCS is quite complex. Not only does it involve a large number of DCS blocks but a great deal of engineering must be put into properly scaling each block and ensuring bumpless transfer from manual to auto.

If, after commissioning, it was discovered that an equipment constraint was being frequently encountered then we would want to include this constraint in the controller. In the DCS case this is likely to greatly increase complexity and would require major re-engineering. With a MVC package the addition would be much simpler.

It is difficult in the DCS to make provision for one of the controllers to be disabled without the other(s). While this might be overcome by only permitting all-or-nothing operation, one controller may saturate. This will change the apparent process gain of the other(s). A properly configured MVC package will handle this situation routinely.

8.7 MVC Principles

It is not the intention here to reproduce the detailed theory of model predictive control (MPC) or how it used for multivariable control (MVC). This has become an almost obligatory section in modern control texts. There are also numerous papers, marketing material and training courses. In this book its description is limited to its general principles; focus instead is on how to apply it and monitor its performance.

MVC packages differ from PID type controllers in a number of aspects. Firstly PID type controllers require a SP. MVC requires only HI and LO limits to be set for each *controlled variable* (CV). The HI limit can be set to the LO limit if a true SP is required. The MVC will exploit the range between LO and HI in two ways – either to avoid a violating a constraint elsewhere on the process or, if all constraints are satisfied, reduce the overall operating cost.

MVC also permits HI/LO constraints to be placed on *manipulated variables* (MV). While this is possible with PID controllers, this will cause a deviation from SP. A MVC will use other MVs to satisfy a constraint if one has reached its limit.

Some MVC packages permit hard and soft constraints. Soft constraints are adhered to if possible but will be violated if this is the only way of avoiding violating a hard constraint. Other packages permit weighting to be applied to constraints so that the engineer can specify which should be violated first if the MVC cannot identify a feasible solution.

Basic PID controls must wait for a disturbance to be measured before responding, whereas MVC predicts future deviations and takes corrective action to avoid future violation of constraints. There are two fundamentally different ways this is done. Some packages use high order Laplace transforms, others use a time series. Time series comprise a linear function of previous values of the MV (and sometimes also CV). In the function CV_n is the predicted next value of the CV; MV_{n-1} is the current value of the MV, MV_{n-2} the previous value etc. The coefficients a_1, a_2 etc. are determined by regression analysis of step test data. Several such functions are in common use in proprietary MVC packages.

The *Finite Impulse Response (FIR)* for a SISO process is given by

$$CV_n = \sum_{i=1}^{n} a_i MV_{n-i} \quad \text{or} \quad \Delta CV_n = \sum_{i=1}^{n} a_i \Delta MV_{n-i} \quad (8.34)$$

The *Finite Step Response (FSR)* is given by

$$CV_n = \sum_{i=1}^{n} a_i \Delta MV_{n-i} \quad (8.35)$$

Dynamic matrix control uses

$$CV_n = \sum_{i=1}^{n-1} a_i \Delta MV_{n-i} + a_n MV_{n-N} \quad (8.36)$$

Also used is

$$CV_n = \sum_{i=1}^{n} a_i MV_{n-i} + \sum_{i=1}^{n} b_i CV_{n-i} \quad (8.37)$$

These equations are then used to predict the effect of **future** changes in the MV. The *prediction horizon* is the number (M) of sample periods used in predicting the value of the CV. The *control horizon* is the number (N) of control moves (ΔMV) that are calculated into the future. In matrix form Equation (8.35), for example, therefore becomes

$$\begin{pmatrix} CV_1 \\ CV_2 \\ CV_3 \\ \\ CV_M \end{pmatrix} = \begin{pmatrix} a_1 & 0 & 0 & & 0 \\ a_2 & a_1 & 0 & & 0 \\ a_3 & a_2 & a_1 & & 0 \\ & & & & \\ a_M & a_{M-1} & a_{M-2} & & a_{M-N+1} \end{pmatrix} \begin{pmatrix} \Delta MV_0 \\ \Delta MV_1 \\ \Delta MV_2 \\ \\ \Delta MV_N \end{pmatrix} \quad (8.38)$$

Future moves are calculated to minimise the sum of the squares of the predicted deviations from target over the control horizon. While the next N moves are calculated, only the first (ΔMV_0) is implemented. At the next cycle the controller recalculates a new set of control moves. This will account for any prediction errors or unmeasured load disturbances.

All MVC packages include at least one parameter that permits the engineer to establish the compromise between fast approach to target and MV movement. Such an approach was adopted in the PID tuning methods described in Chapter 3. Some MVC packages permit the engineer to define *move suppression* which penalises large changes to the MV. Others include a term similar to λ used in the Lambda tuning method and internal model control. This permits the engineer to define a trajectory for the target.

While the techniques covered in Chapter 7 provide similar functionality for SISO controllers these generally need to be custom configured in the DCS. MVC includes the feature as standard. Further the equations above predict the value of a CV based on the changes in a **single** MV. MVC sums the effect of all the MVs. Like other predictors the actual CV is compared to what was predicted and a bias term updated to compensate for any inaccuracy.

PID controllers are either in manual or automatic mode. MVC can be selectively switched to manual. Any MV can be *dropped*, in which case its SP can be changed by the process operator. When calculating future moves the controller effectively sets the HI/LO limits on this MV to its current SP. Some MVs may be categorised as *critical*, dropping one of these disables the whole MVC. Any CV can similarly be dropped, in which case the MVC ignores any constraints placed on it. Some MVC packages support the definition of subcontrollers. These allow part of the overall MVC to be disabled while keeping the remainder in service. Even if disabled the MVC continues to operate in *warm* mode. While it makes no change to any MV, it continues to update its CV predictions so that it operates correctly when recommissioned. If prediction has not been possible for a period, for example because a system failure prevented data collection, the controller will *initialise* – setting all historical values of the MVs to the current value.

MVC supports the addition of disturbance variables (DV) or *feedforward variables*. These are variables which cannot be manipulated by the controller but affect the CV. They are included in the CV prediction and so effectively add feedforward control. Similarly any MV dropped is treated as a DV so that any changes made to it by the process operator are included in the prediction of the CV. Feedforward control can of course be added to PID type controllers but requires configuration of the DCS. MVC includes the feature as standard for all its CVs.

Basic controllers implement a fixed strategy. MVC permit more CVs than MVs and will select which CVs to control based on objective coefficients specified by the engineer. The MVC package will employ a *linear program* (*LP*) or similar algorithm to select the least costly (most profitable) strategy. Objective coefficients are applied to each MV and, in some packages, also to each CV.

PID controllers normally operate in dimensionless form with all inputs and outputs scaled as fraction of instrument range. This is not the case for MVC; its process gain matrix is in engineering units consistent with the units of MVs and CVs.

Although powerful, MVC packages should be applied intelligently. There can be a tendency to assume they can resolve almost any control problem. They can be costly. The first installation has to bear not only the licence cost but also the cost of the platform on which it is to run and engineer training. While the cost is often justifiable, consideration should first be given to less costly solutions – even if they do not capture all the benefits. To make this evaluation it is important to understand all the interactions before proceeding to the design stage.

Most MVC packages assume linear process behaviour. Much can be done outside the package to linearise variables by applying signal conditioning. The MVC includes a feedforward function; however this is a **bias** feedforward. If it makes engineering sense to apply ratio feedforward then the MVC cannot achieve the same performance as a DCS-based ratio algorithm. If the feedforward DV is feed rate then, if it varies by more than $\pm 20\,\%$, the use of ratio feedforward will ensure the process gains in the MVC matrix remain constant. The ratio could then become an MV of the MVC if required.

At the current level of technology MVC packages do not scan frequently enough to handle very fast processes. For example it would be unwise to apply one to compressor antisurge control unless there is a backup fast scanning system to recover from surge.

The intelligent design would normally be a mixture of 'traditional' DCS-based techniques and the MVC.

8.8 Parallel Coordinates

Parallel coordinates is two-dimensional graphical method for representing multiple dimensional space. In the example shown in Figure 8.13, a point in seven-dimensional space is represented by the coordinates (x_1, x_2, x_3, x_4, x_5, x_6, x_7). Since we cannot visualise space of more than three dimensions, the value of each coordinate is plotted on vertical parallel axes. The points are then joined by straight lines.

The technique is well-suited to predicting the behaviour of a multivariable controller, even before steptesting has been started. Plant history databases comprise a number of instrument tag names with measurements collected at regular intervals. If we imagine the data arranged in a matrix so that each column corresponds to either a MV or a CV in the proposed controller and each row is a time stamped snapshot of the value of each parameter. To this we add a column in which we place the value of the proposed MVC objective function (C) derived from the values in the same row (where P are the objective coefficients for the m CVs, and Q the objective coefficients for the n MVs), i.e.

$$C = \sum_{i=1}^{m} P_i CV_i + \sum_{j=1}^{n} Q_j MV_j \qquad (8.39)$$

Each row in the database is then plotted as a line on the parallel coordinates chart. The result will initially appear very confused with a large number of lines superimposed. The next step is to add the HI/LO constraints on each vertical axis. If a line violates any constraint on any axis then the whole line is deleted. The lines remaining will each represent an occasion in the past when all the process conditions satisfied the constraints. The final step is to choose the line for which the value on the cost axis is the lowest. Since this axis is the MVC cost function, the line with the lowest value will represent the operation that the MVC would select.

Provided that the process has at some stage operated close to the optimum (as defined by the MVC) then the chosen data set will give some idea of the operating strategy that the MVC will implement. If different from the established operating strategy, this approach

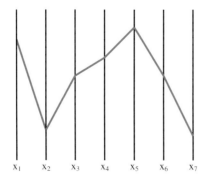

Figure 8.13 Parallel coordinates

188 Process Control

gives an early opportunity to explore why. Any difference should be seen as an opportunity to adopt a more profitable way of operating the process, rather than an error that should be corrected by adjusting the individual objective coefficients.

There are a number of commercially available packages that will allow data to be imported from the process history database and provide end-user tools to simplify its manipulation. Some include enhancements (Reference 2) such as automatically arranging the sequence of the axes, optimising the spacing and filtering out highly interdependent variables. These help support other applications of the technique, of which the main one in the process industry is the diagnosis of operating problems.

8.9 Enhanced Operator Displays

A major challenge with MVC is presenting its actions to both process operators and control engineers in a form that is readily understandable. Particularly with large controllers it is often difficult to diagnose why the controller is adopting a particular strategy. This can lead to the operator disabling the controller, or partly disabling it by tightening the MV constraints. Assuming there is no problem with the controller, such actions result in lost profit improvement opportunities.

While there is yet to be developed an entirely satisfactory solution to this problem, some ideas have been applied successfully. One is the use of a *radar plot*. This is similar to parallel coordinates except that the axes are arranged radially. Only a limited number of CVs and MVs are practicable – perhaps a maximum of around 12, so only the more important variables are included.

Figure 8.14 shows a typical plot. The LO and HI limits are each plotted as a continuous polygon (shown as dashed lines), as is the current operation or predicted steady state (shown

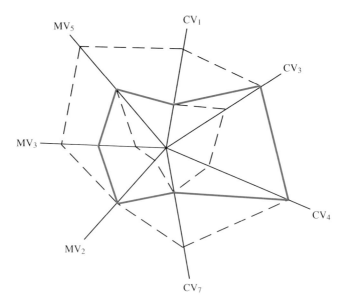

Figure 8.14 Radar plot

as the solid line). A specimen operation might also be included for reference. The human mind would appear to better at recognising shapes rather than tables of numbers. With the use of colour to distinguish multiple plots, the change in shape is often readily recognised as normal or abnormal.

Another graphical approach is the *heat map*. Each critical MV and CV is represented by a horizontal bar. Each bar is divided into small vertical slices. The extreme right hand slice of the bar represents the current situation. As a variable approaches its constraint this slice changes to a colour between pale yellow and bright red representing how the close the variable is to its constraint. At regular intervals the slices all move to the left, with the one on the extreme left being discarded.

This diagram is helpful in showing how variables move in and out of constraint. Again the mind can recall patterns that are known to represent normal or abnormal behaviour.

8.10 MVC Performance Monitoring

The vendors of MVC packages offer increasingly sophisticated tools for monitoring the performance of their applications. The licence fees for such tools can be substantial and many of the functions included may not be seen as valuable by the engineer using them. The purpose of this section of the book is to present a number of ways in which performance might be monitored. Each technique has been applied somewhere, although not all at the same site. It is anticipated that the engineer will identify those that are valuable, decide whether to implement them and assess whether a proprietary package meets the needs.

This section describes a layered approach. At the top are simplistic overview tools primarily for management reporting. Below these the engineer can 'drill down' into increasing levels of detail to diagnose problems. If the engineer is to build the tools, such as in a spreadsheet package, they will need to retrieve information from the process data historian. In addition to the more conventional process measurements held in this database the tools will need the following:

- on/off status for each MVC
- the value of each MVC objective function
- the status of each MV and CV (i.e. have they been excluded or 'dropped')
- upper and lower limits set in the controller for each MV and CV, including hard and soft limits if applicable
- identification of the limiting constraints (not all MVC provide this so it may be necessary to build additional logic into the monitoring tools to flag those close to limits)
- the economic weighting factors (some MVC permit these only on MV, some also on CV)
- either the unbiased prediction for each CV or the value of the bias used in the prediction.

These requirements do increase the load on the historian but most of them compress well. Apart from the last item, the others change comparatively rarely. If the system supports data compression the incremental load will be very small.

Some sites find it important to maintain a high profile for advanced control to retain senior management's attention and commitment. In other sites the management demand regular reports. Performance can be condensed into a single number, i.e. the total benefits captured.

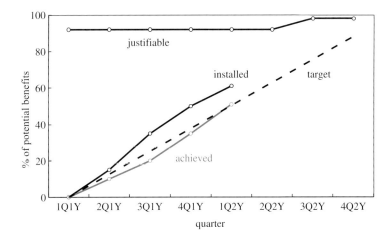

Figure 8.15 Reporting benefits captured

However it is important to remove from this number any changes outside the control of the site – for example changes in feed and product prices. A better approach is to report benefits captured as a fraction of the maximum achievable, as shown in Figure 8.15.

A portion of the available benefits may exist but there is insufficient return on investment to justify the cost of doing so. Rather than exclude them completely the chart includes a 'justifiable' value which is the maximum that could be captured. Future technological developments may bring down costs, and so what is not justifiable today should be occasionally reassessed rather than forgotten.

The 'installed' trend shows what could be captured if every installed control application is working 100 % of the time. The gap between this and the justifiable value represents what new applications are required. The gap between the 'installed' trend and the 'achieved' trend represents what is being lost by applications not being fully utilised.

A quarterly management report including these trends supported by a few summary points can do much to facilitate improvements. Attention can be drawn to manpower shortages on the implementation team or application support. Critical unreliable instrumentation can be identified to support the case for replacement or prioritised maintenance.

While the chart deliberately excludes financial data, there is no reason why the recipient should not be made aware of to what 100 % corresponds. An annual update of the true value of each application should be completed and if this causes any change in any of the indices they can either be back-calculated for previous years or an explanation included in the report.

Reporting benefits as a fraction of what is achievable permits plant-to-plant or site-to-site comparisons. There are also consulting organisations that can provide comparison with competitors in a form such as that shown in Figure 8.16.

The example shown on the chart is capturing about 27 % of the available benefits. In terms of its competition, of the sites surveyed, around 51 % are doing better.

On sites with multiple production units, the next level of detail is a breakdown by unit. Figure 8.17 shows an example.

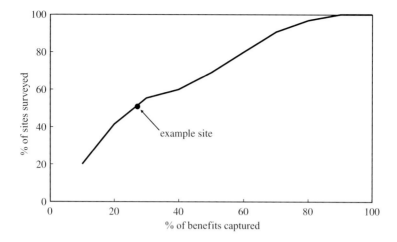

Figure 8.16   Competitive positioning

For this the contribution that each unit makes is represented as a percentage of the total site benefits. Some form of service factor is required to quantify the uptime of each application. This might simply be the percentage of the time that the application is switched on. However it is possible to constrain a MVC so that it makes no changes and still be on 100 % of the time. If required, a more complex, but more realistic, definition of service factor can be used. This might be based on the proportion of MVs not against a constraint.

The next level of detail is monitoring each MVC. It is possible simply to trend the controller's objective function. However, experience shows that these can be quite noisy and show discontinuities as constraints change. But the main concern with the MVC is that it is over-constrained by the operator. It is common for the operators to periodically close

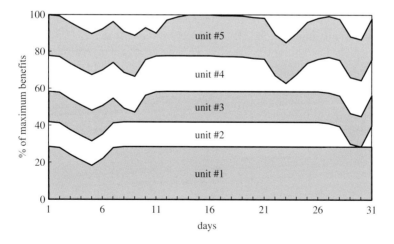

Figure 8.17   Breakdown by unit

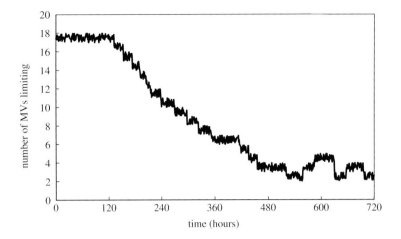

Figure 8.18   Managing MV constraints

the gap between the HI and LO limits on the MVs. This might be done temporarily for a good reason, because of some problem with the controller or the operator understanding of its actions, but the constraints are rarely relaxed again without some intervention by the engineer.

Figure 8.18 shows a trend over a period of about a month where efforts were made to remove as many MV constraints as possible. Some MVC generate engineer-accessible flags to identify whether a MV is limiting. If not then the engineer has to develop some simple logic checking whether each MV is close to a constraint. The flag is set to 1 if the MV is limiting and 0 if not. Any MV 'dropped' is treated as limiting. The flags are totalled and the result historised. At the beginning the controller was virtually disabled – able to manipulate only two MVs. This situation was reversed within the month.

Figure 8.19 shows one of the detailed trends used in support of the exercise. It shows, for MV1, the actual value and the HI and LO constraints. If the MVC supports hard and soft constraints then both should be trended. The chart is useful in determining why the total number of constraining MVs has changed, and from the time of the constraint change, identifying who made the change and why.

Similar trending can be developed for CVs – although these tend to be less used. But detailed monitoring of the value of each CV is worthwhile. The MVC internally calculates a predicted value for each CV from the MVs, using the dynamic models (G) obtained by steptesting.

$$CV_i = \sum_{j=1}^{m} G_{ij} MV_j + bias_i \qquad (8.40)$$

Comparison is made between the predicted value and the actual value; the bias term is then adjusted to bring the two in line. The bias will always be nonzero since it is not true that the CV will be zero when all the MVs are zero. However a large variation in the bias indicates

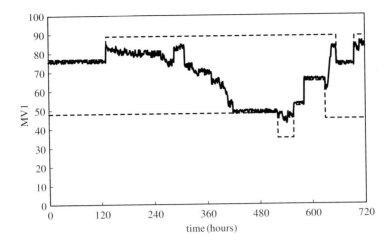

Figure 8.19   Trend of a single MV

a poor model. In order to monitor this it must be possible to retrieve either the bias term or *unbiased CV* from the MVC. Figure 8.20 shows trends of data collected from a MVC for both the unbiased and actual CV.

To the eye the bias (the difference between the two values) appears constant. However by trending the standard deviation of the bias we see in Figure 8.21 that this is not the case. Some event took place, approximately halfway through the collection period, to cause degradation in the accuracy of the CV prediction.

While this trend is an effective detection tool it probably is not practical in this form. It is not immediately obvious whether the reduction in accuracy is sufficient to warrant attention. To address this we instead monitor the performance parameter (ϕ).

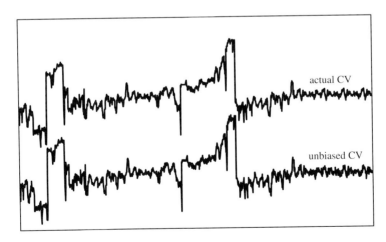

Figure 8.20   Assessing CV prediction error

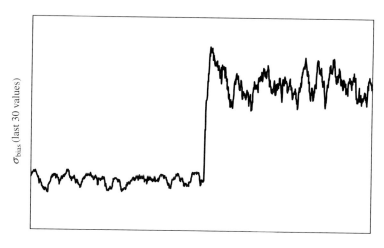

Figure 8.21 Trending the standard deviation of the bias

$$\phi = 1 - \frac{\sigma_{bias}^2}{\sigma_{CV}^2} \qquad (8.41)$$

If the prediction is perfect then ϕ will have the value 1, since the bias will have remained constant and its standard deviation therefore zero. As the variation in the bias approaches the variation in the CV the prediction becomes increasingly valueless. To understand this, let us assume that the prediction of the unbiased CV is that it is always constant. The standard deviation of the bias will therefore be the same as the standard deviation of the actual CV. The controller is effectively ignoring the prediction and ϕ will be zero. As ϕ falls below zero the prediction is so poor that it is creating disturbances greater than the natural disturbances in the CV. Using such a predicted CV is worse than assuming a constant CV and taking no corrective action. Most MVC projects are justified on the basis that the standard deviation in limiting CV will be halved. For this to be achievable ϕ cannot be less than 0.75.

Most MVCs have a large number of CVs and it would be unreasonable to expect the engineer to check all of the trends at frequent intervals. However it is possible to generate an overall performance parameter, for example by trending the number of CVs which fail to meet the required performance. If this trend moves away from zero for significant periods then examination of the individual trends would identify the culprit(s).

Once a poor prediction has been detected we still have the problem of determining the cause. The prediction includes several dynamic models, any one of which may be the source of the inaccuracy. Further it could be caused by the absence of a model. By looking for correlations between ϕ and each of the MVs, the suspect model can be identified. This may simply be by eye – looking at trends of both. Or it may involve the use software, such as a spreadsheet package, to search for correlations between the prediction bias and each of the MVs.

If real process economics are used in the MVC, then a wide range of other monitoring opportunities are created. These include:

• checking that the MVC has truly optimised the process and not simply automated the existing operating strategy

- quantifying the lost opportunity if the operator over-constrains the MVs
- determining the value of debottlenecking projects
- calculating the benefit actually captured by the MVC.

Full details of how these techniques can be developed are included in Chapter 12 as a worked example on a simple distillation column.

References

1. Bristol, E.H. (1966) On a new measure of interactions for multivariable process control. *IEEE Transactions of Automatic Control*, **11**, 133.
2. Yang, J., Peng, W., Ward, M.O. and Rundensteiner, E.A. (2003) Interactive Hierarchical Dimension Ordering, Spacing and Filtering for Exploration of High Dimensional Datasets. Worcester Polytechnic Institute.

9
Inferentials and Analysers

Accurate property measurement is key to the capture of many of the benefits of process control. Money can be made by more closely approaching product quality specifications. Process conditions can be continuously optimised provided good product quality control is in place.

Property measurement falls into two basic categories.

The first are mathematical techniques where basic process measurements of flow, temperature and pressure are used to infer a property. Often also called *soft sensors* or *virtual analysers* they are used mainly to predict product quality but may used for any parameter that cannot be measured directly – such as column flooding, catalyst activity, rate of coking etc.

The second is the use of on-stream analysers to directly measure product quality. It is not the intent of this book to cover any detail of how such analysers operate or how they should be installed or managed. Instead this chapter will focus on the use of their measurements in control strategies.

9.1 Inferential Properties

Even if a reliable on-stream analyser exists it is usually still worthwhile to develop an inferential. Since the inferential is based primarily on basic measurements it will respond much more quickly than the analyser. The analyser could well be located a long way downstream from the point at which the product is produced. And, depending on the design of the sample system and the analytical technique it uses, could introduce additional delays. Figure 9.1 illustrates the benefit of this dynamic advantage on an example process. The two curves are each from an optimally tuned PID controller responding to the same process disturbance. Increasing the sample interval from 30 to 300 seconds resulted in a much larger deviation from SP, sustained for a much longer time.

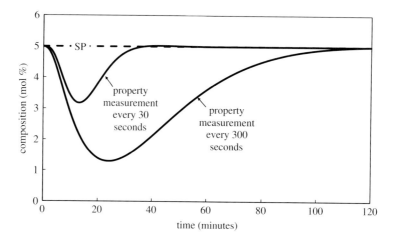

Figure 9.1 Effect of early detection

Figure 9.2 shows the potential economic benefit. Point A represents a typical benchmark with a θ/τ ratio of 4. This might be from a process lag of 5 minutes and a deadtime of 20 minutes – both quite reasonable dynamics for a process such as a distillation column with a chromatograph on the distillate product rundown. In these circumstances an inferential could be expected to reduce the deadtime by at least 10 minutes (point B). Doing so would allow the controller to be tuned more quickly and would result in a reduction by about 33 % in off-spec production.

Inferentials comprise a mathematical function (f) using a number of *independent variables* (x) to predict the value of a *dependent variable* (y).

$$y = f(x_1, x_2, \ldots, x_n) \tag{9.1}$$

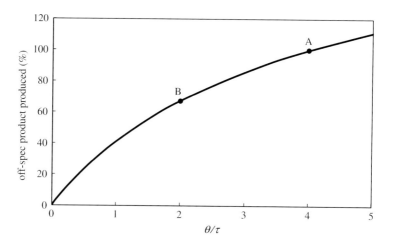

Figure 9.2 Impact of deadtime on off-spec production

They fall into two groups – those derived from *regression analysis* of historical process data and *first-principle* types which rely on engineering calculations. First-principle techniques still require some historical data to calibrate the model and to check its accuracy. While the vendors of first-principle techniques might argue that the volume of data required is less, the key to the success of both techniques is the *quality* of the data. The use of routinely collected data, for example from a plant history database, can often cause inaccuracies in the end result.

Firstly the data may not have been collected at steady state. Not all the variables used in the inferential will have the same process dynamics. Shortly after a disturbance they will all be approaching steady state but to a different degree. The process may be temporarily out of energy or material balance as the inventory of either may be changing. Regression analysis is usually performed on data collected at a fixed interval, possibly averaging several sets of data around the collection time. While any errors introduced will be random and will not necessarily affect the form of the inferential, they will make it difficult to confirm its accuracy. With first principal types, which may use only a few sets of data for calibration, it is more important that data is collected when the process is steady and has been steady for long enough for the deadtime of the dependent variable to expire.

Another potential problem is that of *time-stamping*. The dependent variable is often a laboratory result which may not be available until several hours after the sample was taken. It is therefore necessary to associate it with the operating conditions at the time of the sample. However, sample times are not necessarily reliable. Most sites will sample according to a schedule. However, the true sample time may be very different. It may have been delayed because there was an operating problem at the time or it may be taken early to fit in with the sampler's workload. Often all the samples on a process are scheduled for the same time but they clearly could not all be taken simultaneously. It is a misconception that, if the process is steady, that recording the exact sample time is not important.

Figure 9.3 is based on a hypothetical perfect inferential that agrees exactly with the laboratory result. The process is reasonably steady as seen by the trend of the inferential

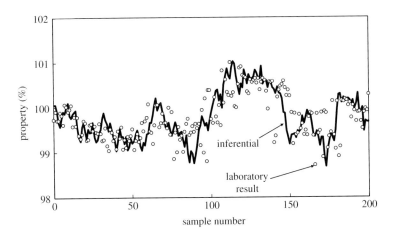

Figure 9.3 Time-stamping error

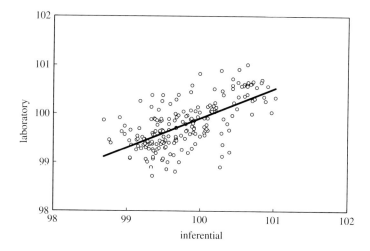

Figure 9.4 Effect of time-stamping error

which varies less than ±1 %. The mismatch of the laboratory samples is caused by introducing, into the sampling time, a random error in the range of ±10 minutes.

Plotting the same information as a scatter chart, Figure 9.4, would suggest that a correlation which we know to be perfect is far from it. The error in prediction is comparable to the variation of the true value. If we were to **develop** an inferential from this information we would have little confidence in its reliability. If we were monitoring the performance of an **existing** inferential then we could be misled into disabling one that is working well. While we could ask the sample taker to record the actual sample time a more reliable approach is to automate this. One approach is to locate a push-button next to the sample point and connect it to the DCS so that it either logs the time when it is pressed or records all the independent variables at the time. Industries, such as pharmaceutical manufacturing where record keeping is of far greater importance, install sample points which record automatically the time that the sample valve is opened. It is also helpful if the LIMS (laboratory information management system) has the facility to record actual rather than scheduled sample time.

Relying on routinely collected data will often not provide sufficient *scatter*. With modern data collection systems it is a trivial exercise to assemble information collected over several years. Even if a laboratory sample is only taken daily, assembling a thousand or more sets of data should present no problem. However, unless the process is required to make multiple grades of the product, each with very different specifications, even without automatic quality control the process operator will have kept the quality very close to target. Any large deviations will usually be due to process upsets and may not provide any reliable steady-state information.

Figure 9.5 shows a typical situation where the development of an inferential, using only routine data, would be unreliable. Inclusion of only a few additional points, collected under test run conditions while moving in stages across a wide operating range, greatly improves accuracy. While regression analysis is generally thought to need 30 or more sets of data,

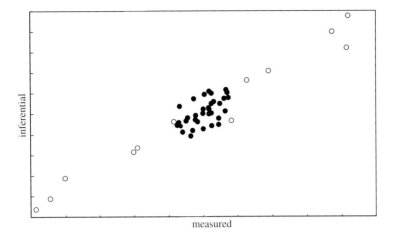

Figure 9.5 Poorly scattered data enhanced with test run results

5 to 10 well scattered, properly collected points will enable a reliable inferential to be developed. Confidence in the calibration of a first-principles model needs a similar amount of data.

It is common practice to collect such additional data during steptesting for an MVC project. Provided steady state is reached then this will be useful to help **validate** an inferential. However it may not be practical to cover all operating scenarios. For example, many inferentials are sensitive to feedstock but all types of feed may not be processed during the steptest phase. Secondly it may prove impossible to develop an inferential from the data collected. It is too late a stage in the project to discover that additional instrumentation will be required.

Simple regression analysis tends to produce inferentials that are arithmetically simple and may therefore be readily built into the DCS using standard features. More complex regressed types, such as neural networks, will require a separate platform and probably some proprietary software. They can therefore be more costly. First-principle models may be provided in *pseudo-code* that the engineer can convert to code appropriate to the DCS. This with its testing and documentation can be time-consuming. Inferentials delivered as 'black boxes' may require less implementation effort but can only be maintained by the supplier.

The decision on whether to use regression or first-principles technology is difficult – particularly if relying solely on the (less than impartial) vendors for information. We have already seen that both techniques require a similar amount of good quality data rather than a large volume of suspect data. Those supplying first-principle models will claim that regression analysis assumes that the input variables are independent and that true independence is unachievable. For example distillation tray temperatures separated by a few trays will track each other closely. While simple least squares regression does include the underlying assumption that the variables are independent, experience shows that it will give good results even if there are cross-correlations. Where there are strongly correlated independents, regression will see little advantage in using both. If there is some advantage to using one over the other, for example because it is a more reliable measurement, then it would be wise to manually exclude the other from the analysis. A good statistical analysis package will identify such cross-correlations and indicate the improvement in accuracy of the inferential that is achieved as each variable is added. There are also many other techniques which do not assume independence. Further, the so-called first-principle techniques can include correlations developed by others by regressing experimental data.

Models based on engineering principles should theoretically adapt more readily to minor process modifications. This would mean that they could be used, unlike regression, without waiting for additional process data to be collected. However they are rarely 'pure' and often include calibration factors. It would be a brave engineer that trusted them implicitly without rechecking the calibration.

Regression analysis is open to abuse if applied without an understanding the process. For example, blindly applying a neural network effectively discards any knowledge of process behaviour. While the resulting inferential may work well, its performance outside the range over which it was trained can be extremely unpredictable. There are examples where this has caused a reversal of the sign of the process gain with respect to the key MV – severely impacting process profitability.

Naively applying linear regression techniques can have a similar impact. With modern spreadsheets and statistical packages it is relatively easy to extract large quantities of data from the process information database and search for all possible correlations. By including a large number of process variables and a wide range of arithmetical transformations (such as powers, logarithms, ratios, cross-products etc.) it will certainly be possible to apparently improve the accuracy of the inferential. However, this is likely to be only a mathematical coincidence. If the inferential includes terms which make no engineering sense (or coefficients which have the wrong sign) then it will fail during a process excursion. However there is also the risk of **excluding** terms that appear at first not to make engineering sense. Chapter 12 gives some examples where nonlinear transformations, ratios and cross-products can make sense, as can coefficients with apparently the wrong sign.

Some first-principle inferentials have a poor reputation as being complex 'black boxes' not fully understood by the engineer and so have fallen into disuse.

If the inferential is to be a CV of a MVC then care needs to be taken with applying regression analysis to derive a **linear** function. Consider the MVC gain matrix shown as Equation (9.2).

$$\begin{pmatrix} K_{11} & K_{12} & . & K_{1n} \\ K_{21} & K_{22} & . & K_{2n} \\ . & . & . & . \\ K_{m1} & K_{m2} & . & K_{mn} \end{pmatrix} \begin{pmatrix} MV_1 \\ MV_2 \\ . \\ MV_n \end{pmatrix} = \begin{pmatrix} CV_1 \\ CV_2 \\ . \\ CV_m \end{pmatrix} \qquad (9.2)$$

The MVC will thus predict CV_1

$$CV_1 = K_{11}MV_1 + K_{12}MV_2 + \ldots + K_{1n}MV_n + bias_1 \qquad (9.3)$$

A linear inferential will have the form

$$y = a_0 + a_1 x_1 + a_2 x_2 + \ldots + a_n x_n \qquad (9.4)$$

If y is used as CV_1 and x_1 is MV_1, x_2 is MV_2 etc. then it is important that a_1 is equal to K_{11}, a_2 is equal to K_{12} etc. Since the inferential's coefficients are derived from regression and the process gains subsequently derived from steptesting, they are unlikely to be exactly the same. If there are other inputs to the inferential that are not included in the gain matrix, then the need for exact agreement will depend on whether those inputs change if an MV is changed.

If the inferential uses process measurements that are physically far apart then a process disturbance may affect one measurement more quickly than another. As a result the inferential may show complex dynamic behaviour, such as inverse response. If used as the measurement of a PID controller then the slow tuning necessary to maintain stability may give very poor control. While, in theory it is possible to dynamically compensate the inputs, the compensation required will depend on the source of the disturbance. One source may cause input 1 to change before input 2, while another may cause the reverse. If the inferential is to be a CV of a MVC, then such packages can handle high order dynamics such as inverse response. However, they too will be prone to the dynamics changing depending on the source of the disturbance. In a regression type inferential it is straightforward to exclude the less critical input if its dynamics are very different and repeat the regression analysis without it. Some accuracy will be sacrificed, but controllability will be greatly improved. In a first-principle model the simplest solution is to assume a constant value for the offending measurement.

So selection is by no means straightforward. The pragmatic approach is to choose the approach that works better in each situation. Regression analysis usually has the lower cost and can be performed by the plant owner using a spreadsheet package or a proprietary development tool. If regression fails to deliver an inferential of sufficient accuracy then a first-principle approach can be explored. This is likely to require a specialist supplier that, if truly convinced of the technology they offer, should be prepared to work on 'no win, no fee' basis. If their product cannot outperform the benchmark established by regression then they would waive their charges.

This leads us to the question of how we assess the accuracy of an inferential.

9.2 Assessing Accuracy

Figure 9.6 shows the method favoured by suppliers of inferentials. Line charts tend to lead one to believe the correlation between the inferential and the actual property is better than it is. Presenting the same data in Figure 9.7 as a scatter plot gives a more precise measure. For example, if the true property is 50 %, the inferential will be between 30 and 70 % – probably far too inaccurate to be of any value.

The other favoured approach is the use of the statistic known as *Pearson* R^2. If there are n sets of data where x is the inferential and y the laboratory result,

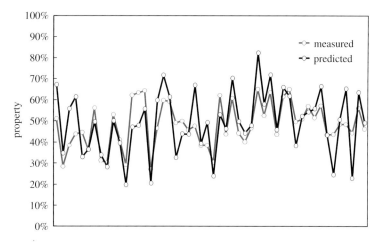

Figure 9.6 Use of line plot to validate inferential

this is defined as

$$R^2 = \frac{\left(\sum_{i=1}^{n}(x_i-\bar{x})(y_i-\bar{y})\right)^2}{\sum_{i=1}^{n}(x_i-\bar{x})^2 \sum_{i}^{n}(y_i-\bar{y})^2} \quad (9.5)$$

A perfect correlation would have a value of 1 for R^2. However a value close to 1 does not necessarily indicate that an inferential is useful. As an illustration, consider the graph shown in Figure 9.8 for the stock price of a process control vendor.

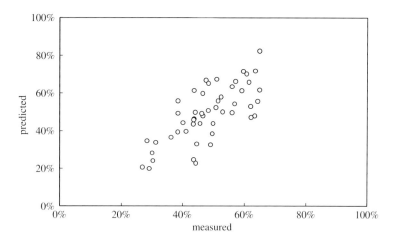

Figure 9.7 Use of scatter plot to validate inferential

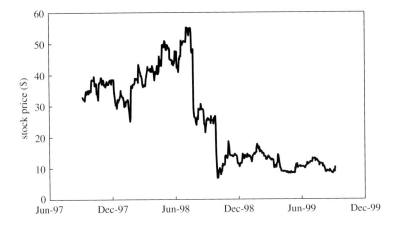

Figure 9.8   Process control vendor stock price

Figure 9.9 shows the performance of an inferential developed by the author. With R^2 of 0.99 one would question why the developer is not a multi-billionaire! The reason is that it failed to predict the large falls in the value of the stock. The three occasions circled undermine completely the usefulness of the prediction. The same is true of an inferential property. If there is no change in the property then, no matter how accurate, the inferential has no value. If it then fails to respond to any significant change it may as well be abandoned.

A better approach is to compare the standard deviation of the prediction error (σ_{error}) against the variation in the actual property (σ_{actual}). We show in Chapter 13 that benefit calculations are usually based on the assumption that the standard deviation of the actual

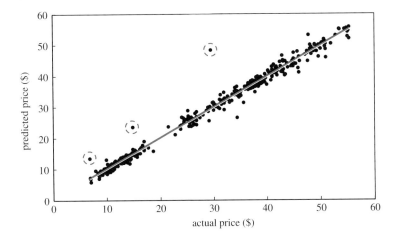

Figure 9.9   Predicting stock price

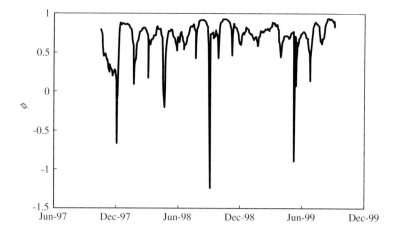

Figure 9.10 Inferential performance parameter

property is halved. If we assume that our control scheme is perfect and the only disturbance comes from the random error in the prediction then, to capture the benefits

$$\sigma_{error} \leq 0.5\sigma_{actual} \qquad (9.6)$$

This can be written in the form of a performance parameter (ϕ)

$$\phi = \left(1 - \frac{\sigma_{error}^2}{\sigma_{actual}^2}\right) \geq 0.75 \qquad (9.7)$$

This parameter will have a value of 1 when the inferential is perfect and 0 when its benefit is zero. But importantly it will be negative if the property controller performance is so bad that process performance would be improved by switching off the controller. Figure 9.10 trends this parameter for the stock price example. It confirms what we know, that the prediction will lose us money on several occasions.

A further limitation of the use of R^2 is that, if there is a perfect relationship between inferential ($PV_{inferential}$) and laboratory result ($PV_{laboratory}$), the value of R^2 will also be 1 for **any** linear function, i.e.

$$PV_{inferential} = a_1 PV_{laboratory} + a_0 \qquad (9.8)$$

So, for example, if a_1 had a value of 3 and a_0 a value of 0, then the inferential would be treble the laboratory result but, according to the statistical test, be working perfectly! The same would apply if a_1 were negative – even though this reverses the sign of the process gain.

To illustrate the difference between R^2 and ϕ, consider the data in Table 9.1. Column 1 is a series of laboratory results. Columns 2, 3 and 4 are the corresponding results from three inferentials derived using different values of a_1 and a_0.

The values of ϕ in columns 2 and 3 of the table confirm, unlike R^2, that the inferential would be so poor that its use would cause control of the property to worsen. In column 4, where only a bias error is introduced, both R^2 and ϕ show that the inferential would be perfect – requiring just a once-off correction for it to be useful.

Table 9.1 Comparison between R^2 and ϕ

$PV_{laboratory}$	$a_1 = 3$, $a_0 = 0$	$a_1 = -1$, $a_0 = 0$	$a_1 = 1$, $a_0 = 5$
4.81	14.43	−4.81	9.81
4.79	14.37	−4.79	9.79
5.25	15.75	−5.25	10.25
5.02	15.06	−5.02	10.02
4.86	14.58	−4.86	9.86
4.96	14.88	−4.96	9.96
5.08	15.24	−5.08	10.08
5.17	15.51	−5.17	10.17
4.98	14.94	−4.98	9.98
4.90	14.70	−4.90	9.90
4.86	14.58	−4.86	9.86
4.98	14.94	−4.98	9.98
4.94	14.82	−4.94	9.94
5.17	15.51	−5.17	10.17
5.01	15.03	−5.01	10.01
5.17	15.51	−5.17	10.17
5.09	15.27	−5.09	10.09
5.16	15.48	−5.16	10.16
4.75	14.25	−4.75	9.75
4.81	14.43	−4.81	9.81
ϕ	−3	−3	1

The parameter (ϕ) can be used both in the development of an inferential and its monitoring. At the development stage we clearly need its value to be greater than 0.75 but, given that this assumes perfect control, in reality it needs to be higher if we are to capture the benefits claimed. A more realistic target is 0.9.

If ϕ is calculated at a high frequency, for example by the use of on-stream analyser measurements, then care must be taken to ensure that the process is at steady state. Because the dynamics of the analyser will be longer than those of the inferential, any change in the inferential will be reflected some time later in the analyser measurement. There will therefore appear to be a transient error, even if both the inferential and analyser are accurate. Alternatively, dynamic compensation can be applied. We cover this later in this chapter.

As a monitoring tool ϕ can be very valuable in the early detection of degradation in the accuracy of an inferential and disabling it before its poor performance does any real harm. However it needs to be used with care. Firstly, if our controller is successful, it will reduce σ_{actual}. Our performance parameter will then fall, misleadingly indicating that the performance of the inferential has degraded. To avoid this we choose a constant value for σ_{actual}, equal to the variation **before** the controller was commissioned. The second issue is that we have to use a number of historical values to calculate σ_{actual} – usually 30. Thus, even if a problem with the inferential is resolved, the performance index will indicate a problem until 30 more laboratory results are taken. While we can reduce the number of historical values used, a better approach would be to treat as outliers the occasion(s) where the inferential is now known to have failed and remove them from the calculation of the index.

Finally we should recognise that a failure may not be due to a problem with the inferential but a problem with the laboratory result. This leads us on to our next topic.

9.3 Laboratory Update of Inferential

With well-integrated information systems it is relatively easy to automatically update the inferential with the latest laboratory result. Any difference between the laboratory result and the value of the inferential at sample time can be used to update the bias term in the inferential calculation.

$$bias_n = bias_{n-1} - K(PV_{inferential} - PV_{laboratory})_{n-1} \quad (9.9)$$

K is a filter parameter set by the engineer to a value between 0 and 1. There is a natural reluctance to set it to 1 since this would accept the full correction immediately; but it may be the laboratory result that is in error. By setting it to a lower value, typically around 0.3, several results will be required for the full correction to be made. In fact, if the error is sustained with K set to 0.3, it will take six updates to eliminate 90 % of the error – given by $100(1 - (1 - 0.3)^6)$. It is possible to optimise the value for K. Using historical data the update can be built into a spreadsheet and K adjusted to minimise the sum of the squares of the error. However in most cases the optimum value of K will be found to be zero!

While, at first glance, updating in this way would seem a good idea, in almost every case it causes the accuracy of the inferential to **degrade**. The laboratory result is subject to error. Most laboratory tests follow a documented standard, which will include estimates of *repeatability* and *reproducibility*. Reproducibility is not of concern here. It relates to the agreement between results obtained from different laboratory instruments, different technicians and different laboratories. Repeatability however is of interest. It relates to agreement between the same sample, analysed by the same technician using the same instrument. It is defined as double the standard deviation of the results, i.e. the 95 % confidence interval. To this must be added many other sources of error such as time-stamping, sample contamination and human error. The variance of the laboratory result is already included in the variance of the inferential error – since this is calculated from the difference between the two. Passing it also through the bias update increases the variance of the inferential error by the factor $(1 + K^2)$. Hence the standard deviation (σ_{error}) will increase by the square root of this factor.

The problem is that we need to distinguish between *bias error* and *random error*. What we have described so far are random errors. A bias error is a systematic difference between the true value and its measurement. It is unlikely to exist in the laboratory result but can arise in the inferential. A change of feedstock may cause a bias error to arise. In the oil refining industry, for example, it is common to have *cold property* specifications on fuels – freeze point, cloud point, pour point etc. These are controlled by changing operating conditions on the process but are also affected by the *paraffinicity* of the crude oil from which the product is derived. Changing the type of crude being processed can therefore cause a bias error in the inferential. In the chemical industry it is common to infer quality of a product based on operating conditions in the reactor in which it is produced. However, as the catalyst activity declines over time, a bias error will accumulate in the inferential.

Table 9.2 CUSUM calculation

Sample	Inferential	Laboratory	Error	CUSUM
1	5.08	4.81	0.27	0.27
2	4.97	4.79	0.18	0.45
3	4.93	5.25	−0.32	0.13
4	5.05	5.02	0.03	0.16
5	5.20	4.86	0.34	0.50
6	5.55	4.96	0.59	1.09
7	5.22	5.08	0.14	1.23
8	5.52	5.17	0.35	1.58
9	5.56	4.98	0.58	2.16
10	5.56	4.90	0.67	2.82
11	5.64	4.86	0.78	3.61
12	4.80	4.98	−0.18	3.43
13	5.16	4.94	0.23	3.65
15	4.95	5.17	−0.22	3.43
16	4.93	5.01	−0.09	3.35
17	4.95	5.17	−0.22	3.13
18	5.17	5.09	0.08	3.21
19	5.17	5.16	0.01	3.22
20	5.16	4.75	0.41	3.63

The best solution to a bias error is to eliminate it at source. If we can achieve this then we can abandon completely any automatic updating of the inferential. In our example it may be possible to detect the change of feedstock or possibly rely on an operator to enter the change in the DCS. There are techniques for compensating for changes in catalyst activity, for example by including in the inferential a parameter representing the total volume of feed processed – maybe weighted by a measure of severity.

If we do need to update automatically then we need to separate the bias error from the total error. The *CUSUM technique* offers an effective solution. In this case CUSUM is the cumulative sum of differences between the inferential and the laboratory result. Table 9.2 presents an example calculation.

Provided the results are in the correct sequence, there is no need for the sample interval to be fixed. Thus if samples are taken at irregular intervals, such as repeat tests, they may still be included. Figure 9.11 shows the CUSUM trend. If the error were 100 % random, the trend would be noisy but horizontal and no bias update is required. If a bias error is present then the gradient of the CUSUM trend is the amount by which the inferential is **overestimating** and so the amount by which the bias should be **reduced**. In our example this value is 0.49. Since it already includes several historical values the correction can applied immediately with confidence.

There is no need to ever reset the CUSUM to zero. However is important to record that a correction has been made so that subsequent estimates of the CUSUM gradient do not include values collected before the correction.

If the error has even a small random component, the performance index (ϕ) will always **reduce** if automatic updating is implemented. The index measures **only** random error. If there was no random error then the index would have a value of 1 – no matter how large the

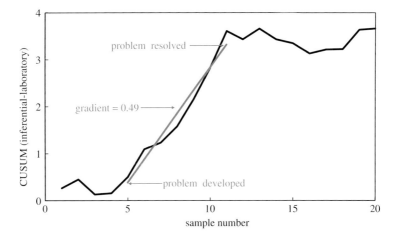

Figure 9.11 CUSUM trend

bias error. The advantage of the CUSUM approach is that it reduces the impact of the random error and so its effect on ϕ is much reduced.

Whether the correction should be fully automated is debatable. It is certainly important to **monitor** random and bias errors frequently but automatic correction is not a substitute for a poor inferential. Its inclusion may disguise a problem. Indeed this is exactly the situation with our predicted stock values. The prediction was quite simply yesterday's value with automatic updating based on today's value.

9.4 Analyser Update of Inferential

Automatic updating using an on-stream analyser measurement is quite different from updating with laboratory results. Analysers can have a reputation of poor reliability but we describe later in the chapter techniques that prevent spurious measurements from disturbing the process or being used to update an inferential. With this measurement validation in place analysers are far less prone to random errors than the laboratory. Secondly analysers provide measurements far more frequently and so the delay introduced by filtering will be far less.

However we have already mentioned the problem that the process dynamics introduce. We could resolve this by only permitting updates when the process is at steady state and has been so long enough for the analyser to respond to any changes on the process. However processes rarely reach a true steady state and updates are likely to be fairly infrequent. Instead we can install the configuration shown in Figure 9.12.

We apply dynamic compensation in the form of a deadtime/lead-lag algorithm. This is tuned in exactly the same way as described in Chapter 6 covering bias feedforward. By performing open loop steps on the MV we obtain the dynamics of both the inferential and

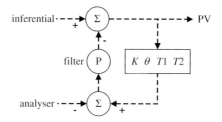

Figure 9.12 Use of analyser to update inferential

the on-stream analyser. Applying Equation (6.19) we get

$$K = -\frac{(K_p)_{analyser}}{(K_p)_{inferential}} \qquad (9.10)$$

The process gain of the analyser and the inferential should be the same and so K should be 1. If the test shows that this is not the case the problem should be resolved before commissioning analyser update – indeed before using the inferential in a controller.

From Equation (6.21))

$$\theta = \theta_{analyser} - \theta_{inferential} \quad T1 = \tau_{inferential} \quad T2 = \tau_{analyser} \qquad (9.11)$$

The analyser deadtime should be significantly larger than that of the inferential – otherwise the inferential serves little purpose – except perhaps as a back-up in the event of analyser failure. So θ will be positive. If not the case, the dynamic compensation should be applied to the analyser measurement.

If the analyser is discontinuous and its sample interval greater than the time it takes the process to reach steady state, then it may not show significant lag. As $T2$ should not be set to zero (because of the effect on the $T1/T2$ ratio) then it is wise only to include the deadtime compensation – by removing the lead-lag or setting $T1$ equal to $T2$.

The way in which the dynamic compensation operates is shown in Figure 9.13. The inferential is shifted (to curve A) by the delay θ. This compensates for the difference in deadtime between the analyser and the inferential K_p. The lead term ($T1$) cancels out the lag in the inferential and the lag term ($T2$) replaces it with the lag of the analyser, changing the output (to curve B). This now closely matches the analyser.

The correction term is the difference between the dynamically compensated inferential and the analyser measurement. The dynamic compensation assumes first order behaviour and so is unlikely to be exact. Further there will be inaccuracies in estimating the values of the time constants. This will cause an apparent error in the inferential but, providing it has the same process gain as the analyser, will be transient. Rather than correct for them instantly a small exponential filter (a lag) is included in the bias update. If the analyser is discontinuous then, between measurements, an error will exist. Again this is transient and will disappear at the next measurement. A substantially heavier filter will be required (with P set to around 0.98). Or, to avoid this, updating could be configured to occur only when the analyser generates a new value.

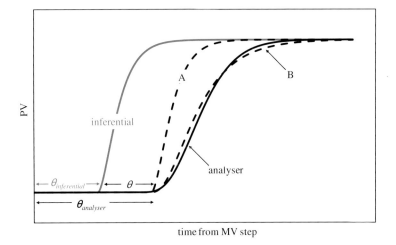

Figure 9.13   Effect of dynamic compensation

The bias used by the inferential should be monitored. Since the updating forces the inferential and the analyser to always agree at steady state, a problem with either measurement will be not be obvious. An increase in the standard deviation of the bias will indicate a problem caused by random error.

9.5   Monitoring On-stream Analysers

Many of the monitoring techniques suggested for inferentials can be applied to analysers. For example the performance index (ϕ) can be used to identify excessive random error between analyser and laboratory. The CUSUM can be used to check for a bias error which can arise particularly if the analysis method does not exactly match the laboratory technique.

However before applying such techniques we should first try to minimise the sources of error. For example locating the check sample point close to the analyser will minimise the time difference between taking the analyser reading and taking the check sample. For discontinuous analysers with a long sample interval an external indicator showing that a new sample is being taken can be used by the sampler so that the check sample is taken at the same time.

If in addition to calibration samples, routine samples are taken of the same product then these should ideally be taken from the same point. If the previously suggested push button or automatic detection is installed then the analyser can also be checked against accurately time-stamped laboratory samples.

Analyser sample delay should be minimised by locating the analyser as close as possible to the process and installing a *fast loop*. This takes a small stream from a high pressure point in the process, routes it close to the analyser and returns it to the process at a point where the pressure is lower. A common approach is to connect the fast loop between the discharge and suction of a product pump. It is not advisable to install the loop around a variable pressure drop, such a control valve, since the sample deadtime will then vary and cause controller tuning problems. If necessary a fast loop pump can be installed. The analyser sample is taken from the fast loop. The sample should be taken as far upstream as possible, again to reduce delay. Vapour travels faster than liquid so taking a sample while still in the vapour phase, or vaporising it at source, will further reduce delay – but the sample lines will need to be heated and insulated. Otherwise the heavier components will condense before the sample reaches the analyser and so affect the result.

The choice of analyser technology may be a trade-off between accuracy and speed of response. We will see in Chapter 9 that fuel gas heating value can be approximately inferred from its specific gravity or derived accurately from a full chromatographic analysis. SG analysers can be installed to give almost no delay, while chromatographs will delay the measurement by many minutes.

Analyser *sample conditioning* should be designed to ensure the sample is 'clean' and in the same condition as that when processed by a laboratory instrument. These recommendations and many others are covered by specialists (Reference 1).

There will inevitably be a difference between analyser and laboratory. Organisations have adopted a variety of approaches to resolving this. Placing responsibility for the accuracy of both devices under the laboratory manager prevents long debate about which result is correct. Moving towards the exclusive use of analysers for product certification raises their profile and the level of management attention given to their maintenance.

Close monitoring permits poorly performing analysers to be identified and the evidence provided to justify their improvement or replacement. It also provides evidence to dubious process operators that a previously suspect analyser is now reliable. In addition to such historical monitoring it is important to check the performance in real time of an analyser being used closed loop. A single undetected failure can result in costs greater than the annual benefit of improved quality control. Process operators and the plant manager will remember, for a long time, the incident of a whole batch of off-spec product that had to be reprocessed or downgraded. This can damage the reputation of all analysers and it takes far more to establish a good reputation than it does to destroy one.

PV validation is a technique which can be applied to any measurement but is of particular important to analysers. A number of checks can be made and automatic control or inferential updating disabled if any of these fail. Firstly the analyser may itself generate alarms. As an addition, sensibly set high/low checks on the measurement will flag a measurement that has moved outside its normal range. This is usually a standard feature within the DCS. If there is an inferential we can use the maximum expected deviation from the analyser to set the high/low checks. The DCS might also offer rate-of-change checking. A measurement moving faster than the process dynamics permit would also be declared invalid. We need also to check for a low rate of change or 'frozen' value. This can occur with failure of discontinuous analysers employing sample-and-hold. While a low rate of change check would detect this it is also likely to generate spurious alarms if the process is particularly steady. A better approach is a *timeout* check. Most discontinuous analysers provide a read-now contact that can be connected as digital input to the DCS. This is used to initiate a countdown timer to a value slightly higher than the sample interval. If this timer reaches zero the analyser is assumed to have failed.

By configuring a tag for each analyser, set equal to 1 if the measurement is valid and to 0 when not, we can historise this tag and use it to trend analyser availability and to average it as required. We can also set up similar tags to monitor the time that each analyser is in automatic control. This information then forms the basis of analyser performance reporting.

We also have to consider what action is taken on restoration of a valid measurement. If the analyser has been out of service for some time then the best approach is to generate a message to the process operator that it can now be restored to automatic control. If the outage is brief then automatic recommissioning might be considered, ensuring that correct initialisation is triggered to ensure the process is not 'bumped' by the measurement being different from that last used by the controller.

Reference

1. Clevett, K.J. (1999) *Process Analyzer Technology*. Chichester: John Wiley & Sons, Ltd.

10

Combustion Control

This chapter confines itself to boilers and other fired process heaters that burn liquid or gaseous fuels, or a mixture of both. Fuel gas in particular can be major source of process disturbances – particularly if its pressure or composition can vary. In the case of mixed firing it may not be possible to manipulate the flow of all the fuels, for example because one may be a by-product, from another part of the process, that cannot be economically stored.

10.1 Fuel Gas Flow Correction

Assuming gas flow is measured using a conventional orifice plate type of flow meter; we covered in Chapter 5 the correction necessary if working in units of **normal volumetric** flow, i.e. measured at **standard conditions**.

$$F_{true} = F_{measured} \sqrt{\frac{MW_{cal}}{MW} \times \frac{P}{P_{cal}} \times \frac{T_{cal}}{T}} \qquad (10.1)$$

Or, if working in **mass** flow units

$$F_{true} = F_{measured} \sqrt{\frac{MW}{MW_{cal}} \times \frac{P}{P_{cal}} \times \frac{T_{cal}}{T}} \qquad (10.2)$$

However, we also mentioned that special attention is required when applying these formulae to fuels. This is because a change in molecular weight not only affects meter calibration but also the heating value of the gas. If Equation (10.1) is used to condition the measurement of a flow controller then an increase in the molecular weight will cause the PV to fall and the flow controller to compensate by opening the control valve. Heating value generally increases with molecular weight and so, to maintain a constant fired duty, we need the control valve to close. The addition of compensation has worsened the impact of the disturbance caused by the change in fuel composition.

Process Control: A Practical Approach   Myke King
© 2011 John Wiley & Sons, Ltd

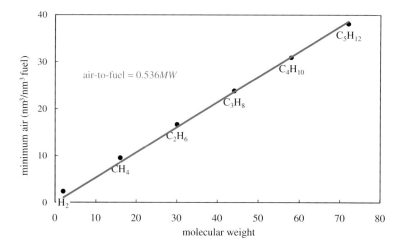

Figure 10.1 Combustion air requirement

It also presents a potential safety hazard. It is common to ratio combustion air flow to the fuel flow measurement. Thus the increasing fuel gas MW would cause a reduction in air flow. As Figure 10.1 shows, using some common fuels as examples, this is opposite to what is required. There is thus the danger of combustion becoming sub-stoichiometric. The resulting loss of heater efficiency would cause the outlet temperature to fall and the controller to increase fuel further.

Before incorporating heating value into the controller we need to ensure we use the correct definition. *Gross heating value (GHV)* is the heat released per unit of fuel if any water produced by combustion is condensed as so also releases its heat of vaporisation. *Net heating value (NHV)* is a lower value because it is based on the water remaining as vapour. Both can be quoted on a volumetric or on a weight basis. The combustion products of most fired heaters leave as flue gas and so we will use NHV. As an energy-saving measure condensing heaters are likely to become more common. The only effect on the control schemes covered in this chapter will be a change of coefficient.

On-stream analysers measuring molecular weight are normally marketed as *densitometers* and so we will base the control design on *specific gravity (SG)*, defined as

$$SG = \frac{MW}{MW_{air}} \tag{10.3}$$

The molecular weight of dry air (MW_{air}) is generally assumed to be 28.96, as derived from Table 10.1.

Figure 10.2 shows the relationship between NHV (on a weight basis) and SG for a number of gases commonly found in fuel gas.

Provided the fuel gas comprises mainly hydrocarbons then its heating value (on a weight basis) is largely independent of composition. Thus calibrating the gas flow meter to measure mass flow and correcting using Equation (10.2) will result in negligible disturbance to the heater as the heating value changes.

Table 10.1 Molecular weight of air

gas	vol %	MW
O_2	20.95	32.00
N_2	78.09	28.01
CO_2	0.03	44.01
Ar	0.93	39.94
air	100.00	28.96

However if hydrogen is present in a significant proportion then this approach will fail because of its very different heating value. Gas compositions are generally quoted on a volume (or molar) basis. Figure 10.3 shows the relationship between weight percentage and volume percentage for hydrogen mixed with gases commonly in fuel. It shows that (on a volume basis) the hydrogen content needs to be substantial before moving away from simple mass flow control of fuel gas.

This is confirmed by Figure 10.4 which shows that hydrogen does not have a significant impact on heating value until its content exceeds around 40 vol %. Above this value we have to adopt a different approach.

Figure 10.5 shows the relationship between NHV, now on a **normal volume** basis, and SG. On this basis the heating value of hydrocarbons varies with SG but in way which can be inferred from

$$NHV = aSG + b \qquad (10.4)$$

The coefficients can be derived theoretically based on the NHV and SG of pure gases. In the engineering units used for the graph, a is 56.14 and b is 5.78. If NHV is measured in units of BTU/SCF, then these coefficients change to 1506 and 139 respectively. However these values should be used only as a guide. The true value will depend on what other components are in the fuel. The presence of inerts such as N_2 and CO_2 will change the relationship, as

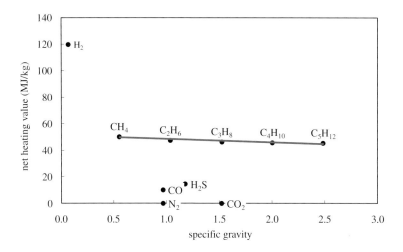

Figure 10.2 Net heating value on a weight basis

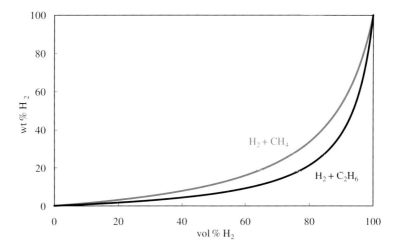

Figure 10.3 Conversion of volume percentage to weight percentage hydrogen

will any nonhydrocarbon fuels such as CO and H_2S. Provided the concentration of these components is small or varies little we can still predict NHV from SG but we need to develop the correlation from real process data. Figure 10.6 shows some typical laboratory data routinely collected from a site's fuel gas system over several months.

At first glance the correlation would appear to be poor with one point (ringed) showing a very large deviation. However a review of the analysis of this sample, shown in Table 10.2, reveals a common problem with sampling.

Poor procedures have resulted is the sample being contaminated with air, as indicated by the O_2 content of 2.70 %. We can remove this component from the analysis provided we also remove the associated N_2. We know from Table 10.1 that the N_2 concentration in air is

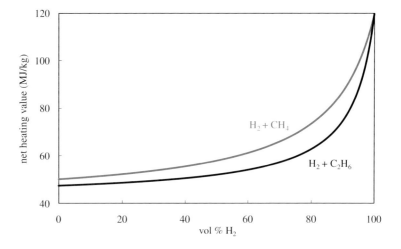

Figure 10.4 Impact of hydrogen on heating value

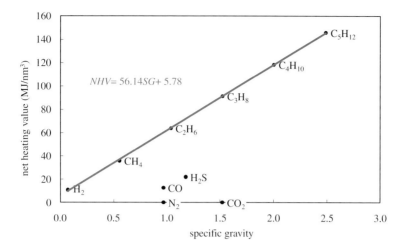

Figure 10.5 Gas heating value on a normal volume basis

3.73 times that of O_2 and so we reduce the N_2 by 10.06 %. The remaining 1.84 % N_2 is that genuinely in the fuel gas. Applying this correction to every sample gives the very reliable correlation shown in Figure 10.6.

Once we have values for *a* and *b* we can modify the signal conditioning so that the flow is measured in energy units, for example MJ/hr or BTU/hr. By combining Equations (10.1), (10.3) and (10.4) we get

$$F_{true} = F_{measured} \sqrt{\frac{SG_{cal}}{SG} \times \frac{P}{P_{cal}} \times \frac{T_{cal}}{T}} \times (aSG + b) \qquad (10.5)$$

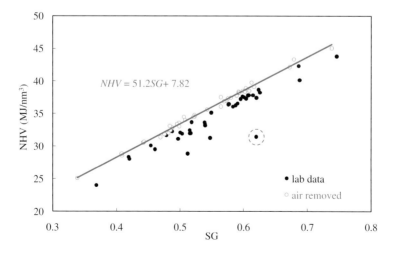

Figure 10.6 Predicting NHV from plant data

Table 10.2 Correcting fuel gas analysis

Gas	Original analysis				Corrected analysis			
	MW	SG	NHV	mol %	MW	SG	NHV	mol %
H_2	2.02	0.07	10.8	46.90	2.02	0.07	10.8	46.90
CH_4	16.04	0.55	35.8	12.10	16.04	0.55	35.8	12.10
C_2H_6	30.07	1.04	63.7	10.10	30.07	1.04	63.7	10.10
C_2H_4	28.05	0.97	59.0	1.70	28.05	0.97	59.0	1.70
C_3H_8	44.10	1.52	91.2	6.70	44.10	1.52	91.2	6.70
C_3H_6	42.08	1.45	85.9	1.20	42.08	1.45	85.9	1.20
C_4H_{10}	58.12	2.01	118.4	3.80	58.12	2.01	118.4	3.80
C_4H_8	56.11	1.94	113.0	0.70	56.11	1.94	113.0	0.70
C_5H_{12}	72.15	2.49	145.3	1.10	72.15	2.49	145.3	1.10
C_6H_{14}	86.18	2.97	172.0	0.30	86.18	2.97	172.0	0.30
O_2	32.00	1.10	0.0	2.70	32.00	1.10	0.0	0.00
N_2	28.01	0.97	0.0	11.90	28.01	0.97	0.0	1.84
CO	28.01	0.97	12.6	0.50	28.01	0.97	12.6	0.50
CO_2	44.01	1.52	0.0	0.20	44.01	1.52	0.0	0.20
H_2S	34.08	1.18	21.9	0.03	34.08	1.18	21.9	0.03
total	17.96	0.62	31.5	99.93	16.37	0.57	36.1	87.16

We can see that, if the molecular weight increases, the PV of the flow controller will now increase. The controller will respond by now closing the control valve.

Some plant owners prefer not to replace the conventional flow measurement with one recording in energy units. They argue that the process operator should be able to see the measurement in its original units. This can be displayed separately or, instead of conditioning the PV, we can apply the reciprocal of the function to the SP.

10.2 Measuring NHV

The technique described above begs the question as to why infer the NHV from SG instead of measuring it directly? Firstly we need the SG measurement in any case for the orifice flow correction. Secondly there is a large cost advantage. Densitometers are a much lower cost instrument than *calorimeters* or others that can be used, such as chromatographs. Further their installation costs considerably less. They are mounted on the pipework itself, much like a flow meter, and do not require an analyser house. Finally there is a dynamic advantage. The residence time in the fuel gas system can be just a few seconds. Densitometers give an almost immediate indication of any change. Any delay could result in the heater outlet temperature detecting the disturbance, and taking corrective action, before the analyser responds. The now belated correction for the composition change would then cause a second disturbance.

We similarly need to ensure that the composition change is not dealt with too early. This can occur if the fuel gas supply pipework is long and the analyser located well upstream of the heater. Before embarking on analyser installation it is important to calculate the residence time between the proposed sample point and the heater. If this is significant then it is possible to delay the measurement in the DCS by the use of a deadtime algorithm. However, in this case, the variability of the residence time should also be checked. If the supply is dedicated to the heater then this involves simply checking the maximum and minimum firing rates for that heater. However, if there are several heaters on site, it is common for there to be fuel gas header supplying all of them. The impact of the demands of the other heaters has on residence time then needs to be taken into account.

Theoretically it is possible to automatically adapt the tuning of the deadtime algorithm based on measured gas flows but this is complex and prone to error if there are multiple fuel gas consumers and producers. It is likely to be more practical to locate the analyser close enough to the heater so that dynamic compensation is not required.

If the residence times for all the heaters are short then it may be possible to locate the analyser on the shared header so that it may be used in the firing control of all the heaters. Under such circumstances greater attention should be given to the integrity of the whole system. A failure which causes the analyser to generate a low, but still believable, measurement would cause the fuel gas consumed by all heaters to rise simultaneously – potentially causing a major pressure disturbance in the fuel header. Given the relatively low cost of densitometers, it is practical to install two close together and cross-check their measurements. Any significant difference between their readings causes the scheme to switch to use the last good value rather than the current value. This provides for *graceful degradation* of the controller. However, when the fault is cleared, each controller using the value should be reinitialised. The new reading could be quite different from the last good reading and recommissioning would otherwise 'bump' the process.

The choice of densitometer is important. Firstly, in order to avoid any measurement delay, it should be of the probe type and not one involving a sample withdrawal system. Secondly, remembering we are using it to infer MW, it should measure density at standard conditions not at stream conditions. While we can of course convert from one to the other in the DCS, this requires measurement of temperature and pressure at the sample point. It is more cost effective for this to be done within the analyser.

If the fuel contains significant concentrations of gases, other than hydrogen and hydrocarbons, whose concentration can vary quickly then the inferred NHV may be unreliable. For example, if there was a large increase in the N_2 or CO_2 content of the gas in Table 10.2, the SG would increase but the NHV would reduce. The correction proposed in Equation (10.5) would therefore cause the flow controller to reduce gas flow, when we need it to increase it. Such a problem would have been apparent when developing the correlation. A possible solution is the use of a *Wobbe Index* analyser, where

$$\text{Wobbe Index} = \frac{NHV}{\sqrt{SG}} \qquad (10.6)$$

The analyser measures the NHV (on a volumetric basis) by continuously withdrawing and burning a sample of the gas. One approach is to record the amount of air necessary to consume it fully. The air flow is adjusted to maintain a constant temperature of the combustion products. Figure 10.7 shows that there is strong correlation between the minimum air requirement and the NHV of common fuel gas components – including inerts and nonhydrocarbons. While not perfect, and not measuring the **minimum** air

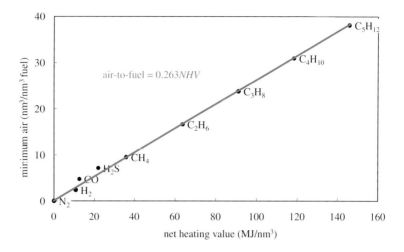

Figure 10.7 Stoichiometric air requirement

required, the analyser can be calibrated to give a measurement accurate enough for control. Another approach is to fix the air flow and then measure the residual oxygen in the combustion products. Both approaches also include a measurement of SG for use in the calculation of Wobbe Index. Some analysers offer the option of also generating a measurement of air demand, known as the *Combustion Air Requirement Index (CARI)*.

The energy flow calculation then becomes

$$F_{true} = F_{measured} \sqrt{SG_{cal} \times \frac{P}{P_{cal}} \times \frac{T_{cal}}{T}} \times \text{Wobbe Index} \qquad (10.7)$$

Wobbe Index analysers are slower than densitometers but do give a continuous measurement. Whether the measurement is fast enough, however, should be checked on a per case basis.

Another possible type of analyser is a chromatograph. This would provide a full analysis of the fuel gas. Using known molecular weights and heating values for the components it is possible to accurately calculate the combined properties. Many chromatographs support this feature or the calculations could be located in the DCS. However a full component analysis would take several minutes, by which time the feedback controller on the heater will have already corrected for the disturbance. In this case a chromatograph would provide an accurate measurement suitable for accounting and monitoring purposes but would be entirely unsuitable for control.

10.3 Dual Firing

Dual fuel firing was cited in Chapter 6 as an example of bias feedforward control. Here we will expand on this technique. The general control problem is illustrated in Figure 10.8.

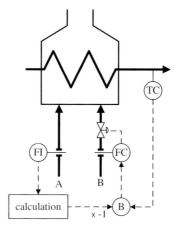

Figure 10.8 Dual firing example

In this example fuel A is a gas over which we have no control. Fuel B is a liquid and its flow may be manipulated to control the heater outlet temperature. The scheme includes a bias feedforward scheme so that changes in fuel A are immediately compensated for by adjusting the flow of fuel B. For this to succeed we have to convert the units of measure of fuel A to be consistent with those of fuel B. By including the heating value of fuel B (NHV_b) in Equation (10.5) we get

$$F_a = F_{measured} \sqrt{\frac{SG_{cal}}{SG}} \times \frac{P}{P_{cal}} \times \frac{T_{cal}}{T} \times \frac{(aSG+b)}{NHV_b} \qquad (10.8)$$

Depending on the choice of units the flow of fuel A (F_a) will now be in TFOE (tons of fuel oil equivalent). In the oil industry the barrel is commonly used as a measure of volume and so BFOE might be used. The output of the temperature controller can be thought of as the total duty demand (in fuel B units) from which is subtracted that delivered by fuel A.

10.4 Inlet Temperature Feedforward

In Chapter 6 we used a fired heater as the example of an application of feedforward control on feed rate. The inclusion of fuel gas pressure, temperature and SG in the duty controller can also be thought of as feedforward schemes. Another potential source of disturbances is heater inlet temperature. Feed to the heater is often preheated by heat exchange with streams in other parts of the process. Any change in the flow or enthalpy of these streams can therefore cause a disturbance. This is likely to be reflected in a similar size change in the outlet temperature. And this change may be propagated again back to heater inlet via the heat integration. Figure 10.9 shows the addition of a suitable bias feedforward scheme.

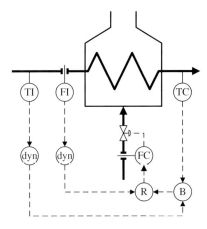

Figure 10.9 Inlet temperature feedforward

From Equation (6.19) we know that the gain in the bias feedforward is given by

$$K = -\frac{(K_p)_d}{(K_p)_m} \tag{10.9}$$

From Chapter 2, we know that $(K_p)_m$ varies inversely with feed rate. If we were to configure the output of the bias algorithm to manipulate the fuel flow directly, rather than the fuel-to-feed ratio, then we would need to include adaptive tuning to automatically adjust K to keep it in proportion to feed rate.

By definition

$$(K_p)_d = \frac{\Delta PV}{\Delta DV} \quad \text{and} \quad (K_p)_m = \frac{\Delta PV}{\Delta MV} \tag{10.10}$$

Thus

$$K = -\frac{\Delta MV}{\Delta DV} \tag{10.11}$$

For simplicity we assume that feed specific heat (c_p) and heater efficiency (η) are constant. If ΔT is the change in inlet temperature then the required change in fuel flow (ΔF energy units) is given by the heat balance

$$F_{feed}.c_p.\Delta T = \Delta F.\eta \tag{10.12}$$

Combining with Equation (10.11) confirms the dependence of K on the feed flow (F_{feed}).

$$K = -\frac{\Delta MV}{\Delta DV} = -\frac{\Delta F}{\Delta T} = -\frac{F_{feed} c_p}{\eta} \tag{10.13}$$

This is not a problem, of course, if feed rate varies little. But if we can retain the feedforward ratio, the MV becomes the ratio SP. By dividing Equation (10.13) by F_{feed} we get

$$K = -\frac{\Delta R}{\Delta T} = -\frac{c_p}{\eta} \qquad (10.14)$$

As it did with feedback control (Chapter 6) the use of the ratio obviates the need, as feed rate changes, to adjust the tuning of the feedforward scheme.

We may choose not to have the flow of the manipulated fuel to be conditioned to energy units if, for example, its properties change little. We then need to include the *NHV* in the calculation of *K*, ensuring we use engineering units consistent with the units of flow.

$$K = -\frac{\Delta R}{\Delta T} = -\frac{c_p}{\eta.NHV} \qquad (10.15)$$

While we can predict the value for *K*, we still need to perform plant tests to obtain the process dynamics required for calculation of the tuning for the dynamic compensation. The procedure is covered in Chapter 6. We need to be careful with the units of *K*. It is usual for bias algorithms, unlike PID controllers, to operate in engineering units and so *K* should be determined in engineering units. Some model identification packages however work in fraction of range. If this is the case then, if *K* is determined using Equation (10.9), it should be converted to engineering units by multiplying by

$$\frac{MV_{range}}{DV_{range}} \qquad (10.16)$$

The result can then be checked for consistency with that obtained from Equation (10.14) or (10.15).

On some heaters, performing the necessary step tests may be impractical. To analyse the results we need to change the inlet temperature with the outlet temperature controller in manual mode. Introducing disturbances to the inlet temperature may not be straightforward and it would be inadvisable to leave the outlet temperature controller in manual mode for long periods waiting for a natural disturbance. A better approach would be to commission the feedforward controller with the value derived theoretically for *K* and dynamic compensation estimated assuming the process dynamics of inlet temperature changes are the same as those for feed rate changes. Careful monitoring of control performance would allow the tuning of the dynamic compensation to be modified as necessary.

10.5 Fuel Pressure Control

Some fired heaters have a pressure, rather than flow, controller on the fuel, as shown in Figure 10.10.

This scheme is used to ensure that burner pressure is kept within limits. Too high a pressure can result in the fuel velocity exceeding the flame velocity so that the flame separates from the burner tip and is potentially extinguished. Too low a pressure on oil burners can result in poor atomisation of the fuel and thus poor combustion. By installing SP limits in the pressure controller, the heater is protected. If the high SP limit is reached the

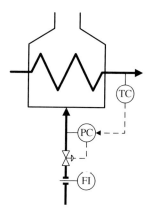

Figure 10.10 Fuel pressure control

operator will be expected to put additional burners into service, and to take some out of service if the low pressure limit is reached.

However, this type of scheme prevents the application of many of the techniques covered in this chapter. It would be difficult to devise a feedforward scheme to deal with disturbances in feed rate, heater inlet temperature or fuel heating value. A better solution is illustrated in Figure 10.11.

This has both a flow controller and a pressure controller. The flow controller permits the addition of the compensation and feedforward schemes. The pressure controller provides burner protection. In this case their outputs are compared in a low signal selector. This provides protection against too high a pressure. If the pressure exceeds the maximum, as entered as SP in the pressure controller, the controller output will reduce to close the valve

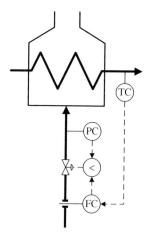

Figure 10.11 Fuel pressure override

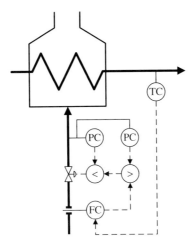

Figure 10.12 Fuel low and high pressure override

and so override the flow controller. The temperature will then not reach its SP and the operator will need to put additional burners into service to relieve the constraint.

If protection against too low a pressure is required then a high signal selector is used. If protection against both low and high pressures is required then many DCS support a middle signal selector. If not, then low and high signal selectors can be configured in series as shown in Figure 10.12.

Another scheme which makes duty control difficult is that often installed on high viscosity fuel oil systems. To prevent pipework blockages, such fuel needs to be kept above a minimum temperature. Should its flow drop, heat losses from the pipework can result in the temperature falling below this minimum. To ensure a flow is maintained, even if a heater is shutdown, fuel is circulated around the site via a heated storage tank. The pipework passes alongside every heater and each burner on the heater can have its own take-off. Since it is not practical to measure the flow to an individual burner, flow meters are installed on the supply to and return from the heater. Fuel consumption is then determined by the difference between these measurements. However, because consumption is small compared to the circulating flow, the calculation is very prone to measurement error. For example if the supply flow is 100 %, measured to ±2 %, and the return flow is 95 %, also measured to ±2 %, then the calculated consumption could vary by a factor of nine, i.e. from 1 % to 9 %. Such a measurement cannot be used as a DV.

10.6 Combustion Air Control

Figure 10.13 is a schematic of a typical combustion air system. Ambient air is routed to the *air preheater* by the *forced draught fan*. Control of air flow may be means of a variable speed drive on the forced draught fan or by some form of throttling of fan discharge. Measurement of flow usually is by pitot tube or annubar. Air then divides to each of the burners; adjustment to individual burners is performed manually by manipulation of *air registers*. The *stack*

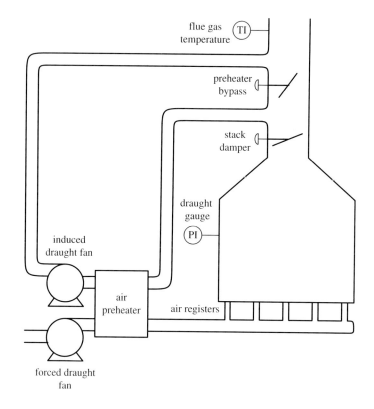

Figure 10.13 Fired heater combustion air schematic

damper is adjusted manually to ensure the firebox pressure (measured by the *draught gauge*) remains negative – in order to avoid the risk of flame exiting through inspection openings etc. The *preheater bypass* will generally be closed but can be opened as necessary should the flue gas temperature fall below dewpoint. On noncondensing heaters any water condensed is likely to be corrosive due to acid gas products produced by the combustion of trace amounts of sulphur compounds such as H_2S. The *induced draught fan* returns the flue gases to the stack.

Alternative configurations are possible which omit either of the fans. If both fans are omitted then recovery of heat from the flue gas is not possible using the type of air preheater shown. Control of air flow on such *natural draught heaters* is by manipulation of the stack damper, although only if this does not violate the need to keep firebox pressure negative.

We clearly need to ensure that our control strategy ensures sufficient air is supplied to ensure complete combustion of the fuel. Leaving unburnt fuel in the flue gas is potentially hazardous; the heater firebox operates under slightly negative pressure. Any *tramp air* which enters through leaks above the combustion zone could potentially cause an explosion in the convection section. Incomplete combustion can result in soot particles thus producing black smoke with the resulting impact on the environment and fouling of any flue gas heat recovery or treatment systems. Once initiated, the problems associated with incomplete combustion can escalate quickly. Because the heater efficiency drops sharply the heater outlet temperature will fall; the controller will respond by increasing the fuel flow.

However, it is undesirable to operate with excess air. Since air enters the heater at ambient conditions and leaves at flue gas temperature any unnecessary air increases the amount of fuel that is required to achieve the desired outlet temperature. Ideally we would like to maintain a stoichiometric air-to-fuel mixture. In practice we need to provide a small amount of excess air to compensate for incomplete mixing and to ensure that full combustion has taken place before the products of combustion leave the firebox.

Figure 10.14 shows the effect that varying the excess air has on flue gas composition, using methane as the example. By measuring flue gas composition we can assess the level of excess air. The most practical measurement is O_2. As with fuel gas properties we need an analyser that responds quickly. Oxygen analysers are available as probe types that can be inserted directly into the stack. Provided there is excess air, oxygen varies approximately linearly with air flow making tuning of the controller straightforward.

The relationship between O_2 and excess air can be developed by examining the chemical equations of combustion. If we first consider pure hydrogen then

$$H_2 + 0.5\,O_2 \rightarrow H_2O \tag{10.17}$$

From Table 10.1 we see that air is 20.95 % O_2. For simplicity let us assume the remainder is N_2. Then, for 10 % excess air

$$H_2 + 0.55\,O_2 + 0.55\left(\frac{79.05}{20.95}\right)N_2 \rightarrow H_2O + 0.05\,O_2 + 0.55\left(\frac{79.05}{20.95}\right)N_2 \tag{10.18}$$

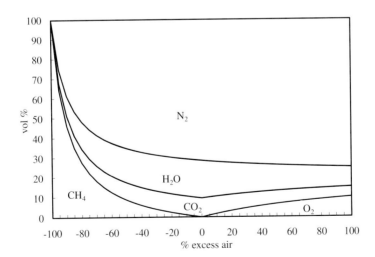

Figure 10.14 Flue gas composition for combustion of methane

On a *dry basis*, where the water remains as a vapour, the molar concentration of O_2 in the flue gas is given by

$$100 \times \frac{0.05}{1+0.05+0.55\left(\frac{79.05}{20.95}\right)} = 1.60\% \qquad (10.19)$$

On a *wet basis*, where the water is condensed to liquid, it becomes

$$100 \times \frac{0.05}{0.05+0.55\left(\frac{79.05}{20.95}\right)} = 2.35\% \qquad (10.20)$$

We can repeat this exercise for pure carbon

$$C + O_2 \rightarrow CO_2 \qquad (10.21)$$

$$C + 1.1\,O_2 + 1.1\left(\frac{79.05}{20.95}\right)N_2 \rightarrow CO_2 + 0.1\,O_2 + 1.1\left(\frac{79.05}{20.95}\right)N_2 \qquad (10.22)$$

Since there is no water product, the wet and dry analyses are the same. The molar concentration of O_2 is thus given by

$$100 \times \frac{0.1}{1+0.1+1.1\left(\frac{79.05}{20.95}\right)} = 1.90\% \qquad (10.23)$$

Figure 10.15 shows how the relationship between flue gas O_2 and excess air varies with fuel composition. This can be issue on multifuel heaters. A low molecular weight fuel, such as a hydrogen/methane mixture, with have a H:C ratio of around 6. For high molecular weight fuel, such as fuel oil, this ratio will be around 2. On a dry basis switching from gas to

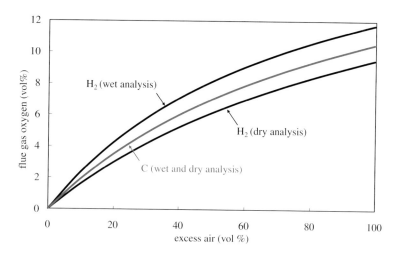

Figure 10.15 Impact of fuel type on flue gas oxygen content

oil will require operation at higher flue gas O_2 to give the same excess air. This, plus the fact that air-to-fuel mixing is less efficient with liquid fuels, is likely to require the operator to occasionally adjust the SP of the O_2 controller.

The main limitation of using O_2 to infer excess air is that it gives no indication of how substoichiometric is the air-to-fuel ratio. Thus the controller will respond at the same speed no matter how large the air shortage.

Not shown in Figure 10.14 is CO (carbon monoxide). As the air-to-fuel ratio approaches the stoichiometric mixture, small amounts of CO will be detectable in the flue gas. This will increase markedly as air rate falls below the minimum required. The CO measurement cannot be used standalone because, like O_2, it only indicates over part of the operating range – showing zero no matter how much excess air is supplied.

If CO and O_2 are both present in significant quantities then this may indicate the presence of tramp air. The combustion is substoichiometric and air is entering the heater after the combustion zone. It indicates the need to seal leaks in the heater casing. If tramp air is not the problem then simultaneous high readings can be built into the analyser validity check.

Many heaters cannot normally operate at the level where CO is detectable, but for those operating at 1 % O_2 or lower, CO can be used to condition the O_2 measurement.

$$PV = O_2 - K.CO \tag{10.24}$$

The coefficient (K) is determined so that the process gain remains approximately constant over the whole operating range. By step testing at higher air rates we can obtain

$$(K_p)_{O_2} = \frac{\Delta O_2}{\Delta(\text{air/fuel ratio})} \tag{10.25}$$

By step testing (carefully!) at minimum air rate we can obtain

$$(K_p)_{CO} = \frac{\Delta CO}{\Delta(\text{air/fuel ratio})} \tag{10.26}$$

Both process gains should be in engineering units. The coefficient (K) is then given by

$$K = \frac{(K_p)_{O_2}}{(K_p)_{CO}} \tag{10.27}$$

Alternatively it may be possible to use historical data to plot the lines shown in Figure 10.16 and determine K from the gradients of these lines.

Figure 10.17 shows how the conditioned measurement varies with excess air. Apart from the unavoidable change in process gain when both analysers show a measurement, we have an approximately linear relationship. Importantly this extends over the whole operating range, giving apparently negative O_2 measurements when there is insufficient air. As an enhancement the value of K can be increased slightly so that the controller will respond more quickly when CO is detected. A cautious approach is advisable to ensure the controller does not go unstable under these circumstances.

The mechanics of implementing automatic control of flue gas O_2 can be complex. There may be problems with the O_2 measurement itself. Poor mixing of the flue gas may mean the measurement is not representative. Tramp air will cause a false measurement of what is

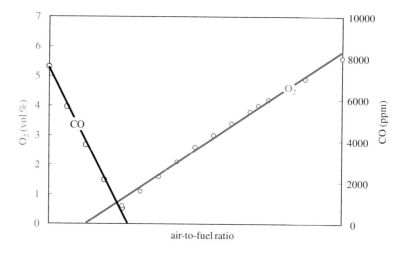

Figure 10.16 Effect of air on flue gas analysis

happening in the combustion zone. Some heaters have multiple cells with their flue gas routed to a common duct. An analyser located here would be of little use for control.

There can also be problems with manipulation of air flow. Natural draught heaters have no fan and so air flow is adjusted by manipulating the position of a stack damper. The relationship between air flow and damper position can be highly nonlinear. The damper is in a potentially corrosive and dirty environment and prone to mechanical failure. It can often exhibit stiction or hysteresis. On forced draught heaters there is the option of throttling the discharge of the fan but this can have many of the problems common to stack dampers. Manipulating the speed of an electrically driven fan, for example by using a *variable frequency drive* (VFD), is an expensive option and can also be nonlinear.

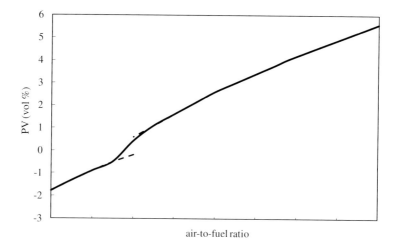

Figure 10.17 Conditioned oxygen measurement

There are also potential operating problems. We have to be sure the control does not increase the probability of substoichiometric combustion or positive firebox pressure.

These are not necessarily reasons for not progressing with improved control. The point is made to demonstrate that implementation may be costly – particularly if retrofitted to an existing heater. Before embarking on implementation we need to ensure the economic payback makes it worthwhile.

The savings will depend on fuel type, flue gas temperature and how close the operation already is to minimum excess air. Figure 10.18 shows the effect that fuel type has on the potential savings if the flue gas temperature is 400 °C (around 750 °F).

It is worth noting the penalty for going substoichiometric. This wastes about 10 times more fuel than the equivalent amount of excess air. Should operation at lower O_2 levels increase the probability of this occurring then this cost, combined will all the other resultant operational problems, could well exceed the annual benefits captured by the controller.

Flue gas temperature also affects the potential for cost saving. If the temperature of the flue gas is reduced by the installation of some form of heat recovery system then the fuel wasted by excess air will be reduced. Figure 10.19 shows the effect, using ethane (C_2H_6) as the fuel example.

Figure 10.20 shows in principle how O_2 control might be added to our dual fired heater example. It assumes some form of flow control is feasible on combustion air, on this occasion via a variable speed drive on the fan. Rather than directly manipulating this flow an air-to-fuel ratio has been installed. This offers several advantages. Firstly it will help maintain the excess air constant during times when the analyser is out of service. Secondly, as covered in Chapter 6, it will obviate the need to retune the controller as feed rate changes. This is particularly important on fired boilers which frequently have a turndown ratio of around 4:1. Without the ratio, a fourfold change in process gain would present a tuning problem.

In this example we have used the fuel **demand** signal as our measure of fuel flow. One could argue that this is desirable on increasing demand since it is an earlier indication of the

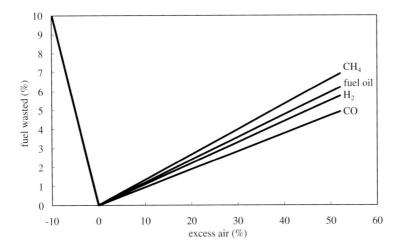

Figure 10.18 Effect of fuel type on fuel wastage

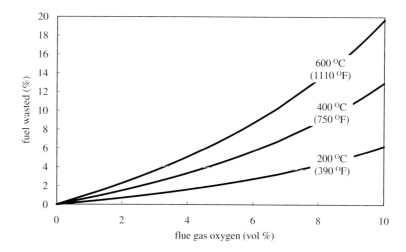

Figure 10.19 Effect of flue gas temperature on fuel wastage

need to increase air flow than that given by the actual fuel flow measurement. However on a decreasing fuel demand this is not the case. The simple ratio also presupposes that there is adequate air for the additional fuel; if the fan is at maximum capacity the temperature controller would still increase fuel. The scheme also assumes that the dynamics of air flow control are the same as those of fuel flow. Due to this size of the actuator it is likely that air flow will increase more slowly than fuel, possibly causing a transient shortage of air enough to take the mixture substoichiometric.

To overcome these potential problems, a *cross-limiting* approach may be adopted. This is shown in Figure 10.21. It is also known as the *lead-lag* scheme (not to be confused with the

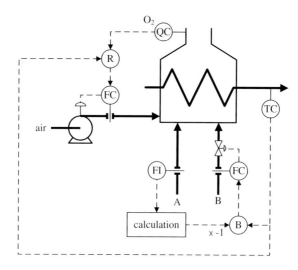

Figure 10.20 Principle of flue gas oxygen control

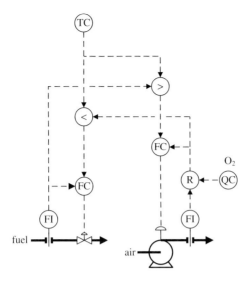

Figure 10.21 Cross-limiting control

lead-lag control algorithm). It gets its name because the air leads the fuel on increasing demand but lags it on decreasing demand.

The ratio is configured as fuel-to-air rather than the otherwise more usual air-to-fuel. It therefore converts the air flow measurement into units of fuel flow. The result is the flow of fuel that can be consumed at the desired excess air rate. This value is used in two places. First it provides the measurement to the air flow controller. Secondly it provides the input into a low signal override (<) on the output of the temperature controller. If the value is less than the fuel demand it prevents the fuel flow from being increased above the value determined from the air flow. However, the fuel demand is also fed, through a high signal selector (>), as SP (in fuel flow units) to the air flow controller. Because it is increasing it is passed through by this selector. As the air flow controller responds to this increase, the override on the temperature controller is relaxed and the fuel SP is permitted to increase. Thus, on increasing demand, fuel will not be increased until sufficient combustion air is delivered.

Conversely, on decreasing demand, the controller output is passed by the low signal selector to the SP of fuel flow controller. The fuel flow measurement provides an input to the high signal selector. If the fuel flow controller does not respond then this will override the output sent to the air flow controller. Thus air is not permitted to reduce until the fuel flow reduces.

Some plant owners take the view that the complexity of the scheme creates a hazard because of potential misunderstanding of its operation by the process operator. One source of confusion is the use of the reciprocal of the air-to-fuel ratio. Another is the air flow controller operating in equivalent fuel units. It is possible to reverse the scheme so that the ratio is on the fuel side, which resolves these issues but then means that the fuel flow control operates in equivalent air units. It is possible to reconfigure the scheme so that both ratios are used and both flow controllers work in their own units, as shown in Figure 10.22.

If the air flow controller is switched to manual mode then the temperature controller may not be permitted to increase fuel, for example if heater feed rate is increased. This will result

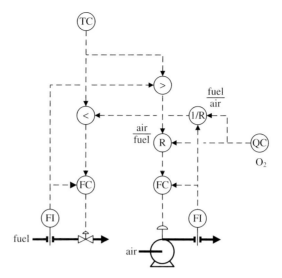

Figure 10.22 Cross-limiting control (alternative)

in the temperature falling below SP. While arguably safer, the operator could well decide to switch the fuel flow controller to manual so that the problem can be corrected.

From an implementation aspect care has to be taken with scaling and initialisation. However there are a large number of the schemes in place throughout industry operating successfully. Some plant owners mandate it as a standard, as do some suppliers of fuel.

10.7 Boiler Control

Boiler control is similar to process heater control, except that boiler duty is manipulated by a steam header pressure controller. Of the schemes outlined so far in this chapter all would be applicable with the exception of the feedforward schemes on feed rate and inlet temperature. The flow of boiler feed water is changed in response to a change in steam drum level, as we saw in Chapter 4, and so follows rather than leads an increase in firing. Feedforward on water flow therefore brings no dynamic advantage. It might be possible to ratio fuel to steam **demand** but even if available as a flow measurement, the dynamic advantage is likely to be small since header pressure will respond quickly to any imbalance between supply and demand. The common advantage of ratio feedforward, i.e. not having the retune the controller as feed rate changes, does not apply to integrating processes. Header pressure control is such a process.

If there is more than one boiler supplying the header, it is common to operate one or more as *baseload* boilers, where the duty is fixed and the remainder as *swing* boilers – used to control the steam pressure. Whether a boiler is in baseload or swing mode affects its process dynamics. Thus, if an individual boiler can be switched between these modes, we must pay particular attention to the control design.

For example, consider the result of a step test performed to obtain the process dynamics for flue gas oxygen control. The purpose of the test is to obtain the process dynamics of flue gas oxygen with respects to changes in air flow. If we change the air rate to a baseload boiler then we change the efficiency of that boiler and hence its steam production. This will in turn disturb the header pressure and the pressure controller will take corrective action by adjusting the firing on the swing boiler. If we repeat the step test with the boiler now in swing mode, the pressure controller will change the duty of the boiler on which we are conducting the test. The change made to firing will affect the flue gas oxygen and we will therefore obtain a different result for the process gain. If we were to use the results of the first test to tune the controller then there is a danger that it will go unstable when the boiler is switched to swing mode.

Of course we could automatically switch tuning with mode changes, but the air-to-fuel ratio scheme already proposed resolves the problem. With this ratio in place the result of testing in swing mode will be the same as that in baseload mode. In swing mode, where the header pressure control adjusts the duty, the air flow will be maintained in proportion and the change in flue gas oxygen will be unaffected.

If the boilers are of different designs then the dynamic relationship between header pressure and boiler firing will vary – depending on which is selected as swing. Similarly the dynamics will change if the number of swing boilers is changed. Different tuning will then be required in the header pressure controller as its MV is changed.

10.8 Fired Heater Pass Balancing

A popular strategy is balancing passes within multi-pass fired heaters. Such heaters will have flow controllers on each pass. The concept is to keep the total flow through the heater at the required value but adjust the distribution of the flow between the passes. The simplest objective is to maintain equal pass flows. Probably the most common is to equalise the individual pass outlet temperatures in the belief that heater efficiency is maximised. Other strategies are possible if the heater is a bottleneck. For example if it is the main hydraulic constraint then equalising the positions of the pass flow control valves, so that no one flow controller saturates before the others, will maximise capacity. If the main constraint is tube skin (tube metal) temperature then, provided this can be measured reliably, these can be equalised. This would permit either the combined heater outlet temperature (or the total feed rate) to be increased. The run length of heaters in coking service is limited by the most coked pass. Balancing the rate of coking of each pass can be exploited by increasing the run length or operating at a higher severity.

In the case where we believe there are efficiency benefits we can predict these. The value of balancing pass outlet temperatures depends on the nonlinear relationship between fired duty and temperature. In the radiant section of the heater this is governed by *Stefan's Law* which states that rate of heat radiated is proportional to the fourth power of the absolute temperature. Assuming our heater operates with an outlet temperature of 300 °C we can use this relationship to construct the curve in Figure 10.23. If the passes are imbalanced so that half are 10 °C hotter than the mean then our chart shows an additional 7.2 % firing is required. For those passes 10 °C below the mean, 6.8 % less firing is required. So by exactly balancing the temperatures the overall saving will be 0.2 % of total firing. Much of this saving might be achieved by occasional manual attention to the distribution but, even if not,

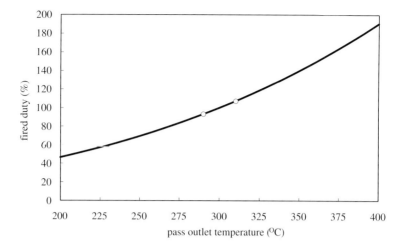

Figure 10.23 Effect of pass outlet temperature on fuel demand

it is unlikely that such an improvement would be measurable. If this is the only source of benefit then it is unlikely to justify the cost of implementation.

However, pass temperature balancing can be very lucrative in situations where there is an economic incentive to operate at the highest possible combined heater outlet temperature. If limited by the metallurgy of the passes, then raising all the pass outlet temperatures to the limit would, in this example, allow the combined temperature to be increased by 10 °C.

Balancing of heaters in coking service should be approached with care. A coked pass will be less efficient in transferring heat from the firebox to the fluid in the pass. Its pass outlet temperature will then be lower. If the objective of pass balancing it to equalise pass temperatures then it will reduce the flow through the pass in order to increase the temperature. This will increase the residence time and accelerate coking. Left unchecked this will shorten run length.

If the heater is a hydraulic constraint then we can use historical data to quantify the relationship between pass flow and valve position. By comparing average valve position to the maximum permitted we can then estimate the overall increase in capacity utilisation.

In the case of a coking constraint (and skin temperature constraint if caused by coking) we can perform a calculation similar to that for efficiency savings. Coking is a chemical reaction governed by the *Arrhenius Equation* which states that rate of reaction (k) is governed by the activation energy (E), the universal gas constant (R) and the absolute temperature (T).

$$k \propto e^{-E/RT} \tag{10.28}$$

The relationship is shown in Figure 10.24. By applying this to the same heater those passes 10 °C hotter than the average will coke 19.5 % more quickly. Those 10 °C cooler will coke 16.8 % more slowly. We could interpret this as the potential to achieve a reduction of 1.3 % in the overall rate of coking. And we could exploit this by increasing the heater outlet temperature by 0.7 °C.

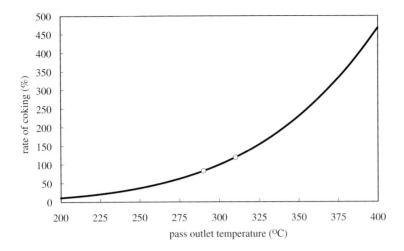

Figure 10.24   Effect of pass outlet temperature on rate of coking

Or we might exploit it by decoking less often. However our calculation assumes that the period that each pass operates above average temperature is the same as the time it spends below. If one pass is consistently the hottest then it will limit run length. Balancing would then increase run length by 24 %, i.e. 100/(100 – 19.5).

Of course we need some measurement of rate of coking for each pass. This might be skin temperature but these measurements are prone to failure. It may be possible to develop an inferential for rate of coking based on measurements such pressure, temperature, flow and (if injected) steam flow.

Before embarking on pass balancing it is advisable to check for interactions between pass flow controllers. Depending on the geometry of the pipework and the pressure differential, increasing the SP of one pass flow starves the others and their corrective action interacts with the first and causes an oscillatory response. The easiest way of dealing with this is to use different gains in each of the flow controllers.

Instrument calibration should also be checked. Often in a difficult service it is common for errors to occur in the flow measurements – for example orifice tappings or the orifice itself may become partially blocked. An apparent maldistribution of flows may not be real – particularly if control valve positions (and/or pass outlet temperatures) are approximately equal.

In terms of the techniques available, there are three basic approaches. Although the schemes described are designed to balance temperatures they can be modified to meet other objectives. The first approach is to use standard algorithms in the DCS as shown in Figure 10.25.

Although, for simplicity, only a two-pass heater is illustrated the technique is applicable to any number of passes. One is selected as master. For each of the other passes the difference between its outlet temperature and that of the master is calculated. Each of these is the PV of a PID controller that has SP of 0. The master flow controller manipulates the master pass control valve directly and each of the others via a bias. Each bias is set by each of

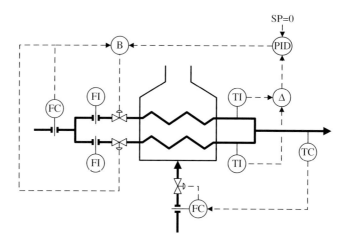

Figure 10.25 Heater pass balancing.

the temperature difference controllers. Upper and lower limits are set on the biases to prevent too large a flow imbalance. Indeed, since we have removed the individual pass flow controllers, it is important to monitor the flow imbalance and alarm any significant maldistribution. This might be an early sign of excessive coking in one pass.

An alternative approach is to custom code the technique. We define a predicted outlet temperature (T_{out}) based on each of the n pass flows (F_i) and outlet temperatures (T_i).

$$T_{out} = \frac{\sum_{i=1}^{n} F_i T_i}{\sum_{i=1}^{n} F_i} \tag{10.29}$$

Using the heater inlet temperature (T_{in}), we predict the pass flows (F_i^*) necessary to balance the heater.

$$F_i^* = \frac{T_i - T_{in}}{T_{out} - T_{in}} F_i \tag{10.30}$$

The use of the derived outlet temperature rather than that measured ensures there is no change in the total flow through the heater.

$$\sum_{i=1}^{n} F_i^* = \sum_{i=1}^{n} F_i \tag{10.31}$$

The advantage of the custom code approach is that a wide range of other checks can be readily added and help improve process integrity. For example validity check can be made on all flows, temperatures and control valve positions. Constraints such as maximum flow imbalance, maximum skin temperature etc. can be included. Logic can be included to handle out of the ordinary situations, such as instrument failure, and the application degraded gracefully rather than an all-or-nothing approach.

The final approach is to make use of a proprietary MVC. For each pass a CV is defined as the pass outlet temperature less the combined outlet temperature. Similarly a CV is defined for each pass as the pass flow less the average pass flow. Each pass flow is included as a MV. If exact temperature balancing is required then the upper and lower limits on the temperature differences are set to 0. The same approach can be used if flow balancing is required. By setting slightly wider limits temperatures can be balanced provided this does not require excessive flow imbalances.

The MVC approach has the advantage that pass balancing may be included in the same controller as that for the rest of the process. So, for example, if the only constraint on increasing feed rate is a pass valve position then heater balancing would automatically be relaxed to allow the increase.

11
Compressor Control

Compressors fall into one of two fundamental types – positive displacement and turbo-machines. Positive displacement machines can be either rotary or reciprocating. They both trap the gas in a cylinder and then force it into a smaller volume and so increase its pressure. Turbo-machines impart velocity to the gas and its momentum carries it into a narrowing space and so its pressure increases. Turbo-machines can be either axial (in which the flow is parallel to the shaft) or centrifugal (in which the flow is at right angles to the shaft). Multistage turbo-machines, with intercooling, are common.

11.1 Polytropic Head

Compressor performance is quoted in terms of *polytropic head*. This is the work done on the gas and its definition is developed from basic gas laws. Firstly *Boyle's Law* states that the volume (V) occupied by a gas is inversely proportional to its absolute pressure (P).

$$V \propto \frac{1}{P} \tag{11.1}$$

Charles' Law states that that volume is directly proportional to absolute temperature (T).

$$V \propto T \tag{11.2}$$

These laws are combined into the *Ideal Gas Law*.

$$PV = RT \tag{11.3}$$

R is the *Universal Gas Constant*. It has a value of 8.314 kJ/kg-mole/K (1.9859 BTU/lb-mole/R). The law, in its normal form, is written on molar basis. On a weight basis we include the molecular weight (MW).

$$PV = \frac{RT}{MW} \tag{11.4}$$

Process Control: A Practical Approach Myke King
© 2011 John Wiley & Sons, Ltd

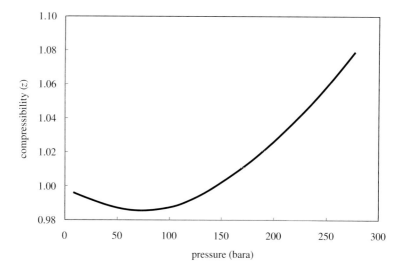

Figure 11.1 Compressibility of air

To accommodate the non-ideal behaviour of the gas, *compressibility (z)* is introduced into the equation.

$$PV = \frac{zRT}{MW} \tag{11.5}$$

Compressibility is determined experimentally for each gas. As an example, that for air is included as Figure 11.1.

The law governing *isentropic* compression can be written as

$$PV^\gamma = \text{constant.} \tag{11.6}$$

The *adiabatic index* (γ) is defined as the ratio of the specific heat of the gas measured at constant pressure (c_p) to that measured at constant volume (c_v).

$$\gamma = \frac{c_p}{c_v} \tag{11.7}$$

The value of γ is available for most gases in data books. For *monatomic* gases, i.e. those with one atom in their molecule such as He and Ar, it is 1.67. For *diatomic* gases, such as H_2, O_2, N_2 and CO, it is typically 1.40. For *triatomic* gases, such as H_2S and SO_2, it is 1.33.

For compression to be described as isentropic (i.e. no change in entropy) firstly it has to be *reversible*. This means that the work done compressing the gas can be fully recovered by decompressing it. This is equivalent to the compressed gas entering the discharge of the compressor, being used to drive the machine backwards which then generates as much energy as was used to compress the gas. Due to losses in the machine, such as those caused by overcoming friction, reversibility is unachievable. Secondly the process has to be *adiabatic*, i.e. no heat must enter or leave the system. Again, due to heat lost to the atmosphere or to other compressor cooling systems, this condition will not be met. We describe a process where entropy changes as *polytropic*.

We can write Equations (11.3) and (11.6) for the suction (s) and discharge (d) of the compressor

$$\frac{P_s V_s}{T_s} = \frac{P_d V_d}{T_d} \qquad (11.8)$$

$$P_s V_s^{\gamma} = P_d V_d^{\gamma} \qquad (11.9)$$

Combining these equations gives, for isentropic compression

$$\frac{P_d}{P_s} = \left(\frac{V_s}{V_d}\right)^{\gamma} = \left(\frac{T_d}{T_s}\right)^{\frac{\gamma}{\gamma-1}} \qquad (11.10)$$

For polytropic compression we can rewrite this as

$$\frac{P_d}{P_s} = \left(\frac{V_s}{V_d}\right)^{n} = \left(\frac{T_d}{T_s}\right)^{\frac{n}{n-1}} \quad n \neq \gamma \qquad (11.11)$$

Polytropic efficiency (η_p) describes how close compression is to isentropic and is defined as

$$\eta_p = \frac{\left(\frac{n}{n-1}\right)}{\left(\frac{\gamma}{\gamma-1}\right)} \qquad (11.12)$$

Polytropic head (H_p) is the work done on the gas; by definition this is given by

$$H_p = \int_s^d V \, dP \qquad (11.13)$$

Modifying Equation 11.9 for polytropic compression

$$PV^n = P_s V_s^n \quad \text{or} \quad V = P_s^{\frac{1}{n}} V_s P^{\frac{-1}{n}} \qquad (11.14)$$

Substituting for V in Equation 11.13

$$H_p = P_s^{\frac{1}{n}} V_s \int_s^d P^{\frac{-1}{n}} \, dP \qquad (11.15)$$

Thus

$$H_p = P_s^{\frac{1}{n}} V_s \frac{n}{n-1} \left[P_d^{\frac{n-1}{n}} - P_s^{\frac{n-1}{n}} \right] = P_s V_s \frac{n}{n-1} \left[\left(\frac{P_d}{P_s}\right)^{\frac{n-1}{n}} - 1 \right] \qquad (11.16)$$

Substituting from Equations 11.12 and 11.5

$$H_p = \eta_p \frac{\gamma}{\gamma-1} \frac{z_s RT_s}{MW} \left[\left(\frac{P_d}{P_s}\right)^{\frac{\gamma-1}{\eta_p \gamma}} - 1 \right] \qquad (11.17)$$

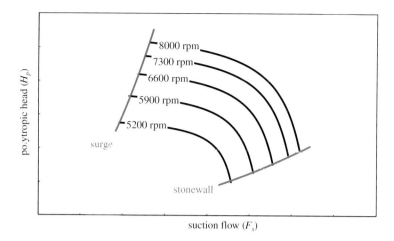

Figure 11.2   Compressor performance

Remember that pressure (P) and temperature (T) are on an absolute basis. Polytropic head (H_p) will have units such J/kg or BTU/lb.

Figure 11.2 shows a typical set of compressor performance curves. The performance curve for a centrifugal compressor, unlike that for a pump, terminates before the flow reaches zero. At too low a flow the compressor will *surge*. The impeller discharge pressure temporarily falls below that in the discharge pipework causing a transient flow reversal. This causes large and rapid fluctuations in flow and pressure. It can also be extremely noisy, although much of the noise may arise from the check (non-return) valve in the discharge pipework opening and closing rapidly. On some machines the resulting vibration can damage the compressor and/or its gearbox very quickly. Others tolerate the condition for longer.

The upper limit of the performance curve is *stonewall*. This arises when the speed of the gas, relative to the impeller, approaches the speed of sound (at conditions within the machine). Since gas cannot travel faster than this, the maximum capacity of the machine has been reached. No damage to the machine is likely under these conditions.

11.2   Flow Control (Turbo-Machines)

Compressor performance curves should strictly be plots of polytropic head (H_p) against suction flow measured in **actual** volumetric units (F_s). However, it is common for discharge pressure, or the ratio of discharge to suction pressure, to replace polytropic head. In order to simplify the description of how possible flow controls operate, we will use the curves in this form. These approximations assume suction conditions and molecular weight remain constant. Should either change then the performance curve, drawn on this basis, will move.

Figure 11.3 shows a simplified process diagram showing the compressor delivering gas to a downstream process. This pressure drop through the process increases in proportion with the square of the flow (measured in actual volumetric units). Assuming the pressure at the

Compressor Control 247

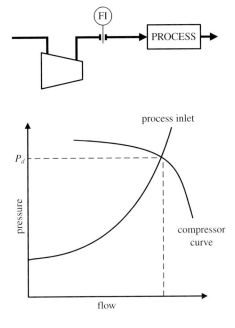

Figure 11.3 Process diagram

exit of the process is constant then we can add the *process curve* to the compressor performance curve. Since the flow through the compressor must equal the flow through the process, and the compressor discharge pressure must be the same as the process inlet pressure, the compressor will operate where the two lines cross.

In order to adjust the flow, we have to cause either the process curve or the compressor curve to move. Alternatively we have to remove the condition that discharge and process inlet pressures are equal, or that the flows are equal.

Figure 11.4 shows the first of several possible schemes. By placing a control valve between the compressor and the process the pressures are no longer equal. Or, if we think of the valve as now part of the process, the process curve will now move as we change the valve opening.

Because of the nonlinear behaviour of both the compressor and the process the relationship between flow and valve Δp is highly nonlinear. An equal percentage valve or some other form of controller output conditioning (as described in Chapter 5) can be used to avoid problems in tuning the flow controller.

As might be expected, because we are expending energy to raise the pressure of the gas only to partially reduce it again across the valve, the scheme is not energy efficient. Its range is limited since, as we close the valve, the process curve approaches the end of the compressor curve and surge occurs.

Figure 11.5 shows how the control valve can be relocated to the suction of the compressor. Because we have plotted the compressor curve in terms of discharge pressure, rather than polytropic head, the curve moves when we change suction pressure. Because of the lower pressures involved this is more energy efficient than discharge throttling. It also has a greater range because the surge point reduces as suction pressure is reduced.

248 Process Control

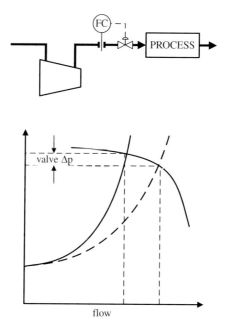

Figure 11.4   Throttling discharge

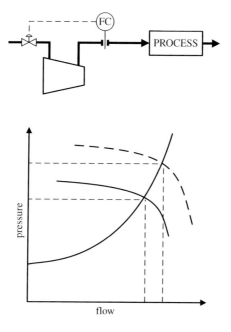

Figure 11.5   Throttling suction

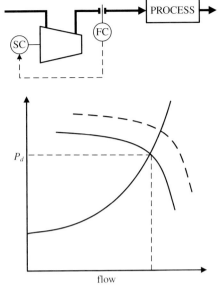

Figure 11.6 Adjusting speed

One concern is that it is possible for the suction pressure to fall below atmospheric pressure and any leaks in the pipework would permit air to enter the compressor. If compressing a flammable gas this could cause an explosion inside the machine.

If the compressor is of the variable speed type then we can use this to control the flow, as shown in Figure 11.6. By changing the speed we move from one compressor curve to another. This is energy efficient and, because the surge point moves, can operate over a wide range. However variable frequency drives (VFD) for large electric motors are costly. Variable speed steam turbine drivers have a mixed reputation. Most success has been had with gas turbine drivers.

Figure 11.7 shows the use of inlet guide-vanes. These convert the inlet gas's forward momentum into rotational momentum. The angle of the guide-vanes is adjustable; conventionally negative angles give *pre-rotation* and positive angles produce *counter-rotation*. Pre-rotation increases the compressor efficiency permitting it to deliver a greater flow at the same discharge pressure. They are therefore also energy efficient. But, at steep angles, the guide-vanes effectively behave like suction throttling. Because the guide vanes are inside the compressor, adjusting them changes the geometry of the machine itself. Thus another family of compressor curves will exist for each guide-vane angle – each with a different surge point. Mechanically guide-vanes are more complex and more costly to maintain.

By partially recycling gas through the compressor we remove the condition that the flow through the machine must equal the flow though the process. The scheme shown in Figure 11.8 offers the greatest range since the surge point is never approached. It is, however, costly to install and operate. In reality the system is more complex than drawn; for example the recycle must be cooled. It is the least energy efficient approach. It can, however,

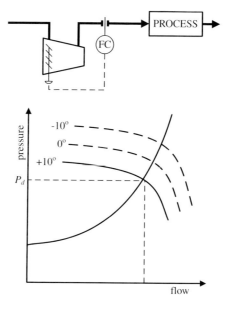

Figure 11.7 Inlet guide vanes

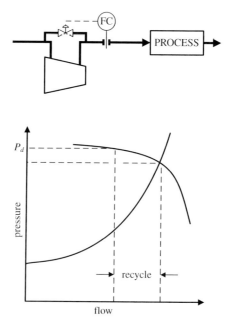

Figure 11.8 Recycle manipulation

be used to increase the range of the other schemes described, by recycling just enough to avoid surge.

11.3 Flow Control (Reciprocating Machines)

Many of the schemes described for turbo-machines can be applied to reciprocating machines. Suction throttling reduces the suction pressure and so less gas (on a weight basis) enters the cylinder and thus less gas will be discharged. But discharge throttling has little effect on flow. Once in the cylinder the machine will deliver a fixed mass of gas at each stroke. Throttling simply means that the machine has to work harder to overcome the restriction. Speed control is an energy efficient means of controlling flow. Recycling is effective but, like turbo-machines, is costly.

A further option is *cylinder loading* which alters the effective compression ratio. Reciprocating machines follow the cycle shown in Figure 11.9. At point A the piston is at *top dead centre* (*TDC*), having just finished discharging gas. The machine then begins the suction stoke. At point B the inlet valve opens and then closes at point C when the piston reaches *bottom dead centre* (*BDC*). It then begins the compression stroke with the exhaust valve opening at D and then closing at A.

Cylinder loading changes the point in the cycle at which the inlet valve closes. If set at 75 % load, on reaching BDC the valve remains open. The piston begins the compression stroke but no actual compression takes places. Instead gas leaves via the inlet valve until it closes at point C_{75}. At this point true compression begins but on a gas volume that is 75 % of maximum. The load can similarly be reduced to 50 % or 25 % of maximum. Cylinder loading is normally set manually local to the machine. However, there are examples of it being automated to provide true flow control.

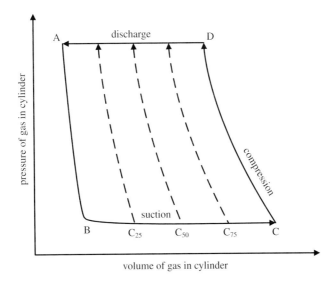

Figure 11.9 Cylinder loading

11.4 Anti-Surge Control

Surge is not a problem with positive displacement machines. On turbo-machines it can be avoided by recycling but this is costly. However the cost of repair and lost production that can arise from surging a compressor may be greater. The objective is to minimise recycling without jeopardising the machine. To achieve this we need to be able to predict, using the available instrumentation, that surge is about to occur.

The complexity of surge avoidance depends much on the machine and its duty. For example a machine running at a fixed speed and compressing a gas of constant composition might require only minimum flow protection, as shown in Figure 11.10.

The flow measurement is normally located on the suction side, since compressor curves are normally presented in terms of actual suction flow. However, providing suction and discharge pressures are constant, the scheme will operate equally effectively if the flow measurement is downstream of the compressor. However it relies on there being a significant change in flow as the surge point is approached, i.e. the compressor curve is reasonably 'flat' in this area. Should this not be the case then the scheme shown in Figure 11.11 is preferable. This will cause the recycle to open if the discharge pressure exceeds SP. Here pressure must change significantly as surge is approached and so the scheme is better for 'steep' compressor curves.

For variable speed machines surge is a line, not a single point. The next group of schemes are based on developing an equation for this line – either from the compressor manufacturer's data or from plant testing.

The surge line can usually be represented by a quadratic function

$$H_p = f(F_s^2) \tag{11.18}$$

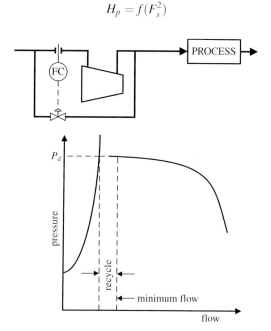

Figure 11.10 Minimum flow protection

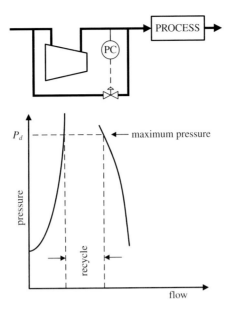

Figure 11.11   Maximum pressure protection

Figure 11.12 shows the compressor curves now redrawn by plotting against F_s^2 instead of F_s. The surge line is now approximately straight. A second line has been added, building in a safety margin of 15 % over the minimum flow.

F_s is measured by an orifice plate type meter and is related to the pressure drop (dp) across the orifice.

$$F_s = c_d \frac{\pi d^2}{4} \sqrt{\frac{2dp}{\rho}} \tag{11.19}$$

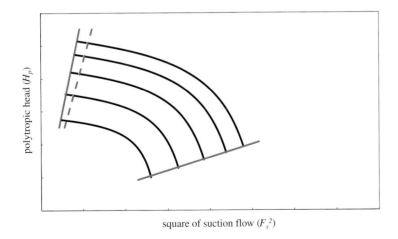

Figure 11.12   Linearised surge line

If we assume the discharge coefficient (c_d), the pipe diameter (d) and the fluid density (ρ) are all constant then building in the 15 % margin gives

$$(1.15F_s)^2 \propto 1.15^2 dp \qquad (11.20)$$

Or, if m is gradient of the minimum flow line and c the intercept on the polytropic head axis, then

$$H_p = m.(1.15^2 dp) + c \qquad (11.21)$$

To apply this technique we need to be able to measure H_p and dp. If we have a suction flow meter then the measurement of dp is readily available. We need only to compensate for any square root extraction that may be in the control system (see Chapter 5). However the measurement of H_p is not so straightforward. Equation 11.17 includes values which we cannot measure – such as polytropic efficiency. There are a number of possible approaches which use a parameter related to H_p. A common example is the pressure rise ($P_d - P_s$).

By plotting ($P_d - P_s$) against $1.15^2 dp$ we can determine m and c but we need a number of points collected close to surge. It may be possible to develop a correlation from information provided by the compressor manufacturer. Or, if the machine has surged on previous occasions, the process data may exist in the plant history database. Failing either of these approaches it would be necessary to obtain data by testing close to the surge point – a process that needs to be managed carefully!

Historically this type of scheme has been implemented as shown in Figure 11.13. Unusually the controller SP is a process measurement – in this case ($P_d - P_s$). The controller PV is the value that ($P_d - P_s$) would have if the compressor were operating at the minimum flow. The controller, by manipulating the recycle, will attempt to eliminate the difference between the PV and the SP. If recycle is needed to avoid surge then, by doing so, this will ensure that the compressor operates at the minimum flow. If recycle is not needed then the controller will fully close the recycle valve leaving an offset between the PV and the SP.

While effective, the rather unconventional approach requires a slightly different approach to implementation. Firstly, when obtaining the process dynamics for tuning

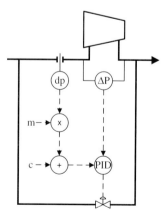

Figure 11.13 Implementation of anti-surge control

purposes by stepping the recycle valve, both the SP and the PV will change. Model identification therefore should be based on the difference PV – SP. Secondly the controller cannot initialise using PV tracking. If switched to automatic when the compressor is operating above minimum flow, with the recycle valve closed, the controller will do nothing. If the flow is below minimum then the controller will immediately take corrective action – as required. However if the recycle is open unnecessarily, then the corrective action taken by the controller can cause a process upset. Lastly, the unconventional approach is a disadvantage for operator understanding and does not permit any adjustment by the operator of how closely surge is approached.

An alternative approach is to rearrange the equation of the minimum flow line

$$(P_d - P_s) = m.dp + c \tag{11.22}$$

Substituting for dp in Equation 11.19

$$F_{min} = c_d \frac{\pi d^2}{4} \sqrt{\frac{2\left(\frac{P_d - P_s - c}{m}\right)}{\rho}} \tag{11.23}$$

This is then a measure of the minimum suction flow. This value could be used as the SP of a minimum flow type antisurge controller. Or it could be subtracted from the measured flow to give the margin to surge in flow units

$$F_{margin} = F_s - F_{min} \tag{11.24}$$

This would then be the PV of the antisurge controller with the operator entering the required margin as SP.

It should be emphasised that the use of the pressure rise across the machine is just one possible approximation to polytropic head. Pressure rise is a special case of the function $(aP_s + bP_d + c)$, where the coefficients a, b and c are chosen for each machine. Another approach is to use the pressure ratio (P_d/P_s) – remembering that, unlike the pressure rise, it is important here to convert to absolute pressures. Other schemes use the speed of the machine or its power consumption. There is also a range of schemes which use some nonlinear function of dp rather than just its measurement. The most effective solution is machine-specific and each should be evaluated before building the controller.

There is however a more rigorous approach. By combining Equations 11.11 and 11.12 we get

$$\left(\frac{P_d}{P_s}\right)^{\frac{\gamma-1}{\eta_p \gamma}} = \frac{T_d}{T_s} \tag{11.25}$$

Rearranging we get

$$\eta_p \frac{\gamma}{\gamma - 1} = \frac{\log\left(\frac{P_d}{P_s}\right)}{\log\left(\frac{T_d}{T_s}\right)} \tag{11.26}$$

Substituting Equations 11.25 and 11.26 into Equation 11.17 we get an alternative definition of polytropic head

$$H_p = \frac{\log\left(\frac{P_d}{P_s}\right)}{\log\left(\frac{T_d}{T_s}\right)} \frac{z_s R T_s}{MW} \left[\frac{T_d}{T_s} - 1\right] \tag{11.27}$$

By writing Equation 11.19 for compressor suction we get

$$F_s^2 \propto \frac{dp}{\rho_s} \tag{11.28}$$

The gas density (ρ_s) is inversely proportional to specific volume (V). So, from Equation 11.5, we get

$$\rho_s \propto \frac{MW \cdot P_s}{z_s T_s} \tag{11.29}$$

Replacing ρ_s in Equation 11.28 gives

$$F_s^2 \propto \frac{dp \cdot z_s T_s}{MW \cdot P_s} \tag{11.30}$$

We assume that the surge line can be represented by a quadratic, i.e. H_p varies linearly with F_s^2 and so, since H_p is zero when F_s^2 is zero, the gradient of the surge line is given by

$$\frac{H_p}{F_s^2} \propto \frac{\log\left(\frac{P_d}{P_s}\right)}{\log\left(\frac{T_d}{T_s}\right)} \frac{P_s}{dp} \left[\frac{T_d}{T_s} - 1\right] \tag{11.31}$$

Terms that are difficult to measure, such as polytropic efficiency (η_p), compressibility (z_s) and molecular weight (MW), have now been eliminated. Provided there are measurements of suction flow; suction and discharge pressure; and suction and discharge temperature, then we can calculate a parameter which is proportional to the slope of the surge line. This parameter is used as the PV of the antisurge controller. The value of this parameter at which the compressor surges can be predicted by using data collected historically or by taking the machine into surge. Remembering to move the machine away from surge the gradient of this line is **reduced**. Thus to incorporate a safety margin the SP of the controller is set to 15 % (say) **less** than the point at which surge is known to occur.

This scheme should prove more robust as gas composition and other operating conditions vary. It is thus often used in proprietary compressor control systems.

It is important that the controllers in all the antisurge schemes described have anti-reset windup. Unless the required gas flow is **always** below the surge point the controller will mostly be operating with the recycle valve fully shut, i.e. it will be saturated. The valve needs to open quickly when surge is approached and any windup could result in the compressor surging before the valve opens. For the same reason it is also common for quick opening valves to be used. These present a tuning challenge; the nonlinearity they introduce means that the process gain is highest when they initially open. The antisurge controller needs to be tuned therefore for operation close to this condition; otherwise it may become unstable. Obtaining the process dynamics from plant testing also needs to be approached with care since the tests should ideally be performed with the machine close to surge.

In addition to antisurge control, it is common to have a surge recovery scheme. This relies on the detection of rapid changes in pressure or flow and then overrides the antisurge scheme to fully open the recycle valve. It can also increase the safety margin used by the antisurge controller so that, when the antisurge scheme is permitted to take back control, the same situation should not recur.

The antisurge controller will interact with other schemes on the compressor. For example, it opening the recycle valve will cause the discharge pressure to drop and hence also the flow to the downstream process. The flow (or pressure) controller will then to take corrective action. To avoid this interaction propagating this controller should be tuned to act significantly slower than the antisurge scheme.

In multistage machines, the antisurge scheme on the first stage will temporarily starve the next stage(s) of gas – possibly causing them to approach surge. Feedforward compensation may be required to avoid significant disturbances.

Special consideration must be given to compressors operating in parallel. Even so-called identical machines will have slightly different compressor curves. The machines must operate with the same inlet pressures and the same outlet pressures. If operating on a flat section of the curves, equalising the pressure rise may result in very different flows through

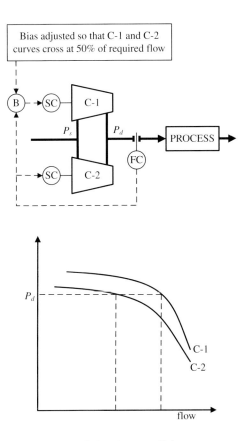

Figure 11.14 Balancing parallel compressors

each machine. It is possible that one may approach surge and the other stonewall. It is also possible that the load will swing between the machines – particularly during process disturbances or compressor start-up. It is important that the flow control strategy balances the machines – for example by applying a bias to one of the speed controllers, so that both machines operate at roughly the same distance from surge. Figure 11.14 shows a possible scheme.

12
Distillation Control

Figure 12.1 shows an example of the internal arrangement of a distillation column. Liquid from the *downcomer* flows across the *tray*. Vapour from the tray below passes through the liquid – exchanging less volatile components for more volatile ones from the liquid. Trays are numbered but there is no convention as whether they are counted from the top or the bottom of the column. In this book we number them from the bottom.

When counting trays in a column we generally refer to actual trays. We can also count *theoretical trays* (also known as *theoretical stages*). Actual trays are not 100 % efficient, theoretical trays are. We can convert from actual to theoretical trays by multiplying by the tray efficiency. There are columns where actual trays are replaced by packing; then we would refer to theoretical trays per unit height of packing.

Although not used in this book, other texts refer to the section of the column above the feed tray as the *rectifying* or the *enriching* section. In this section the vapour flow is greater than the liquid flow. The lower section is referred to as the *stripping* or *exhausting* section. Here the vapour flow is smaller than the liquid flow.

A typical external arrangement is shown in Figure 12.2. Feed enters the column on the *feed tray*. Depending on its enthalpy it will split between vapour and liquid. Vapour joins that rising from the *reboiler* and travels up the column to the condenser. A *total condenser*, as the name suggests, condenses all the vapour. A *partial condenser* does not, resulting in the need to withdraw a gaseous product from the *reflux drum*. This drum, also known as the *reflux accumulator* or *overheads drum* collects the condensed liquid. Part is pumped back to the column as *reflux*, the remainder leaves as *distillate* – also known as *overheads*. Whether this needs to be pumped will depend on the downstream pressure. The liquid portion of the feed joins the reflux and leaves the base of the column. Part of it is evaporated by the *reboiler*; the remainder leaves as *bottoms*. Again the need for pumping will depend on the downstream pressure.

Figure 12.3 describes the column's operating envelop. *Blowing* can occur when the internal vapour rate is high. Instead of the vapour breaking up into bubbles to produce foam on each tray, its velocity is such that it produces 'tubes' of vapour through the liquid on the

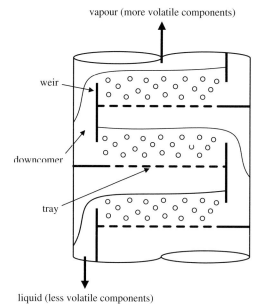

Figure 12.1 Column internals

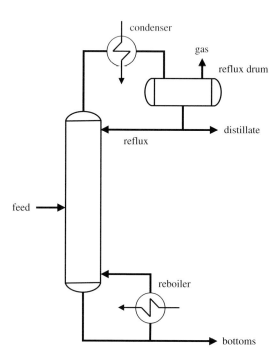

Figure 12.2 Basic column

Distillation Control 261

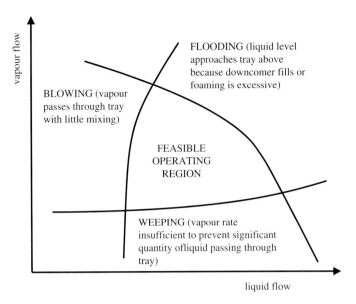

Figure 12.3 Operating envelop

tray. The surface area of vapour/liquid contact is substantially reduced and the tray efficiency falls.

Conversely, *weeping* can occur when the vapour rate is very low. Without sufficient vapour passing through the tray, liquid instead weeps through the tray. It effectively bypasses the tray and again the tray efficiency falls.

Blowing and weeping are unusual in conventional distillation columns. To produce more vapour we increase the reboiler duty but, in order to maintain the heat balance, we increase the condenser duty. Depending on the configuration of the controls, the additional liquid formed in the reflux drum will be returned to the column. Thus vapour and liquid rates change almost in proportion – avoiding the blowing and weeping constraints. However there are columns where the two flows are not related. For example, reflux may be provided from some external source – as it is on gas scrubbing columns.

Flooding is the most common of the hydraulic constraints likely to be encountered. There are two mechanisms that cause it. The first, *downcomer flooding*, arises if the maximum internal liquid rate is exceeded. Liquid flows through the downcomer under gravity. The level of liquid built up in the downcomer is as result of a balance between the pressure drop across it and the head of liquid held above it on the tray. As the flow increases, the pressure drop increases (with the square of the flow) and so the head must increase. Ultimately the level reaches the tray above and the tray ceases to provide any separation.

Jet flooding is caused by excessive vapour rates or foaming. Increasing vapour mixed with the liquid on the tray decreases the density of the foam. Thus a greater depth is required to provide the head necessary for the liquid to pass down the downcomer. Again the level of the foam will ultimately reach the tray above.

The problem is exacerbated by the composition controllers on the column. The drop in tray efficiency will result in a reduction in product purity. The controllers will respond by increasing reboiler duty and/or reflux, thus further increasing the internal traffic. While flooding may have occurred in only a small section of the column, without intervention it will quickly propagate to the whole column.

From a control perspective, accurate detection of flooding is difficult. One approach is to monitor the pressure drop across the column. However this is usually too late an indication; flooding is well established by the time a significant change is detected. A better approach is to measure the pressure drop across small sections of the column known to flood first. However this too can be unreliable; it is common for the column to be operating normally with a high pressure drop one day, and to be flooding at a lower pressure drop on another. Attempts have been made to use tray loading calculations, usually used for column design, as a means of quantifying approach to flooding but these have proved unreliable. Using any of these techniques as an override to restrict the column will overconstrain the operation.

12.1 Key Components

Most distillation processes are multicomponent. A column common to many industries is the *LPG splitter*. The feed composition is shown in Table 12.1. It is primarily a propane/butane mixture which is required to be separated to specified purity targets. While at first glance the feed appears to comprise four components, butane has two isomers (isobutane and n-butane) and pentane has three. Further it is likely that small amounts of unsaturated material will be present, such as C_3H_6, C_4H_8, C_5H_{10} and their isomers. Thus an apparently binary distillation actually involves a substantial number of components.

While the theory clearly exists to allow such columns to be designed, it is complex and its use here will make difficult the explanation of how the control strategies operate. We will therefore simplify the approach by treating the column as a binary separation. We do this by identifying the *key components*. The *light key* (*LK*) is the lightest component that will be found in any quantity in the bottom product. The *heavy key* (*HK*) is the heaviest that will be found in the distillate product.

As an example let us assume that the distillate must be 95 % pure propane and the bottoms butane must contain 10 % propane. Table 12.1 shows the resulting products. While the lightest in the bottoms is ethane, it is present only in a small concentration. The main impurity is propane. Similarly butane is the main impurity in the distillate. We therefore define these as the light and heavy key components. To close the mass balance we lump

Table 12.1 LPG splitter feed and product analysis

Component	Feed	Distillate		Bottoms	
	moles/hr	moles/hr	mol %	moles/hr	mol %
C_2H_6	0.7	0.5	1.3	0.2	0.4
C_3H_8	45.0	39.1	95.0	5.9	10.0
C_4H_{10}	52.7	1.4	3.5	51.3	87.1
C_5H_{12}	1.6	0.1	0.2	1.5	2.5
total	100.0	41.1	100.0	58.9	100.0

together all the components lighter than the light key component as the *light light key* (*LLK*) component and all those heavier than the heavy key component as the *heavy heavy key* (*HHK*) component.

12.2 Relative Volatility

The underlying principle of separating components by distillation is that, when a liquid is partially evaporated, the composition of the vapour produced is different from that of the liquid. One component must be more volatile than the other(s). Ease of separation depends on *relative volatility*.

There are a number of correlations which predict how pure components behave. The most commonly documented is the *Antoine Equation* which predicts the *vapour pressure* of the pure component (P_0) when at the temperature (T).

$$\ln(P_0) = A - \frac{B}{T + C} \qquad (12.1)$$

A, B and C are constants determined experimentally. They are readily available from data books and the Internet. They have engineering units so their numerical value will depend on the units of measure of pressure and temperature. They also change if the Antoine Equation is based on the logarithm to base 10, i.e. $\log_{10}(P_0)$.

Table 12.2 gives values for some common components. These assume pressure is measured in bara and temperature in °C.

As an aside, the Antoine Equation has a number of uses. Normal boiling point is defined as the temperature (T_b) at which the vapour pressure reaches atmospheric pressure (1.01325 bara). Rearranging Equation (12.1)

$$T_b = \frac{B}{A - 0.013163} - C \qquad (12.2)$$

Of course the boiling point can be determined at any pressure – a technique we shall use later to validate pressure compensation of tray temperatures.

Antoine's Law can also be used to determine *Reid Vapour Pressure* (*RVP*). RVP is defined as the vapour pressure at 100 °F (37.8 °C).

$$RVP = e^{A - \frac{B}{37.8 + C}} \qquad (12.3)$$

Table 12.3 shows the values of the properties predicted.

Table 12.2 Antoine coefficients

Component		A	B	C
propene	C_3H_6	9.08250	1807.529	247.00
propane	C_3H_8	9.04654	1850.841	246.99
n-butane	C_4H_{10}	9.05800	2154.897	238.73

Table 12.3 Predicted properties

Component		T_b °C	RVP bara
propene	C_3H_6	−47.7	15.4
propane	C_3H_8	−42.1	12.8
butane	C_4H_{10}	−0.5	3.5

Returning to the use of Antoine in predicting volatilities, the coefficients in Table 12.1 were used to plot the curves in Figure 12.4. As might be expected, propene and propane have very similar vapour pressures and so would be difficult to separate. Propane and butane however would be a relatively easy separation.

To quantify this we will develop a definition of relative volatility from some basic equations of state. Firstly *Raoult's Law* states that the partial pressure of a component i in the vapour (p_i) is proportional to its molar fraction in the liquid (x_i).

$$p_i = (P_0)_i x_i \qquad (12.4)$$

For Raoult's Law to apply, the components need to chemically similar. As an improvement, for non-ideal systems, *Henry's Law* can be used. It uses the experimentally determined *Henry's constant (H)* as the constant of proportionality rather than the vapour pressure (P_0).

$$p_i = H_i x_i \qquad (12.5)$$

Dalton's Law of Partial Pressures states that the partial pressures of the components sum to the total pressure (P).

$$\sum_{i=1}^{n} p_i = P \qquad (12.6)$$

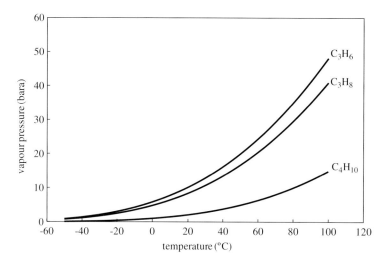

Figure 12.4 Vapour pressure of common components

The *Ideal Gas Law* states that partial pressure is proportional to the mole fraction (y_i) of the ith component in the vapour.

$$p_i = P y_i \qquad (12.7)$$

Combining these laws gives

$$y_i = \frac{(P_0)_i}{P} x_i \quad \text{or} \quad y_i = \frac{H_i}{P} x_i \qquad (12.8)$$

Volatility (α) is defined as the ratio of the mole fraction of the component in the vapour to that in the liquid.

$$\alpha_i = \frac{y_i}{x_i} = \frac{(P_0)_i}{P} \quad \text{or} \quad \frac{H_i}{P} \qquad (12.9)$$

Relative volatility is the ratio of the volatility of one component to that of another.

$$\alpha_{ij} = \frac{\alpha_i}{\alpha_j} = \frac{(P_0)_i}{(P_0)_j} \quad \text{or} \quad \frac{H_i}{H_j} \qquad (12.10)$$

Since we are working with components that are chemically similar we can assume ideal behaviour and use the first definition of α_{ij}. Figure 12.5 shows how relative volatility varies with temperature. A value of 1 for relative volatility indicates that the components cannot be separated by distillation because the composition of the vapour evaporated is identical to that of the liquid. As expected the relative volatility for propene/propane is close to this value. That for propane/butane is far greater.

We will address later, in the section covering optimisation, the variation of relative volatility with temperature. Since liquids in the distillation column are normally at their bubble point and vapours at their dew point, to reduce the temperature at which separation

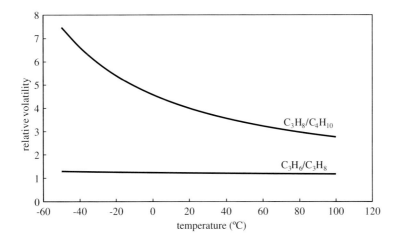

Figure 12.5 Relative volatility of common mixtures

takes place we would reduce pressure. Thus operating at a lower pressure makes separation easier.

12.3 McCabe-Thiele Diagram

The technique usually taught to explore column operation is the *McCabe-Thiele Diagram*. While it's use in industry is about as rare as the use of control theory it does help us understand column operation. Figure 12.6 shows part of the McCabe-Thiele construction for the separation of propane (C_3) from butane (C_4).

For a binary mixture, from Equation (12.9):

$$\alpha_{LK} = \frac{y}{x} \quad \text{and} \quad \alpha_{HK} = \frac{1-y}{1-x} \qquad (12.11)$$

From Equation (12.10) relative volatility is given by

$$\alpha = \frac{\alpha_{LK}}{\alpha_{HK}} = \frac{y(1-x)}{x(1-y)} \qquad (12.12)$$

Rearranging gives the equation for the vapour line

$$y = \frac{\alpha.x}{1 + (\alpha-1)x} \qquad (12.13)$$

If we were to start with a liquid containing 0.5 mole fraction C_3 (point A) and allow it to evaporate partially in a closed container to equilibrium conditions, then the vapour would have the composition at B. If we were to remove the vapour and condense it, it would then contain about 0.75 mole fraction of C_3. The liquid left in the container would of course be substantially richer in C_4. This is one theoretical tray.

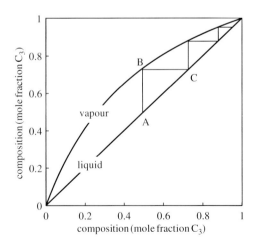

Figure 12.6   Theoretical tray for propane/butane separation

Distillation Control 267

We can see that we only need repeat this process with the condensed vapour three or four more times to obtain a high purity C_3 product. Strictly we should recalculate α at each tray since, as shown by Figure 12.5, relative volatility changes with temperature and temperature reduces as we move up the column.

The method as shown assumes total reflux, i.e. no overhead product is drawn. The reflux ratio (R/D) is therefore infinite. It therefore gives us the minimum number of theoretical trays.

Figure 12.7 shows the molar balance for the basic distillation column on which most of this chapter is based.

The molar feed rate to the column is F. It has *feed quality* of q. This rather confusing term has nothing to do with the composition of the feed but is a measure of its enthalpy. Its definition is given by

$$q = \frac{\text{heat used converting a mole of feed to saturated vapour}}{\text{latent heat of vaporisation of feed}} \qquad (12.14)$$

But perhaps an easier way of understanding it is thinking of it as the fraction of the feed which leaves the feed tray as liquid. Under normal conditions the liquid held on each tray is

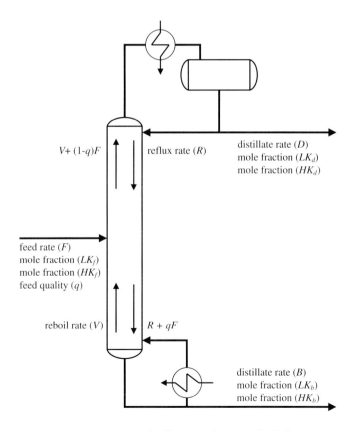

Figure 12.7 Basic distillation column molar balance

at its bubble point and the vapour leaving each tray is at its dew point. While this may not be true in the feed section, heat exchange between liquid and vapour takes place so that, within a few trays of the feed tray, equilibrium is reached.

However, if the feed entering is liquid exactly at its bubble point, no heat exchange is necessary. Thus all of the liquid entering as feed will leave the feed tray as liquid and so q will be 1. Similarly, if the feed entering is vapour exactly at its dew point, all of the feed will leave the feed tray as vapour and q is 0. Values of q between 0 and 1 are possible. Under these circumstances the feed will be a mixture of vapour and liquid at the saturation temperature. In fact q is then a measure of the *wetness* of the mixture.

Values outside the range of 0 to 1 are also possible. It is common for the feed to be liquid below its bubble point. On entering the column, the temperature of this liquid has to be raised to its bubble point. The energy to do this comes from condensing some of the vapour that would otherwise leave the feed tray. Thus the flow of liquid leaving the feed tray will be greater than the flow of feed, so q is greater than 1. Similarly, if the feed is superheated vapour then, on entering the column, it gives up heat to bring its temperature down to dew point. This heat will vaporise some of the liquid that would otherwise have left the feed tray and so q is negative.

Trouton's Law states that, for a pure component

$$\frac{MW.\lambda}{T} = \text{constant} \tag{12.15}$$

MW is the molecular weight, λ the latent heat of vaporisation (per unit mass) and T is the boiling point (absolute temperature). In distillation the boiling points of components will be similar – particularly as they are measured on an absolute basis. This means that the number of moles of liquid being evaporated to vapour is equal to the number of moles of vapour being condensed to provide the energy necessary – no matter what the composition of the vapour and liquid. This is known as *constant molal overflow* and means the liquid flow leaving each tray is the same for all the trays above the feed tray. While it changes to a new value at the feed tray it will be this value for all trays below the feed tray. The same can be said of the vapour flows.

We can therefore choose any tray in the upper section of the column and write the same mass balance. If x is the mole fraction of light key component in the liquid flowing on to a tray above the feed tray, and y is the mole fraction in the vapour leaving the same tray, then the balance around the top section of the column for the light key component, assuming a total condenser, is given by

$$[V + (1-q)F]y = R.x + D.LK_d \tag{12.16}$$

A mass balance round the top section of the column gives

$$V + (1-q)F = R + D \tag{12.17}$$

By combining Equations (12.16) and (12.17) we get the equation of the *top operating line*

$$y = \frac{R}{R+D}x + \frac{D.LK_d}{R+D} \tag{12.18}$$

We can similarly write a balance for the light key component for the section of the column below the feed tray.

$$(R+q.F)x = V.y + B.LK_b \tag{12.19}$$

From the overall mass balance

$$B = F - D \tag{12.20}$$

Rearranging Equation (12.17) gives

$$V = R + D - (1-q)F \tag{12.21}$$

Using Equations (12.20) and (12.21) to eliminate B and V from Equation (12.19), we get the equation of the *bottom operating line*:

$$y = \frac{R + qF}{R + D - (1-q)F} x - \frac{(F-D)LK_b}{R + D - (1-q)F} \tag{12.22}$$

Provided the reflux is returned to the column at its bubble point, i.e. there is no subcooling by the condenser, Equations (12.18) and (12.22) can be plotted on the McCabe-Thiele diagram. These are shown in Figure 12.8; they lie between the vapour and liquid lines. By establishing a realistic reflux we have increased the number of theoretical trays required. We have saved energy but now require a taller column. Column design is therefore a trade-off between operating cost and cost of construction.

The number of *actual* trays will be greater than the theoretical number to allow for inefficiency. In reality the vapour and liquid leaving the tray will not be in equilibrium. To do so would require a very large tray residence time and an uneconomically large column diameter.

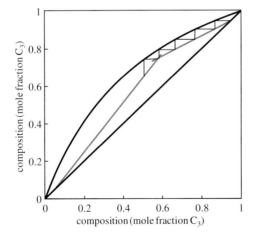

Figure 12.8 Use of reflux for propane/butane separation

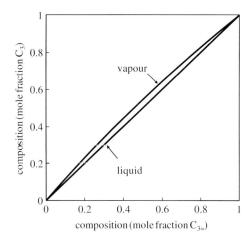

Figure 12.9 Propene/propane separation

If we draw the McCabe-Thiele drawing for the separation of propene and propane (Figure 12.9) we can see that we will need far more theoretical trays.

However the propene/propene mixture is still 'well-behaved'. Figure 12.10 shows the vapour and liquid lines for the separation of ethanol and water. In this case the vapour line crosses the liquid line. Known as an *azeotrope*, it prevents the purity of ethanol exceeding around 95 %. Indeed, if the feed contained a higher proportion of ethanol, the overhead product would be richer in water. The position of the azeotrope depends on pressure, so varying this may offer a solution. If not, then another component can be introduced. For example, if very high purity ethanol is required, the addition of benzene will permit this.

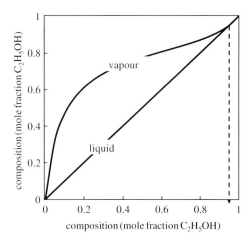

Figure 12.10 Separation showing azeotrope

12.4 Cut and Separation

Cut and *separation* are the key parameters in determining the composition of the distillate and bottoms. We can write a total mass balance for the column in Figure 12.7.

$$F = D + B \tag{12.23}$$

We can also write a mass balance for one of the components, for example LK

$$F.LK_f = D.LK_d + B.LK_b \tag{12.24}$$

Combining these two equations to eliminate B gives

$$\frac{D}{F} = \frac{LK_f - LK_b}{LK_d - LK_b} \tag{12.25}$$

The proportion of feed drawn as distillate is the *distillate cut*. We could equally have developed an expression for *bottoms cut*

$$\frac{B}{F} = \frac{LK_d - LK_f}{LK_d - LK_b} = 1 - \frac{D}{F} \tag{12.26}$$

If we take the example in Table 12.1,

$$\frac{D}{F} = \frac{45 - 10}{95 - 10} = 41\% \tag{12.27}$$

This shows us that 41 % of the feed **must** be drawn as distillate. This is a necessary condition to exactly meet both product composition targets.

Depending on how the composition targets are defined we may need to recalculate cut. For example, defining the specification on distillate in terms of C_4 content (HK_d) is equivalent to setting a target for LK_d only if it is truly a binary distillation. If there is any change in the LLK then it would be necessary to change the cut.

We conclude from this that any control scheme we design should only change the distillate flow if the feed rate changes (to maintain D/F constant), if the feed composition changes or if there is any change in the target composition for either product. If distillate flow is changed for other reasons, such as a change in column pressure or feed enthalpy, then it is certain that at least one product composition will move away from target.

Rearranging Equation (12.25),

$$LK_b = \frac{LK_f - \left(\frac{D}{F}\right)LK_d}{1 - \left(\frac{D}{F}\right)} \tag{12.28}$$

Putting in the values from our example gives,

$$LK_b = \frac{45 - 0.41 LK_d}{1 - 0.41} = 76.5 - 0.697 LK_d \tag{12.29}$$

Using 10% for LK_b and 95% for LK_d is only one solution of this equation. In fact there is a wide range of solutions as shown in Figure 12.11 by the coloured line.

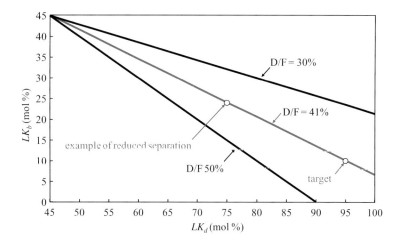

Figure 12.11 Importance of cut

Figure 12.11 shows that, if the cut is wrong, we **cannot** meet both product specifications. For example if D/F is fixed at 30 % we can meet the target for LK_d but not that for LK_b. If it is set at 50 % we can meet LK_b but not that for LK_d. Meeting the required cut is a necessary but not a sufficient condition of meeting the target compositions. For example $LK_d = LK_b = LK_f$ is one solution to Equation (12.29); the cut may be correct but there is no separation. Similarly a value of 75 % for LKd and 24 % for LKb satisfies Equation (12.29). This reduced separation example is shown on Figure 12.11.

We can show diagrammatically in Figure 12.12 the effect of changing separation. In both the target and reduced separation cases the distillate cut is maintained at 0.41.

For multicomponent distillation we can use a *true boiling point (TBP)* curve to represent the same two cases. A TBP curve is obtained in the laboratory, in principle, by slowly

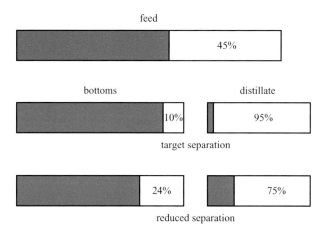

Figure 12.12 Changing separation

Distillation Control 273

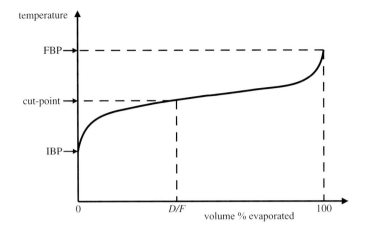

Figure 12.13 TBP curve for column feed

heating the liquid sample and recording the volume evaporated. In practice the test is a little more complex. Originally it was performed as a batch distillation using a column with a very large number of trays and a very high reflux ratio. Nowadays it is done by mass spectrometry.

Figure 12.13 shows the curve for the feed to our case study column. IBP is the *initial boiling point* and FBP the *final boiling point* (or *end point*).

Our cutpoint, currently expressed as a fraction of feed, can now be defined as a *cutpoint temperature*. In concept, any feed material boiling at a temperature below this value will leave the column as distillate and any boiling above this temperature will leave as bottoms. In practice such a perfect separation is impossible.

Figure 12.14 shows the TBP curves for the two products if we meet out target separation. It shows that, as required, 95 % of the distillate boils below the cutpoint – as does 10 % of the

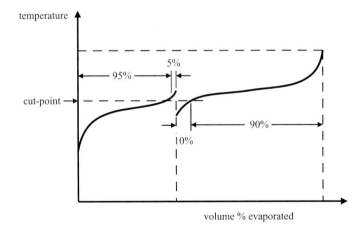

Figure 12.14 Target separation

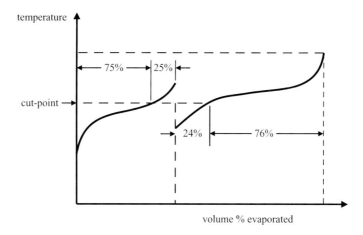

Figure 12.15 Reduced separation

bottoms. As a result, the TBP curves overlap. This overlap is a measure of separation and is defined as the FBP of the distillate less the IBP of the bottoms.

Figure 12.15 shows the reduced separation case where the overlap is far greater.

Table 12.4 shows how the compositions in Table 12.1 are modified by changing to the reduced separation.

In order to quantify separation (S) we will take the definition from the *Fenske Equation*.

$$\alpha^N = S = \frac{\left(\dfrac{LK_d}{HK_d}\right)}{\left(\dfrac{LK_b}{HK_b}\right)} \qquad (12.30)$$

The Fenske Equation is normally used to estimate the minimum number of theoretical trays (N) for a column, i.e. the number necessary to achieve the required separation when operating with total reflux. While of little value in column design it is a parameter used by others in inferential property calculations. We will cover this later in this chapter.

Table 12.4 Reduced separation case

Component	Feed	Distillate		Bottoms	
	moles/hr	moles/hr	mol %	moles/hr	mol %
C_2H_6	0.7	0.5	1.3	0.2	0.4
C_3H_8	45.0	30.9	75.0	14.1	24.0
C_4H_{10}	52.7	9.6	23.5	43.1	73.1
C_5H_{12}	1.6	0.1	0.2	1.5	2.5
total	100.0	41.1	100.0	58.9	100.0

Referring to Table 12.1, we can calculate our target separation.

$$S = \frac{\left(\dfrac{95.0}{3.5}\right)}{\left(\dfrac{10.0}{87.1}\right)} = 236 \tag{12.31}$$

From Table 12.4 we can similarly calculate it for the reduced separation case.

$$S = \frac{\left(\dfrac{75.0}{23.5}\right)}{\left(\dfrac{24.0}{73.1}\right)} = 9.7 \tag{12.32}$$

We see that for a change of about 20 % in product purity the value for S changes by two orders of magnitude. To obtain a more linear relationship, log(S) is commonly used. Indeed the Fenske Equation is often documented in the form

$$N = \frac{\log(S)}{\log(\alpha)} \tag{12.33}$$

For truly binary systems the required S can be calculated from the product purity targets.

$$S = \frac{\left(\dfrac{LK_d}{100 - LK_d}\right)}{\left(\dfrac{100 - HK_b}{HK_b}\right)} = \frac{LK_d HK_b}{(100 - LK_d)(100 - HK_b)} \tag{12.34}$$

Figure 12.16 shows that, despite using log(S), the relationship with purity remains nonlinear.

Alternatively the required S can be calculated from the product *impurity* targets

$$S = \frac{\left(\dfrac{100 - LK_b}{LK_b}\right)}{\left(\dfrac{HK_d}{100 - HK_d}\right)} = \frac{(100 - LK_b)(100 - HK_d)}{LK_b HK_d} \tag{12.35}$$

In this form S the nonlinearity is much less – even though S has been plotted rather than log(S), as illustrated in Figure 12.17.

So far we have assumed that we want to exactly meet the specifications. However purity targets are usually set as inequalities; for example our propane has to be **at least** 95 % pure. It is permitted to produce it at a higher purity and, should there be an economic advantage, we might wish to do.

Let us consider first the situation where only the distillate product has a specification. We are permitted to produce a bottoms product of any composition. For example, if we chose to

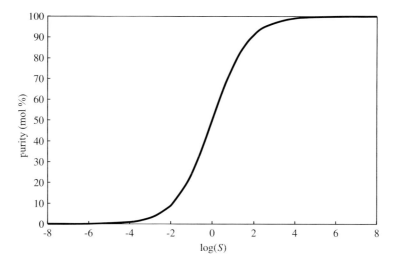

Figure 12.16 Relationship between purity and separation

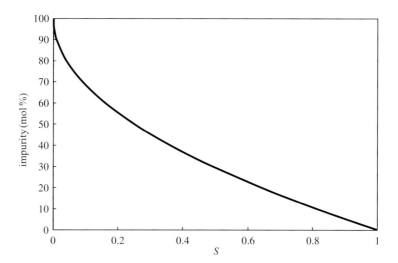

Figure 12.17 Relationship between impurity and separation

produce bottoms containing 20 % propane, then our distillate cut, from Equation (12.25) would be

$$\frac{D}{F} = \frac{45-20}{95-20} = 33\% \quad (12.36)$$

While we are free to choose our own target for LK_b, it is still subject to constraints. For example, it cannot exceed LK_f. The bottoms cannot contain more light key material than the feed. Equation (12.25) would require us to operate with a negative cut! We can however reduce LK_b to zero. While it is not possible to make *both* products 100 % pure, it is possible to make *either* product completely pure – even on columns with relatively poor separation. It may of course be economically disastrous in that doing so might lose very large quantities of the more valuable component in the lower value product.

Thus

$$0 < LK_b < LK_f \qquad (12.37)$$

If we put these constraints into Equation (12.25) we get

$$\frac{45-45}{95-45} < \frac{D}{F} < \frac{45-0}{95-0} \quad \text{or} \quad 0\% < \frac{D}{F} < 47\% \qquad (12.38)$$

The effect that cut has the bottoms composition, while keeping the distillate composition constant, is shown in Figure 12.18.

While it is possible, provided we keep the distillate cut between 0 and 47 %, to meet the distillate 95 % purity target, we have to compensate by adjusting separation. As shown in Figure 12.19, as cut approaches the upper limit, separation approaches infinity.

A similar approach allows us to develop the condition for keeping the bottoms composition at its specification. We now assume that LK_b is fixed at 10 % but we are free to choose any target for LK_d. The distillate cannot contain less light key than the feed and so LK_d is constrained as

$$LK_f < LK_d < 100 \qquad (12.39)$$

If we put these constraints into Equation (12.25) we get

$$\frac{45-10}{100-10} < \frac{D}{F} < \frac{45-10}{45-10} \quad \text{or} \quad 39\% < \frac{D}{F} < 100\% \qquad (12.40)$$

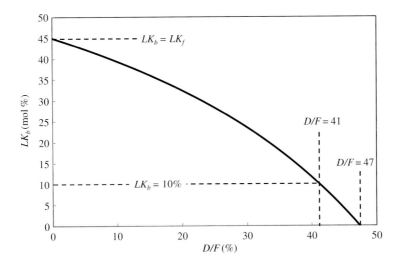

Figure 12.18 Effect of cut on bottoms composition

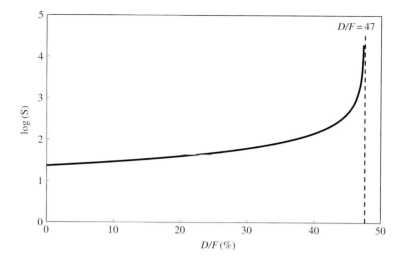

Figure 12.19 Keeping distillate composition at target

The effect that cut has the distillate composition, while keeping the bottoms composition constant, is shown in Figure 12.20. For simplicity we have treated the distillation as truly binary so that

$$HK_f = 100 - LK_f \qquad (12.41)$$

$$HK_d = 100 - LK_d \qquad (12.42)$$

$$HK_b = 100 - LK_b \qquad (12.43)$$

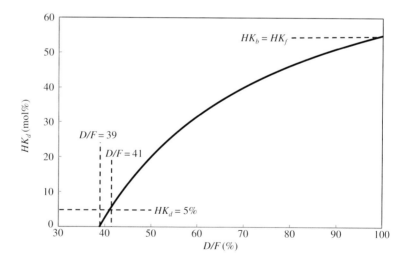

Figure 12.20 Effect of cut on distillate composition

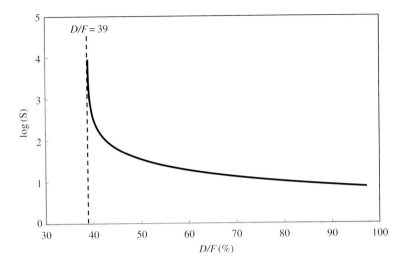

Figure 12.21   Keeping bottoms composition at target

While it is possible, provided we keep the distillate cut between 39 and 100 %, to meet the bottoms 90 % purity target, we have to compensate by adjusting separation. As shown in Figure 12.21, as cut approaches the lower limit, separation again approaches infinity.

We now combine the curves from Figures 12.19 and 12.21 to show the constraints in keeping *both* products within specification. We must operate within the area coloured in Figure 12.22. This confirms that, if we require both product compositions exactly on target

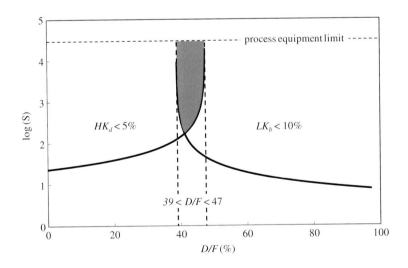

Figure 12.22   Keeping both product compositions at target

we must operate where the lines cross, with a distillate cut of 41 %. This requires the lowest separation and therefore the lowest energy.

If we treat the specifications as constraints rather than absolute targets then we can operate at a distillate cut within the range 39 to 47%. However, as we approach these limits, tray hydraulics or condenser/reboiler capacity will constrain how much separation can be increased. The true feasible range will be considerably narrower.

Where we wish to operate within this feasible space will depend on the process economics. If both products have a similar value then there is no incentive to maximise one at the expense of the other. Profit would therefore be maximum by operating exactly on both specifications and would be reflected as an energy saving. But, for example, if propane were considerably more valuable than butane then we would not wish to leave C_3 material in the bottom product. It would economic to produce butane at purity greater than the specification demands. We would, however, want to operate at the lowest permitted propane purity in order to maximise the amount of C_4 material that is sold at the propane price. We would therefore wish to operate at a higher distillate cut. How large depends on the cost of the additional energy required to recover the additional propane. This falls into the area of optimisation that we cover later in this chapter.

While a range of around 8 % for distillate cut might seem quite broad, this is as a result of the composition targets being relatively low purities. If, for example, both products were required to be better than 99.5 % pure then the feasible range would be given by Equation (12.25) as

$$\frac{45-0.5}{100-0.5} < \frac{D}{F} < \frac{45-0}{99.5-0} \quad \text{or} \quad 44.7\% < \frac{D}{F} < 45.2\% \qquad (12.44)$$

Such purity targets are common for propene destined for polymer production, adding a further challenge to the already difficult separation from propane.

In practice we do not have separation as a MV; we instead adjust it by changing *fractionation*. The terms separation and fractionation are often used interchangeably. This is not strictly correct. Separation is a measure of product composition, while fractionation is a measure of energy used. They are certainly related in that increasing fractionation will increase separation but, as we will see later, separation is influenced by many other parameters.

Figure 12.23 shows the effect that varying the reboiler duty has on the composition of both products. In this case cut has been kept constant which, depending on how the basic controls have been configured, may not be the real situation. It nevertheless shows that, as we change fractionation, both product compositions change.

Similarly, depending on the control configuration, we may not be able to directly manipulate distillate flow rate and keep fractionation constant. However, for the purposes of explanation, this has been the test performed for Figure 12.24. We see that changing it again affects both compositions.

These figures illustrate some of the issues that our control design will later have to address. Firstly if both MVs affect both product compositions, which one do we select for distillate composition control and which for bottoms? Secondly, if we adjust one MV to correct off-spec production, how do we deal with the problem that this will put the other product off grade?

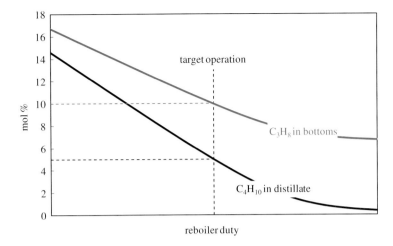

Figure 12.23 Effect of reboiler duty on product compositions

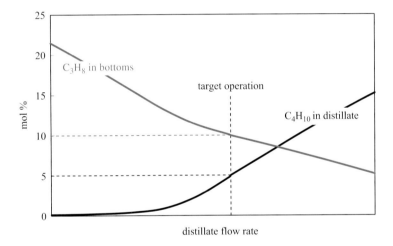

Figure 12.24 Effect of distillate flow rate on product compositions

12.5 Effect of Process Design

Separation is determined by a number of factors – both operating conditions and column design. Figure 12.25 shows the effect of varying reboiler duty while keeping the cut constant. Under these circumstances increasing duty has no affect on product yields. But it does increase the liquid and vapour rates in the column and so improves the purity of both products and hence separation. The relationship is highly nonlinear; as reboiler duty is increased the benefit of the additional duty reduces. The achievable purity is limited by the

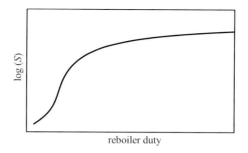

Figure 12.25 Effect of reboiler duty on separation

number of trays although hydraulic limitations within the column will be encountered long before the theoretical limit is reached (Figure 12.3).

Figure 12.26 shows the effect of varying the number of trays, again with the cut held constant. Again we reach the point where the addition of further trays gives no benefit. Separation is limited by the reboiler duty.

Figure 12.27 combines the effect of reboiler duty with that of the number of trays. By fixing the product compositions we can assess the energy saving as the number of trays is increased. As expected from the previous tests the relationship is highly nonlinear. Too few trays mean that the column would be extremely costly to operate. Too many greatly increase construction cost with virtually no impact on energy savings.

In much the same way that separation is affected by the number of trays in the column, it is also affected by their efficiency. Inefficiencies arise because of incomplete mixing on each tray and because it is not practical for the residence time to be large enough for the vapour and liquid to reach equilibrium.

There are two common definitions of tray efficiency (η). The *Fenske tray efficiency* is applied to the whole column

$$\eta = \frac{N_{theoretical}}{N_{actual}} \tag{12.45}$$

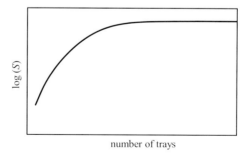

Figure 12.26 Effect of number of trays on separation

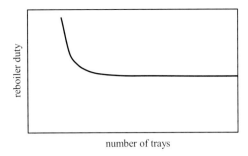

Figure 12.27 Trade-off between number of trays and reboiler duty

We can use Equation (12.45) to redefine the relative volatility for use in the McCabe-Thiele diagram.

$$\alpha_{actual} = (\alpha_{theoretical})^\eta \qquad (12.46)$$

The *Murphree tray efficiency* is defined as the fraction of the theoretical change in composition actually achieved across a tray; for the nth tray

$$\eta = \frac{(y_n)_{actual} - y_{n-1}}{(y_n)_{theoretical} - y_{n-1}} \qquad (12.47)$$

To incorporate this definition into the McCabe-Thiele diagram, we plot an actual vapour line part way between the operating lines and the theoretical vapour line.

Figure 12.28 shows the effect that tray efficiency has on separation.

As we did when we varied the number of trays we can explore what change in reboiler duty is necessary to maintain the compositions constant as tray efficiency changes. This is shown in Figure 12.29. Tray efficiency varies widely depending on the material being processed. Typically a LPG splitter would have a tray efficiency of around 85 %. The chart shows there is little to be gained in energy savings by upgrading to a more efficient tray. However, columns processing much heavier components derived from crude oil, such as bitumen, would have an efficiency of around 20 % and offer the potential for large energy savings.

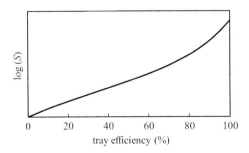

Figure 12.28 Effect of tray efficiency on separation

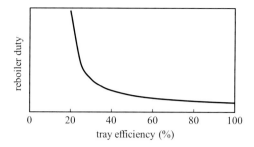

Figure 12.29 Effect of tray efficiency on reboiler duty

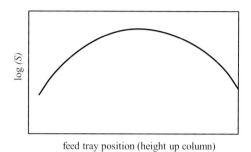

Figure 12.30 Effect of feed tray location on separation

The location of the feed tray also affects separation. Figure 12.30 shows the effect of moving this above and below its optimum location.

Again we can adjust the reboiler duty to compensate for poor positioning of the feed tray. Figure 12.31 shows the result when maintaining constant product compositions. Sensitivity studies also show the effect of changing the feed composition and feed enthalpy. In this case the optimum is fairly 'flat' and so the cost of being a tray or two away from the ideal location is small. It is common to build flexibility into the column design and include the facility to switch feed tray location. Switching is done by operation of manual block valves and is not something that would normally be included in a control strategy.

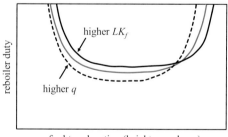

Figure 12.31 Optimising feed tray location

12.6 Basic Controls

There are several issues which need to be addressed when designing the basic control for the column. The first is one of *pairing*. We will see that on our simple column that there are five PVs that we must control – pressure, reflux drum level, column base level, distillate composition and bottoms composition. We normally have available five MVs – distillate flow, bottoms flow, reflux flow, reboiler duty and condenser duty. We need therefore to decide which MV is going to be used to control which PV. Theoretically there are 5!, or 120, possible combinations. While many of these are nonsensical, a large number of feasible schemes are possible.

Our next problem is one of nonlinearity. We have seen already that the relationship between separation and fractionation is highly nonlinear. We can also expect the process dynamics to be difficult. Parts of the process, such as the reflux drum, hold large inventories that will introduce large process lags. Each tray introduces a transport delay – resulting in very large process deadtimes on columns with a large number of trays. Further we have seen already the interactive nature of the process – changing any one the MVs affects all of the PVs.

The process will be subject to multiple disturbances. If fed from an upstream unit the feed rate, feed composition and feed enthalpy can all vary. Disturbances may enter through the reboiler – particularly if the duty is provided through heat integration with another part of the process. Disturbances can enter through the condenser – particularly from *air-fin* types that are subject to sudden changes in ambient temperature or rainfall. We may deliberately vary pressure to optimise the column and we may occasionally change product specifications.

Some of the instrumentation, most notably on-stream analysers, can be very costly. Chromatographs are commonly used which, along with the necessary housing and sampling system, are particularly costly. Other types such as near-infra red (NIR) and nuclear mass resonance (NMR) devices can be more so and involve high ongoing support costs. It may be that we cannot economically justify such instrumentation and need to compromise on the control design.

The basic instrumentation available to us is shown in Figure 12.32. For our case study we have assumed that the condenser uses cooling water and the reboiler uses a heating fluid, for example, steam. We will examine alternatives later. While the drawing does show a gaseous product, this is downgraded to flare and is only intended to be generated when required to avoid overpressurisation.

By most standards the column is generously instrumented – particularly with on-stream analysers. It is not the intention to suggest that all columns should be so endowed. Not all of the schemes we will design for this column are economically justifiable or applicable on all. It is assumed by this stage the control engineer needs no help in installing the necessary flow controllers so these have already been included on the products, reflux and reboiler heating fluid.

12.7 Pressure Control

One of the key requirements of the basic column controllers is to maintain the energy balance. Energy enters the column as feed enthalpy and in the reboiler. It leaves as product enthalpy and in the condenser. If we neglect losses these inputs and outputs must balance.

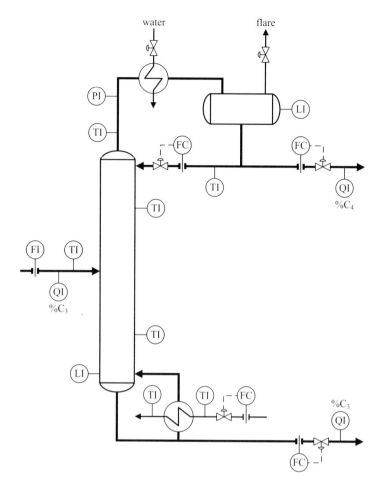

Figure 12.32 Basic instrumentation

While we may have some limited control over feed enthalpy, and maybe some control over product enthalpy, the main source of energy is the reboiler and the main sink is the condenser. If the input energy is greater than the output then more vapour will be produced than condensed and the column pressure will rise. By controlling column pressure we therefore maintain the energy balance.

Pressure makes for good control because it responds quickly to any energy imbalance, but there are also other good reasons for controlling it. Clearly we wish to ensure the unit is safe and that normal process disturbances do not lift relief valves. Pressure affects tray loading. If pressure is reduced, given that the space occupied by the vapour is constant, the vapour velocity must increase to maintain the same molar flow. If operating close to its design limit the disturbance could result in the column hitting the blowing or flooding constraint. Similarly pressure affects dew points and bubble points which change the log mean temperature difference (LMTD) across the condenser and across the reboiler. Again, if operating close to design limits, one of these may become a temporary capacity limit.

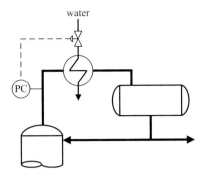

Figure 12.33   Manipulating cooling water

We have seen that pressure affects bubble points and relative volatility; any disturbance to pressure will therefore disturb product composition. While we will show later that there can be advantages in adjusting pressure, we want to do this in managed way, making the changes slowly and taking compensating action to maintain constant composition. We want to avoid deviations from the pressure controller SP.

There is a very wide range of possible process designs for the control of pressure. It is beyond the scope of this book to go into great detail of the process aspects. Its aim is to highlight the issues that can arise with controller design and tuning. The schemes fall into three fundamental groups – those that adjust the rate at which vapour is condensed, those that adjust the rate at which vapour leaves the process and those that adjust the vapour generated.

Taking the first the first of these groups, there are a number of possible methods – including manipulation of coolant rate or temperature, changing the efficiency of the condenser or partially bypassing it.

Figure 12.33 shows one of the most common schemes – manipulation of coolant flow. The water may come from a closed system where heat is removed from the circulating water in a cooling tower. Or it may be a once-through system taking water from the sea or a river. A flow controller on the water is normally not justified. If the water supply pressure is reasonably constant it offers no advantage and, for natural sources of water, may be prone to fouling and corrosion. There is often a **minimum** limit put on the water flow. In fouling services it is important to keep the flow turbulent to keep any solids in suspension. In salt water service it is often necessary to keep the water exit temperature below about 50 °C (around 120 °F), to prevent excessive corrosion of the mild steel tube bundle. The response of the controller can be highly nonlinear but, because of the fast process dynamics, its sluggish response when the process gain is low generally goes unnoticed.

On columns where the temperature of the coolant must be lower than that possible using water a refrigerant must be used. This might be ammonia or propane. Figure 12.34 shows a typical configuration. The shell side of the condenser is partially filled with boiling liquid refrigerant, thus removing heat as its heat of vaporisation. The column pressure controller manipulates the flow of refrigerant vapour returned to the compressor. An alternative configuration is to have the level controller manipulate this flow and the pressure controller manipulate the flow of liquid refrigerant. With either configuration there will be interaction between the controllers – resolved by tuning one controller a little slower than the other.

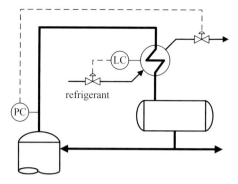

Figure 12.34   Manipulating refrigerant flow

The system can become quite complex if designed to recover energy transferred to the refrigerant. The refrigerant vapour may be used to drive a turbine to provide pumping or compression elsewhere in the process. Or the refrigerant compressor discharge may be cooled by providing reboil energy for the column. As a result, disturbances can be propagated through the refrigerant system back to the column – requiring greater attention to the control design.

It is common for air to be the condenser coolant. Air-fin condensers comprise layers of finned tubes over which air is forced by fans. There are a number of ways in which air flow may be manipulated. As shown in Figure 12.35, the pitch of the fan blades or the fan speed can be adjusted. However there are usually a large number of fans in place. While it may not be necessary to install such mechanisms on all the fans, it does become costly and, like most

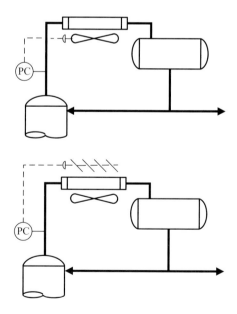

Figure 12.35   Manipulating air flow

mechanical systems, subject to reliability problems. It is also possible to change the number of fans in service. While this can be automated, it must be combined with some other more continuous variable to handle the situation where n fans are insufficient but $n + 1$ fans are too many. Another approach is to manipulate louvers placed in the stream of air exiting the condenser. This too is costly and subject to mechanical problems.

Air-fin condensers can be subject to rapid changes in ambient conditions. A sharp drop in temperature or a rainstorm can cause the reflux to be subcooled. On entering the column the reflux is brought back to its bubble point by condensing some of the rising vapour. This condensed vapour provides additional reflux; thus the *internal reflux* will be greater than the measured external flow. This then disturbs the product compositions.

It is possible to install an internal reflux controller. If ΔT is the temperature drop across the condenser, c_p the specific heat of the reflux material and λ its latent heat of vaporisation then

$$R_{internal} = R_{external} \left[1 + \frac{c_p \Delta T}{\lambda} \right] \qquad (12.48)$$

As shown in Figure 12.36, we can modify the SP of the **external** reflux flow controller to take account of the subcooling and maintain a constant internal reflux flow.

While in principle this scheme appears to be beneficial, it may not perform well. One issue is timing. The reflux drum introduces a large process lag and so its exit temperature will change later than its inlet. Using the condenser outlet temperature would result in the reflux being corrected too early – although it would be possible to lag the measurement of temperature difference. Alternatively the drum exit temperature could be retained and the inlet temperature lagged.

The temperature difference can vary for reasons other than subcooling. For example an increase in the heavy key component in the overhead vapour will cause an increase in vapour temperature. The internal reflux controller will then reduce the reflux flow – the opposite of what is required to deal with the composition change. Using, instead of the overhead temperature, a constant set at a typical value can resolve this. But then the correction for subcooling, although directionally correct, will not be of the correct magnitude.

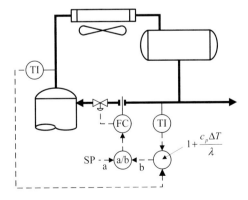

Figure 12.36 Internal reflux control

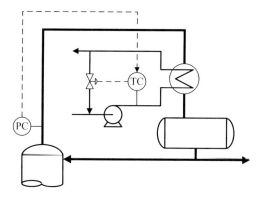

Figure 12.37 Manipulation of coolant temperature

Moving to the next subcategory of schemes, another approach is to manipulate coolant temperature. While not applicable to air-fins it does offer some advantage if condenser fouling is an issue, since it permits a high coolant flow to be maintained – no matter what condenser duty is required. As can be seen in Figure 12.37, it does require additional pumping and so is more costly to implement.

The next subgroup of strategies are techniques which adjust the efficiency of the condenser. The first of these, shown in Figure 12.38, places a valve in the vapour line before the condenser. The pressure drop across the valve lowers the temperature at which the vapour condenses, hence lowering the LMTD across the condenser and therefore the heat it removes.

Another way of reducing the efficiency of the condenser is to reduce the effective surface area used for heat transfer. Figure 12.39 shows how, by placing the control valve under the condenser, liquid can accumulate in the condenser. This *flooded condenser* is less efficient because less heat transfer takes place in the submerged part of the tube bundle. Here sensible heat is removed in subcooling the liquid, in the exposed part of the bundle, heat of vaporisation is removed condensing the vapour. The pressure controller indirectly changes the level of liquid in the condenser.

While the flooded condenser approach offers little advantage if the coolant is liquid, it is beneficial if applied to air-fin condensers. It can replace the potentially unreliable

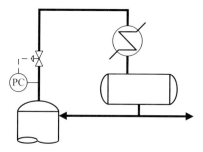

Figure 12.38 Reduction of condensation temperature

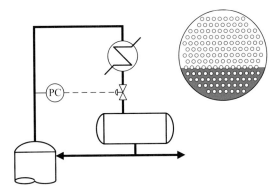

Figure 12.39 Flooded condenser

mechanisms for manipulating air flow and is considerably less costly. With air-fin condensers it is the tube side that is flooded. It is better applied to single pass condensers. It is a myth that the condenser should be inclined to avoid the step changes in surface area that might occur as layers of tubes are exposed. It is unlikely that in the presence of the turbulence of condensation, and of the gradient of the liquid level necessary for flow to occur, that a full layer of tubes is exposed before the liquid begins to drain from the layer below. Plus the condenser is likely to be inclined slightly in any case to ensure it is self-draining. The scheme is shown in Figure 12.40.

Conversion of any existing scheme to flooded condenser is more than just a change in control configuration. It requires a full process design check and is likely to result in changes to relief valves and other safety-related systems. There are also a number of ways in which it can be configured. For example the drum can also be flooded thus avoiding the need to install a valve on its inlet. The pressure controller can then either manipulate either the reflux flow or the distillate flow directly.

If the drum is not flooded then it is better if the condensate enters the drum above the liquid level to avoid interaction between the drum level controller and the column pressure controller. Otherwise a change in liquid level will affect the column pressure. A pressure equalising line between the column and the drum, although not essential, will keep the

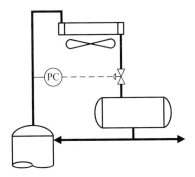

Figure 12.40 Flooded air-fin condenser

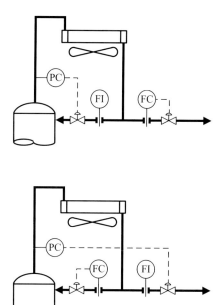

Figure 12.41 Drum-less flooded condenser

drum pressure constant. This is particularly beneficial if a vapour product is taken from the drum, since pressure variations would disturb the composition of both this and the liquid distillate product.

The drum can also be omitted. This gives a considerable cost saving on new plant. It also eliminates a major source of process lag thus allowing the composition controllers to be tuned to act much more quickly. Its disadvantage is that, in the event of a process upset, there is very little liquid inventory.

The two possible drum-less schemes are shown as Figure 12.41. The decision as to whether to control the liquid level (in the condenser) by manipulating reflux or distillate flow is part of a much more wide-ranging consideration that will be covered in the next section. However controlling pressure by manipulating the reflux flow helps considerably with disturbances to the energy balance, for example those caused by changes in ambient conditions around the air-fin condenser. We saw earlier in this chapter that distillate cut should be kept constant during such disturbances. If for example there is a rainstorm, the condenser duty will increase – condensing more vapour and reducing column pressure. The pressure controller will resolve this relatively quickly by reducing the reflux to build level in the condenser. In doing so it helps compensate for the drop in reflux temperature and so maintain a more constant internal reflux. With the cut kept constant by the distillate flow controller, and the fractionation kept approximately constant by the reduction in external reflux, both product compositions will remain relatively unchanged. However, as we shall see later, there are several other factors to consider before selecting this configuration.

Figure 12.42 shows an alternative configuration for a flooded condenser. The control valve is located **before** the condenser. It important that the condensate enters the **drum** below the liquid level and the equalising line between column and drum is **essential**. If the

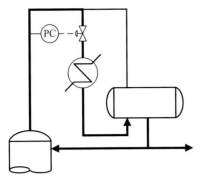

Figure 12.42 Alternative configuration for flooded condenser

pressure in the column rises above SP the pressure controller will open the valve. This increases the pressure in the condenser. Since the pressure in the drum is unaffected the liquid level in the condenser falls until the head of liquid matches the pressure drop across the control valve. This exposes more tubes and therefore brings down the pressure.

From an operator perspective the scheme is conceptually more difficult to understand. It requires a larger, and hence more costly, control valve because it is located in the large diameter vapour line rather than the smaller diameter condensate line. The process design is also more difficult. Since it offers nothing over the alternatives it should not be considered further.

Our last subgroup of techniques which adjust the quantity of vapour condensed is known as the *hot gas bypass* or *hot vapour bypass*. It can be implemented in several ways; the first of which, shown in Figure 12.43, we have already discussed as an option with the basic flooded condenser approach. It includes the addition of what we previously described as an equalising line, but can also be thought of as a vapour bypass. Care is needed in the process design to ensure good mixing between the bypassed vapour and the subcooled condensate. Otherwise the reflux temperature is likely to fluctuate, causing disturbances to the internal reflux.

Figure 12.44 shows the same scheme but with the condenser located below the drum. This has advantages in terms of easier access for maintenance work. Because it is the

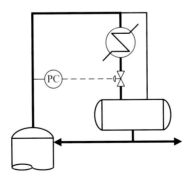

Figure 12.43 Hot gas bypass

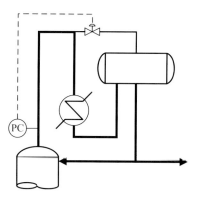

Figure 12.44 Hot gas bypass with condenser below drum

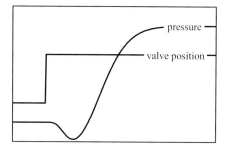

Figure 12.45 Inverse response

smaller line it is tempting to reduce the installation cost by locating the control valve in the bypass rather than in the condensate line.

This however can give problems with inverse response. This is illustrated in Figure 12.45.

When the valve is opened two competing processes take place. The first is one of material transfer from the column to the drum. This has the effect of **reducing** column pressure. The second, because of the bypass being opened, is one of heat transfer reducing the amount of vapour condensed and so **increasing** pressure. Because the dynamics of material transfer are generally faster than those of heat transfer we see the first of these effects. However the drum pressure rises quickly and the material transfer slows. The heat transfer process ultimately prevails and the pressure rises above that at which it was before the control valve moved. The amount of inverse behaviour depends on the relative dynamics of the two processes. On some columns it may not be noticeable; on others it may be severe.

While it is possible to tune a PID controller in this situation, it is necessary to greatly reduce the controller gain – thus making it very slow to respond to disturbances. An alternative solution is to relocate the pressure transmitter from the column to the drum, as shown in Figure 12.46. While the two processes affecting pressure are unchanged, they no longer compete. While a simple, low cost solution – particularly if trying to resolve a problem on an existing unit, its limitation is that it does not control the pressure in the

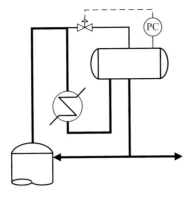

Figure 12.46 Relocating pressure transmitter

column. While the variation is likely to be relatively minor, it will have an effect on product composition.

The ideal solution is to locate control valves in both locations, as shown in Figure 12.47. That in the condensate line is used to control column pressure, while that in the bypass used to control drum pressure. If the condenser is below the drum there must be sufficient difference between the two pressures to overcome the maximum liquid head. If this is not the case then, on high pressure, the column pressure controller will saturate and the pressure will rise until it is sufficient to overcome the head. While not necessarily unsafe, it does mean that full control of pressure will be lost.

Rather than rely on the process operator to maintain sufficient pressure difference, the drum pressure controller may be replaced by a differential pressure controller (dPC), as shown in Figure 12.48. The SP of this controller could be fixed and not adjustable by the operator, or a safe minimum limit configured. It might also include logic that disables column pressure control if the dPC is switched to manual.

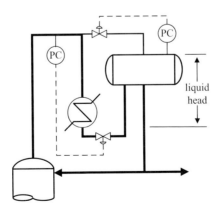

Figure 12.47 Use of two pressure controllers

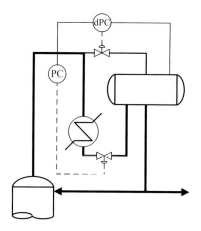

Figure 12.48 Pressure difference controller

The two controllers will interact. If column pressure rises above its SP the controller will open the condensate valve so that more vapour passes through the condenser. But, if the column pressure rises, the measurement of the dPC will also increase and its controller will respond by opening the bypass. This will reduce the flow of vapour to the condenser – the opposite of what is needed. To break the interaction the dPC should be configured to use the SP of the column pressure controller, not its PV.

The next main group of schemes manipulate the vapour flow leaving the process. Clearly these only work if the condenser is not a total condenser. But, if there is a significant flow of vapour, adjusting it has the most direct control on pressure and will have an immediate effect. How we actually manipulate the flow will depend on the vapour handling system. If the vapour is simply routed to a lower pressure system then we need only place a control valve in its line, as shown in Figure 12.49 – for both noncondensing and partial condensing situations.

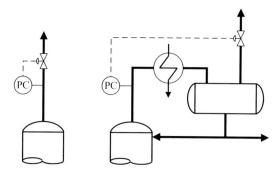

Figure 12.49 Manipulation of vapour flow

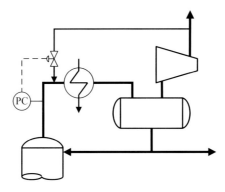

Figure 12.50 Manipulation of compressor spillback

If the overhead vapour is routed to a compressor, then all of the schemes described in Chapter 11 can be applied. These include suction or discharge throttling, manipulating speed or inlet guide vanes and the use of recycle (or *spillback*). In the latter case the spillback can be cooled through a dedicated exchanger or routed back to the condenser inlet, as shown in Figure 12.50.

The pressure in vacuum distillation columns with is similarly controlled by manipulating the spillback around the ejectors, as shown in Figure 12.51. The installation of the pressure transmitter needs special attention. It is important that the impulse line is self-draining back to source; otherwise a liquid head can build up and cause a false pressure measurement. The liquid may also boil and cause a noisy measurement.

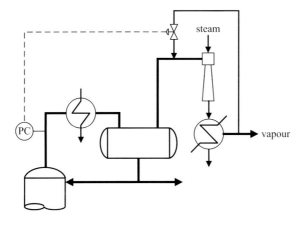

Figure 12.51 Manipulation of ejector spillback

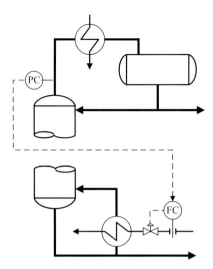

Figure 12.52 Manipulation of vapour generation

Throttling either the inlet or discharge from an ejector, or throttling the motive fluid, is not generally successful because of the impact they have on ejector performance. However it is common for the ejector spillback to be closed and the pressure controller on manual. This is not a reflection on the performance of the scheme. It can be economically very attractive to operate the column at the lowest possible pressure, even if this means the pressure fluctuating somewhat. We will cover later in the chapter techniques for compensating for such fluctuations so that product composition is not affected. And we will also return later to pressure optimisation.

The last of the strategies we might consider for pressure control is manipulation of the rate at which vapour is generated. A typical scheme is shown in Figure 12.52. This scheme should only be applied under special circumstances. While quite feasible is does result in the loss of the main fractionation variable as a means of controlling product composition. But there are columns where tight control of composition is not required; this may be because they are deliberately overfractionated.

The use of split range pressure controllers is common on distillation columns. They enable the operating range to be extended beyond the point where a MV saturates. One example, covered in Chapter 5, dealt with controlling pressure by manipulating vapour production, but also importing a noncondensible stream when completely closing the vapour off-take valve was not enough to raise the pressure.

Another example might be venting vapour if the condenser limit has been reached. In Chapter 5 we explained why it was often better to install two independent controllers, rather than the split range approach. Figure 12.53 shows how this would work on our case study column. The split range scheme, on rising pressure, is configured to first open the cooling water valve. If the controller output reaches 50 % the cooling water valve will be fully open and, if necessary, the valve to flare will begin to open. In the preferred design the controller manipulating the cooling water valve would have a SP slightly lower than that manipulating the flare valve.

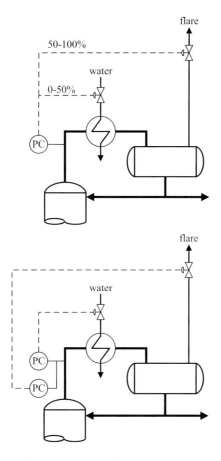

Figure 12.53 Preferred alternative to split-ranging

12.8 Level Control

After maintaining the energy balance, the next prime objective of the basic column controllers is to maintain the material balance. Material enters as feed and leaves as products. If these are not in balance then the inventory in the process will change. This is reflected by changing levels in the reflux drum and/or the column base. By controlling these levels we maintain the material balance across the column.

So far we have used up one of our available MVs (condenser duty) to control one of our PVs (pressure). There remain four MVs from which we can select two to provide control of our two levels. To these we can add feed rate, since on some columns this is available as an MV. Table 12.5 shows that there are in theory 20 potential schemes.

Two of the potential schemes can be rejected immediately, on the grounds that they do not meet the objective of maintaining the material balance across the column. Figure 12.54 shows one of these (see note 1 in Table 12.5). A change in feed rate will cause the column to move out of material balance but, since both product flows are fixed by flow controllers, no

Table 12.5 Potential level control strategies

		secondary of column level controller				
		feed	distillate	reflux	bottoms	reboil
secondary of reflux drum level controller	feed	×	impractical	impractical	impractical	impractical
	distillate	not always available	×	impractical	energy balance scheme (4)	material balance scheme (5)
	reflux	not always available (3)	impractical	×	material balance scheme (6)	violates material balance (1)
	bottoms	impractical	impractical (2)	impractical	×	impractical
	reboil	impractical	impractical	violates material balance	impractical	×

corrective action can be taken. For example, an increase in feed rate will cause the column level to increase. The column level controller will increase the reboiler duty, which will increase column pressure. The pressure controller will then increase condenser duty. The additional condensate will cause the reflux drum level to rise and its level controller will return it to the column, where it will again cause the column level to rise. If left unchecked the column will ultimately fill with liquid. The second scheme rejected again has both product flows fixed – compounded by a rather strange configuration of level controllers.

Many of the schemes are described as 'impractical'. One of them (note 2) is shown as Figure 12.55. Many engineers would reject the scheme instinctively; it simply does not look 'right'. However it would function correctly provided one of the level controllers is configured as reverse acting, i.e. on increasing PV it will reduce its output. In this example the column level controller is reverse-acting. So, if the operator were to reduce its SP, it will reduce the distillate flow. The drum level will then begin to rise and so its controller will increase the bottoms flow – bringing down the column level as required.

Similarly, if the operator were to reduce the SP of the drum level controller, it would increase the bottoms flow. The column level will fall causing the level controller to increase the distillate flow and so bring down the drum level.

What makes the scheme impractical is that each level controller relies on the other to take corrective action. If one of the controllers were switched to manual then the output of the other would eventually saturate. While in some instances it is unavoidable, it is not generally advisable to design such *nested* controllers – particularly mutually nested controllers. All of the schemes described as impractical in Table 12.5 involve unnecessary nesting.

Two of the schemes involve manipulation of the feed rate. On many columns this is not an option since they receive their feed from a process upstream that has its feed rate fixed to meet some other criterion. Even on those which receive feed from tankage, manipulating

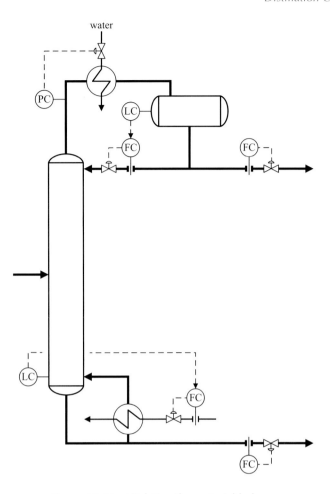

Figure 12.54 Violating the material balance

feed rate may be undesirable. Operating plans will often specify feed rather than product rate. Nevertheless there are occasions where the scheme is used. There are examples of a series of several columns all having level controllers cascaded to each of the feed flow controllers.

Figure 12.56 shows one of the two possible schemes (note 3). It is often used when both products are routed to downstream processes and it is a requirement that the feed to the processes is kept constant. In reality this is probably not achievable. While we can manipulate reboiler duty to provide control of composition, this will only vary separation. To vary the cut at least one of the product flows will need adjustment. One could argue that it would be better to cascade the drum level to the distillate flow. This would permit averaging level control to be applied to make smooth changes to the downstream process flows and release the reflux flow as a MV for composition control.

Of the three schemes remaining, one (note 4) has both product flows as MVs of level controllers, as shown in Figure 12.57. This is known as the *energy balance* scheme. The

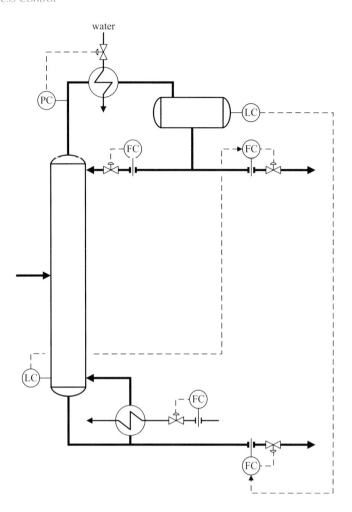

Figure 12.55 Impractical level control configuration

remaining MVs, reboil and reflux, will then be used for composition control. Both MVs affect both cut and separation. So, unlike other configurations, it is not simple to vary cut without changing separation or vice-versa. This does not imply that there is any problem with the scheme; indeed it is the most commonly used configuration.

The two remaining schemes are known as *material balance* schemes. One of product flows is not used for level control and remains available as a MV for composition control. The first of these schemes (note 5) is shown in Figure 12.58.

This scheme is undesirable for a number of reasons. Firstly it can exhibit inverse response; additional reboiler duty causes the vapour volume in the reboiler to increase and so displace some liquid into the column base. The level then initially rises before falling as a result of the increased vaporisation. Secondly the thermal inertia of the reboiler will introduce a large lag, not common in level control. Controller tuning must therefore be relatively slow. It may be that it cannot respond sufficiently quickly to routine changes in

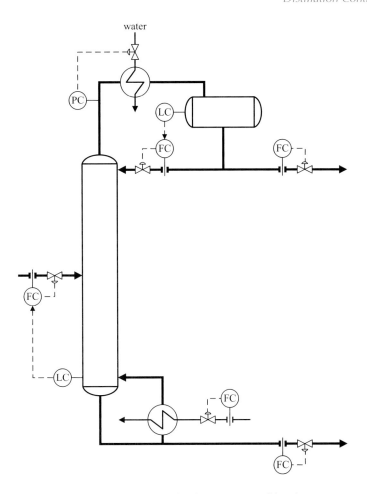

Figure 12.56 Using feed rate to control level.

reflux or bottoms flow. So, to avoid violating level alarms, changes have to be made slowly – thus degrading the performance of the composition controllers that manipulate these variables.

However there are occasions where use of the scheme proves necessary. For example, if the bottoms flow represents only a small part of the feed, manipulating it over its full range may not be sufficient to control column level. The remaining MVs, that will be used for composition control are bottoms flow, which determines cut, and reflux flow, which determines separation.

Figure 12.59 shows the more common version of the material balance scheme (note 6). This scheme does not share the problems with the previous scheme and thus is in common use. The remaining MVs that will be used for composition control are distillate flow, which determines cut, and reboil duty, which determines separation.

The result of the exercise just completed is that, in designing a level control strategy, the choice in the majority of cases is restricted to one of two options. Either we apply the energy

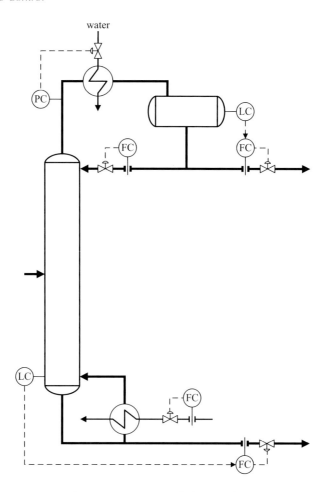

Figure 12.57 Energy balance scheme

balance scheme and cascade the drum level controller to the distillate flow, or we select the preferred material balance scheme and cascade the level controller to the reflux. There are a number of considerations in making this decision.

We have seen that cut is the prime variable in determining product composition. With the material balance scheme in place, cut can only be varied by changing the distillate flow. However, this in itself has no effect on product composition. Changing the distillate flow causes the drum level to deviate from SP and it is the corrective action of this controller changing reflux that changes the composition. To achieve tight control of composition we therefore need tight level control. In the case of the energy balance scheme, we are free to choose tight or averaging level control, since the composition controller will manipulate reflux directly. If the distillate product is routed to a downstream process, averaging level control will exploit the surge capacity of the reflux drum to minimise flow disturbances. Even if the distillate product is routed to tankage, if it is cooled via heat integration, then averaging level control will minimise disturbances to the energy balance.

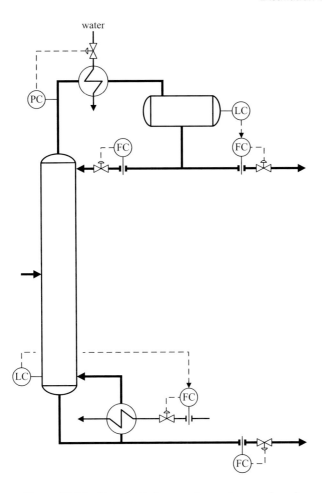

Figure 12.58 Material balance scheme (less preferred)

A similar argument can be applied to the choice of energy balance over the less preferred version of the material balance scheme. In theory we can install averaging tuning in the column level controller. However, if the holdup in the column is small then this will give little advantage. Further, depending on the reboiler configuration, permitting the level to vary may cause problems with its performance. Nevertheless there are occasions where averaging level control will be beneficial.

Another consideration, in making the choice, is the reflux ratio (defined in this book as R/D, not $R/(R+D)$ as it is in some). A large reflux ratio favours the use of the material balance scheme because the relatively small distillate flow may not offer the range necessary to adequately control drum level. Similarly a low reflux ratio ($\ll 1$) favours the use of the energy balance scheme.

The material balance scheme is favoured when it is important to maintain the cut constant. Both versions of the scheme do this. We have seen that this is desirable when the prime disturbances arise from disruption to the energy balance – such as rainstorms on air-

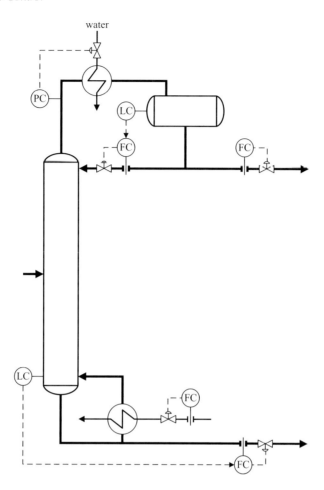

Figure 12.59 Material balance scheme (preferred)

fin condensers, changes to feed enthalpy, changes in column pressure and variation in reboiler duty due to heat integration.

We have seen that we can define cut both in terms of product yield but also in terms of temperature. A tray temperature controller can be configured to manipulate whatever variable remains after the level controllers are configured. And so we can control cut whether we have selected either the energy balance or the material balance scheme. However, there are occasions when tray temperature is insensitive to changes in product composition – for example when separating components with very similar bubble points. Under these circumstances the cut control provided by the material balance scheme is advantageous.

The energy balance scheme is favoured when the main disturbances arise from changes to the material balance. For example, a change in feed composition requires that the cut be changed in order to maintain product compositions on target. The energy balance scheme permits the cut to change. While it is unlikely to change the cut to the correct value it does change it in the right direction and will outperform either material balance scheme. For the same reason, the energy balance scheme will perform better during feed rate changes.

We have seen that the distillate composition controller on the preferred material balance scheme relies on the level controller. If this is switched to manual then the composition control should be automatically disabled and not permitted to be recommissioned until the level is back on automatic. Similarly, on the less preferred version of the material balance scheme, bottoms composition control should not be permitted if the column level controller is switched to manual. Without this precaution the composition controller will ramp its output until it saturates.

A limitation of the material balance control scheme is that if the product flow is not used as a MV for composition control, for example if the controller is switched to manual, it may not be possible to control the composition of the other product. The cut may be such that achieving the target composition is infeasible. This problem can also arise dynamically. If the composition controller manipulating cut is much slower than that manipulating separation then the latter may be need to be tuned to act slow enough for the other to first correct the cut. This is common if the distillate composition is controlled using an on-stream analyser to manipulate the distillate flow. The lag and deadtime imposed by the reflux drum and analyser means that control will be slow compared to the other composition controller which might comprise an inferential manipulating reboil duty.

The material balance scheme is often favoured on high purity columns. We have seen that the permitted range of cut is much smaller under these circumstances. A scheme that keeps tight control of cut should therefore perform better.

On many columns the decision is not clear cut. Here the approach should be to make a preliminary selection of one of the schemes, identify its limitations and attempt to enhance the scheme to deal with these. If this fails then switch to the alternative and enhance this one. For example we might have good reasons to select the material balance scheme but the column is subject to changes in feed rate. Installation of the feedforward scheme shown in Figure 12.60 will maintain a constant D/F ratio and so overcomes this limitation. While not quite as simple as drawn, a full description of feedforward control is presented later in this chapter.

The same column might also be subject to changes in feed composition. If we are in a position to measure this on-stream, or infer it, then the proportion of light key in the feed (LK_f) can be fed forward to the ratio controller target by applying Equation (12.25) – using the target values of LK_b and LK_d. Figure 12.61 shows a simplified version of the scheme.

Another modification can be made to the material balance scheme aimed at overcoming the lag introduced by the reflux drum. This is shown in Figure 12.62.

Any changes to the SP of the distillate flow controller made to correct composition are passed directly to the reflux flow controller SP, rather than wait for the drum level controller. If the level controller is tuned tightly according to the method given in Chapter 4, this would also be achieved by a conventional level controller. However the scheme includes an additional term (K). If K is set to -1, the change in distillate flow SP is passed to the reflux flow SP as an equal and opposite change. As a result there is no change in drum level. However K can be set as required. For example, setting a value less than -1

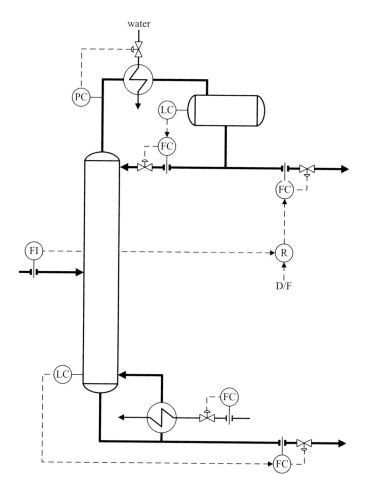

Figure 12.60 Enhancing the material balance scheme to maintain cut

will result in the reflux being changed by more than the change in distillate. The drum level will then change and the controller will take corrective action to bring the reflux back to the correct value. The 'kick' this introduces will help overcome the overall process lag.

The scheme has introduced something similar to a lead-lag algorithm, as described in Chapter 6, where K is effectively the $T1/T2$ ratio. The lag is governed by the speed of response of the drum level controller. Tuning would be by trial-and-error, without an obvious measure of how well the scheme is performing. If such dynamic compensation is justified it would more straightforward to use a conventional lead-lag algorithm and tune it according to the method given in Chapter 6.

The alternative approach of selecting the energy balance scheme and checking for its limitations might show that it does not handle well disturbances to condenser and reboiler duty. We have seen (Figure 12.36) how the use of internal reflux control might be applied to the first of these problems. If the disturbances to the reboiler duty arise from heat

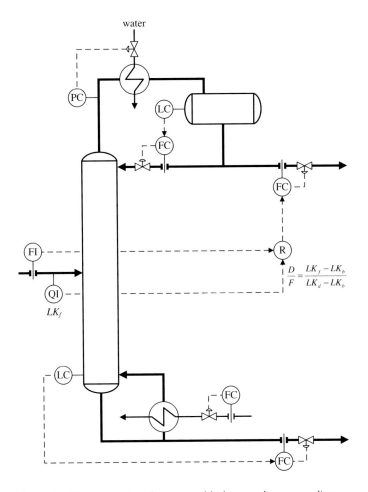

Figure 12.61 Enhancing the material balance scheme to adjust cut

integration we can adopt a similar approach as shown in Figure 12.63. Thinking of the reboiler duty SP in energy units, we use the temperature drop across the reboiler to calculate $c_p\Delta T$. Dividing the duty SP by this value generates the flow SP. Thus, any disturbance to the temperature of the heating fluid will be immediately compensated for by the change in flow. While the lag of the reboiler means that outlet and inlet temperature do not change at the same time, the scheme will still work well. Lagging the inlet temperature to compensate for the reboiler lag would be counter-productive since it would delay the flow correction.

If the reboiler is a fired heater then it may be subject to disturbances in the fuel system. Chapter 10 describes techniques which will keep duty constant.

In addition to the energy balance or material balance options, there are various hybrid schemes – the most well known of which is the simplified *Ryskamp scheme* (Reference 1). In this version of the scheme we add a reflux ratio controller to the energy balance scheme,

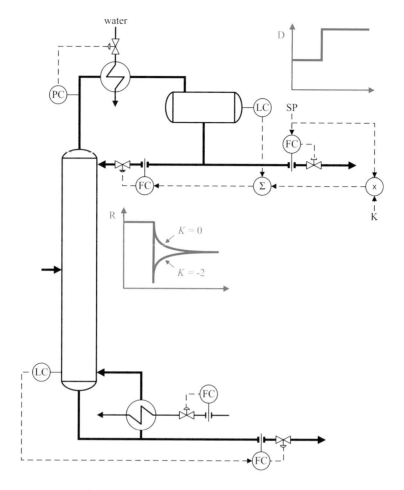

Figure 12.62 Compensating for reflux drum lag

as shown in Figure 12.64. The input to the ratio algorithm in the drum level controller output (or the distillate flow measurement). So, as the level controller takes corrective action, it moves reflux and distillate in proportion – maintaining a constant reflux ratio.

An advantage of the scheme is that it helps break the interaction between the two composition controllers. Figure 12.65 shows, on an example column, the effect of changing reboiler duty with each of the three level control schemes in place. The composition targets for HK_d and LK_b are both 5 %. It confirms that the material balance scheme is better than the energy balance scheme at keeping HK_d constant, but the Ryskamp greatly outperforms both. This means that when the bottoms composition controller adjusts reboiler duty, the disturbance made to the distillate composition is almost negligible. The distillate composition controller would now adjust the reflux ratio. Doing so will still cause a disturbance to the bottoms composition but breaking the interaction in one direction is enough to enable stable composition control.

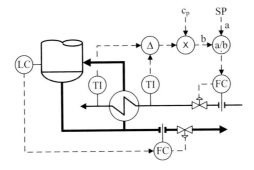

Figure 12.63 Enhancing the energy balance scheme

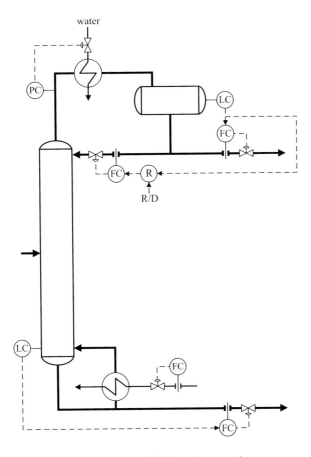

Figure 12.64 Simplified Ryskamp scheme

312 Process Control

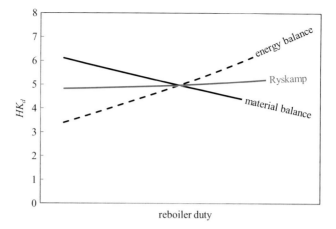

Figure 12.65 Performance of Ryskamp scheme for distillate composition

A second advantage of the Ryskamp scheme is that, compared to the material balance scheme, it makes the bottoms composition more sensitive to changes in reboiler duty, as seen in Figure 12.66. This means that less adjustment of reboiler duty is required – reducing the interaction further.

Of course, because the Ryskamp scheme offers benefits in one operating scenario does not mean it is universally the best scheme. Figure 12.67 shows how each of the schemes perform as feed composition changes. As we expect, the energy balance scheme maintains the distillate composition closer to its target than the material balance scheme. But, despite keeping the reflux ratio constant, the Ryskamp scheme performs poorly.

Figure 12.68 explains why Ryskamp does not handle feed composition changes as well as the energy balance scheme. As we know from Equation (12.25) distillate product rate (D) varies linearly with feed composition (LK_f). Perhaps what is not immediately obvious is

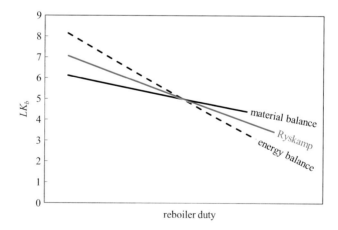

Figure 12.66 Performance of Ryskamp scheme for bottoms composition

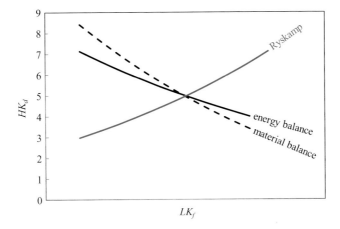

Figure 12.67 Performance of Ryskamp scheme for changes in feed composition

why the reflux passes through a maximum. To explain this, consider a feed that has a composition identical to the bottoms product specification. No energy is required to operate the column; it effectively becomes the pipework necessary for the feed to all be pumped out as bottoms. No reboil and no reflux are required.

Now we consider the other extreme case, where the feed has a composition identical to the distillate product specification. Again no separation is required; we need to provide sufficient reboil duty to fully vaporise the feed, and sufficient condenser duty to return it to liquid in the reflux drum, but no reflux is required.

Figure 12.68 shows that keeping the reflux ratio constant is not what is required. Indeed the value required varies greatly with feed composition – particularly as the light key in the feed gets small.

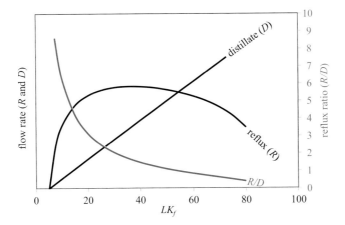

Figure 12.68 Effect of feed composition

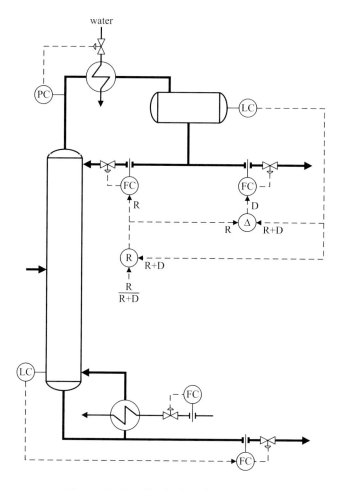

Figure 12.69 Original Ryskamp scheme

Figure 12.69 shows the original version of the Ryskamp scheme. This maintains constant a different definition of reflux ratio, i.e. $R/(R + D)$. This is chosen because it is the slope of the top operating line on the McCabe-Thiele diagram – see Equation (12.18). The output of the level controller is $R + D$. From this is subtracted R to generate the SP for the distillate flow controller. The reflux ratio target is multiplied by the level controller output to generate the SP for the reflux flow controller.

In terms of composition control the performance of this version of the scheme is unlikely to be distinguishable from that shown in Figure 12.64. The additional complexity therefore might not be justified. But while it is still not correct to keep the modified reflux ratio constant as feed composition changes, Figure 12.70 shows a more linear relationship. This would be helpful if feedforward on feed composition was being contemplated.

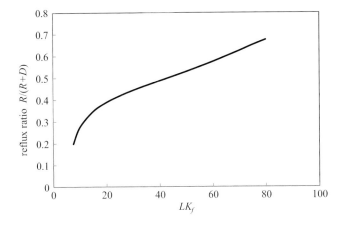

Figure 12.70 Effect of feed composition on the original Ryskamp scheme

12.9 Tray Temperature Control

Although on-stream analysers are readily available for many properties they do have limitations. They can be subject to large delays – either due to the analyser technology itself or the position of its sample point. For example, the sample point for distillate is commonly located on the discharge of the product pump – usually after any product coolers; thus any change in product composition must first pass through the reflux drum before it is detectable by the analyser. Further, once the disturbance has been detected, we have a drum full of off-spec material that must pass through the product system. Secondly, although analyser technology is steadily improving, they are more prone to failure than other instrumentation. Some are also required to be taken out of service regularly for calibration and preventative maintenance. And thirdly, they are usually expensive to install and maintain. There may simply not be sufficient benefit to be had to justify the cost.

On many columns tray temperature control offers a method which, although not as accurate as an analyser, provides a degree of composition control and overcomes these problems. It is not necessarily a replacement for a higher level of composition control; if both are feasible then, as we shall show later, they can operate in conjunction. Tray temperature control works on the principle that liquid on the trays is at its bubble point. Bubble point is related to composition and so fixing the bubble point provides some level of composition control. As we have seen cut can also be expressed as temperature and so controlling tray temperature helps maintain cut.

Ideally the control engineer should be involved in the selection of the tray(s) on which the temperature controller will be installed. It is more often the case that existing temperatures, selected by others, have to be used. Retrofitting new temperatures is usually difficult. Most vessels are stress-relieved after construction. This involves heating the vessel to a temperature of around 650 °C (1200 °F) – depending on the alloy. Doing so avoids stress corrosion cracking when the vessel is in service. The stresses introduced by drilling the vessel and welding an additional thermowell would require this process to be repeated – at

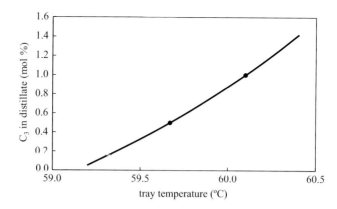

Figure 12.71 Insensitivity of temperature to composition in C3 splitter

least in the area around the weld. With the vessel in place, probably lagged with insulation material, attached to the structure and to pipework, this would be a very costly exercise.

In theory we should control the temperature at the point at which the product is withdrawn from the column. However there are several reasons why this may not be practical. Firstly, the liquid may not be homogeneously at its bubble point. This is likely to be the situation close to the top tray since reflux is often subcooled. It can also occur at the base of the column if the vapour from the reboiler is superheated. Secondly, in pseudo-binary columns, the relationship between composition and bubble point also depends on the proportion of non-key components in the product. The greatest proportion of LLK will occur at the top of the column, and of HHK at the bottom. Thus, if non-key composition varies, these are the regions most prone to inaccuracy. Finally, particularly on high purity columns, the temperature may not be sensitive to changes in composition.

When moving away from the top and bottom of the column, there are other issues to consider. If the components in the feed have very similar bubble points it may not be possible to identify a tray anywhere where the temperature is sufficiently sensitive to changes in composition. Figure 12.71 shows the relationship between bubble point and composition for a *C3 splitter* – a column separating propene from propane. A common requirement is that the propene be 99.5 % pure. The relationship shows that doubling the permitted amount of propane in the product changes the tray temperature by about 0.4 °C (0.7 °F) – a change too small to be measured accurately by conventional instrumentation. In such a case it is likely that accurate composition control can only be achieved with the use of an on-stream analyser.

The relationship between composition and temperature may be nonlinear. While this will not affect temperature control it will give tuning problems with any composition controller that may later be cascaded to the temperature controller. Figure 12.72 shows such a problem arising from a poorly selected tray in our LPG splitter. Considering the design case, where our target is 5 % C_4 in distillate then, provided we are between 4 and 6 % the variation in the slope of the relationship corresponds to an acceptable variation in process gain of about ± 20 %. However if a disturbance were to take the composition down to around 1 %, the process gain falls by a factor of around 10 – meaning that the composition controller would be very slow in returning to target.

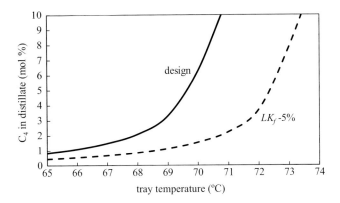

Figure 12.72 Nonlinear relationship between composition and temperature

Similarly the relationship between temperature and its MV may be nonlinear which **will** give tuning problems for the temperature controller. We will address this later when completing the exercise of checking the suitability of a chosen tray temperature.

There is also potentially a problem in locating the temperature too close to the feed tray. On the tray itself it is unlikely that liquid will be at its bubble point. There is also an issue with trays close to the feed tray because the relationship between tray temperature and composition can be sensitive to feed composition. The dashed curve in Figure 12.72 shows another reason why the tray is a poor choice. A 5 % reduction in the LK content of the feed causes a major change in the relationship between distillate composition and tray temperature. If we were controlling the tray at around 70 °C to meet our 5 % target and the feed composition were to change, then our product composition would move to around 1.4 %.

Figure 12.73 shows temperature profiles drawn for two different feed compositions. In both cases the products are exactly at target composition. The profiles show that, if we were to fix the temperature at any point in the column, at least one of the product compositions

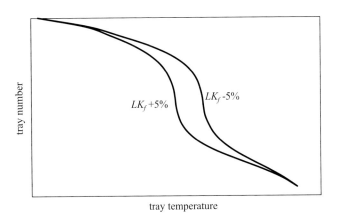

Figure 12.73 Effect of feed composition on temperature profile

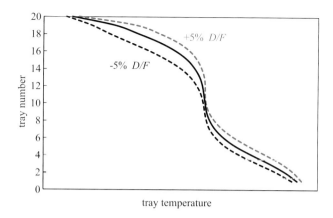

Figure 12.74 Effect of changing cut (with material balance scheme in place)

would change if feed composition changed. As one might expect, the nearer to the feed tray that we control the temperature the more pronounced the problem.

The process of selecting tray temperature(s) for composition control is usually completed using a simulation of the column, rather the real column. Properly it should be done before the column design is frozen. On columns, already built, that have multiple tray temperatures installed, it is possible to execute the plant tests required but they will be very disruptive to the operation.

The column at this stage of the control design will have two MVs remaining for use to control the composition of both products. Which these variables are depends on the choice of level control configuration. If the material balance scheme is in place then, usually, distillate flow and reboil are available or, less usually, bottoms flow and reflux. If the energy balance scheme is in place then reboil and reflux are available.

On a 20 tray column, taking the first of these schemes as an example, Figure 12.74 shows the effect of changing the cut by ±5 %; this may be achieved by changing the distillate flow by ±5 % or, less usually, the bottoms flow by ±5 %. If we are changing the distillate flow then we would likely wish to use this to control the **distillate** composition and so we would select a tray in the upper section of the column.

Similarly Figure 12.75 shows the effect of varying reboil duty by ±5 %. Since the cut is fixed by the distillate (or bottoms) flow controller, increasing reboil duty increases separation without changing product yields. Thus both products approach the bubble point of the pure LK and HK components and the temperature profile **rotates** (anti-clockwise).

If we repeat the test but with the energy balance scheme in place, as shown in Figure 12.76, we see that the temperature profile **translates** with reboil duty – because both cut and separation are changed.

Figure 12.77 shows how the profile changes if we change reflux by ±5 % on a column with the energy balance scheme.

Using the energy balance scheme as an example, the profiles in Figure 12.76 have been plotted in Figure 12.78 as **changes** in temperature. In choosing a suitable tray we look for sensitivity and linearity. Sensitivity is measured by the distance between the two profiles and linearity by the symmetry. We can see that tray 4 shows the greatest sensitivity. In terms

Distillation Control 319

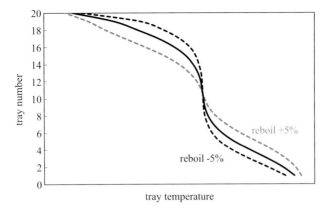

Figure 12.75 Effect of changing reboil duty (with material balance scheme in place)

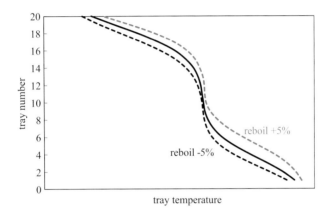

Figure 12.76 Effect of changing reboil duty (with energy balance scheme in place)

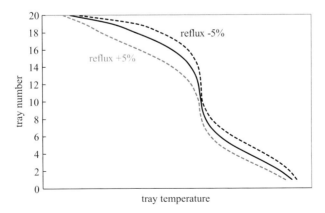

Figure 12.77 Effect of changing reflux (with energy balance scheme in place)

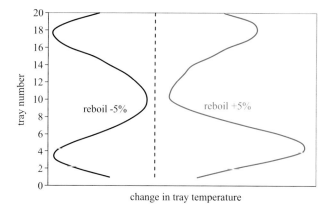

Figure 12.78   Change in temperature due to reboil

of linearity an increase in reboil duty has an effect about 50 % greater than a decrease. While nonlinear, it falls just within our criterion of being able to select a value for process gain which varies by less than 20 %. Had it not, then we might reconsider our choice. For example tray 2, although considerably less sensitive, shows almost exactly linear behaviour. Tray 3 might be considered as a good compromise.

Figure 12.79 shows, in a different form, the sensitivity and linearity of trays 3 and 4. But the curves have also been plotted for a different feed composition. If we were to install a temperature controller on tray 4, initially holding the composition at 5 %, then the change in feed would cause to composition to increase by about 0.8 %. Using tray 3, since it further from the feed tray, the disturbance is reduced to about 0.5 %. Taking the better linearity into account then tray 3 is probably the better choice.

While we have confirmed there is sufficient linearity between composition and temperature our first concern is that we can control temperature by manipulating reboil duty. This check is shown in Figure 12.80 which shows how the temperatures on tray 3 and 4 vary.

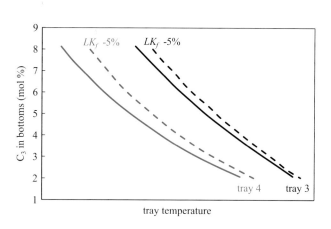

Figure 12.79   Effect of feed composition on lower tray temperature

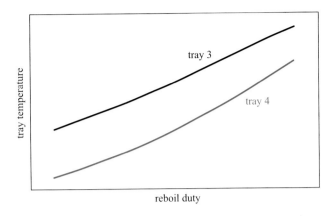

Figure 12.80 Linearity of lower temperature control

While both would work well, tray 3 shows greater linearity while tray 4 shows greater sensitivity.

Having configured the lower temperature controller we can complete a similar exercise for the upper controller. Figure 12.81 shows that changes in temperature, derived from the profiles in Figure 12.77, in response to changes in reflux. Here we can see that tray 18 shows the greatest sensitivity. An increase in reflux causes a change about 40 % greater than a decrease – well within our criterion for linearity. If it were not, then tray 19 shows greater linearity with slight loss of sensitivity but might be rejected because it is too close to the reflux tray above. Tray 17, although of similar sensitivity to tray 18, is well outside our linearity criterion with an increase in reflux causing a temperature change around 75 % larger than a decrease.

Figure 12.82 shows the effect that a change in feed composition would have on composition control. Tray 18 remains the preferred choice, since it is further from the feed tray. Fixing the temperature to initially hold the composition at 5 % would result in it changing by about 0.4 % when the feed composition changes.

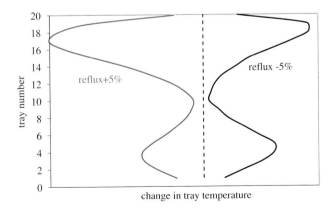

Figure 12.81 Change in temperature due to reflux

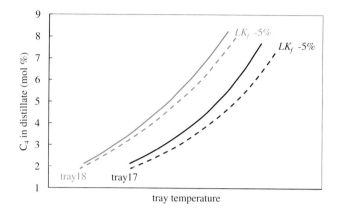

Figure 12.82 Effect of feed composition on upper tray temperature

Figure 12.83 shows that tray 18 is readily controllable by manipulating reflux.

It can be advantageous to use multiple tray temperature measurements in the same controller. There may not a measurement on the required tray but it may be possible to infer a value by interpolating between two or more other measurements. Or a temperature on one tray may become insensitive to composition changes under different operating conditions; incorporating a second measurement that correspondingly becomes more sensitive will maintain composition control.

If T_1 is the temperature measured on tray n_1, T_2 is the temperature measured on tray n_2 and T is the temperature on the required tray n, then the equation of the section of the temperature profile connecting them, assuming a straight line, is

$$\frac{T-T_1}{T_2-T_1} = \frac{n-n_1}{n_2-n_1} \tag{12.49}$$

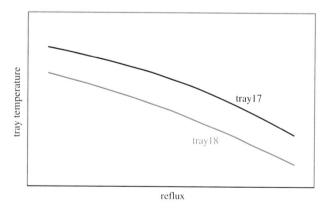

Figure 12.83 Linearity of upper temperature control

Rearranging

$$T = \frac{n_2 - n}{n_2 - n_1} T_1 + \frac{n - n_1}{n_2 - n_1} T_2 \qquad (12.50)$$

This is a weighted average of the two measured temperatures with weighting factors designed to give the temperature on a tray where the required measurement does not exist. If multiple measurements are used to maintain sensitivity then the weighting should be chosen to keep constant the process gain between composition and temperature.

Unusual dynamic behaviour may arise if the tray temperatures are far apart. Any resulting inverse response may require the controller to be so detuned as to counter the advantage of using multiple measurements.

Equation (12.49) can be rearranged

$$n = \frac{T_2 - T}{T_2 - T_1} n_1 + \frac{T - T_1}{T_2 - T_1} n_2 \qquad (12.51)$$

This is useful if the temperature profile is very steep, for example if temperature is insensitive to composition throughout the column. T then becomes the required temperature and n the tray on which this temperature target is currently being met. The value of n is used as the PV of a controller with the required tray as its SP.

At this stage we should emphasise that we are not yet in a position to put **both** temperatures on control. From our understanding of the process we know that the two controllers will interact – possibly to the point of becoming unstable. We will address this later. Until then we will proceed on the basis that only one of the controllers will be in service.

A very common configuration is to place a single temperature controller (TC) in the lower section of the column – manipulating reboiler duty. Figure 12.84 shows the effect of doing so with the energy balance scheme in place. During feed rate changes, with no TC, both product compositions will clearly vary. Figure 12.84 show the impact on the distillate composition if reflux (R) and reboil (V) are held constant (coloured line). It also shows that,

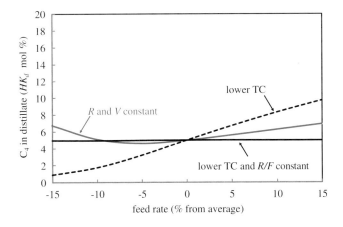

Figure 12.84 Impact of lower temperature control (energy balance scheme)

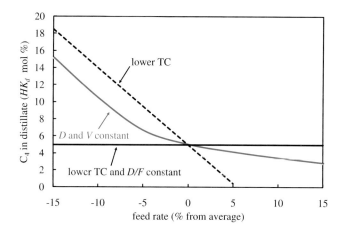

Figure 12.85 Impact of lower temperature control (material balance scheme)

if the TC is commissioned to manipulate the reboiler duty, while this will maintain the bottoms composition almost constant it does so at the cost of worsening the variability of the distillate composition. To compensate for this reflux must be adjusted. Without a second TC in place one possible solution is to maintain reflux in proportion to feed rate.

Figure 12.85 shows that making feed rate changes with the material balance scheme in place gives the same problem and that this can be resolved by keeping distillate flow in proportion to feed.

The conclusion is therefore that implementing temperature control in the lower section of the column causes the distillate composition to vary considerably more when column feed rate changes – no matter what level control strategy is adopted. This can be resolved by the implementation of ratio feedforward to the remaining unused MV.

Figures 12.86 and 12.87 show the same tests but this time with a TC commissioned in the upper section of the column. In the case of the energy balance scheme it manipulates reflux; if

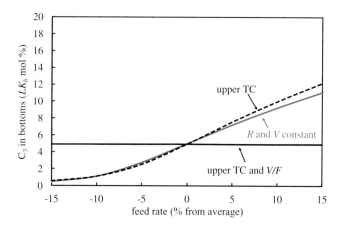

Figure 12.86 Impact of upper temperature control (energy balance scheme)

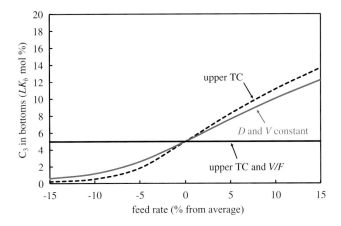

Figure 12.87 Impact of upper temperature control (material balance scheme)

the material balance scheme is in place it manipulates distillate flow. In both cases controlling the distillate composition has a much lesser effect on the bottoms composition. This demonstrates the importance of cut versus separation as a means of controlling composition. The TC in the upper section manipulates distillate flow, either directly or indirectly via reflux, and keeps D/F approximately constant. Not compensating fractionation has less of an impact. However, while the upper TC does not significantly worsen the control of bottoms composition, maintaining a constant V/F ratio is still very beneficial.

Manipulating cut should therefore the first choice if our aim is to control only one of the product compositions. While it is quite feasible to adopt this approach if the TC is close to the top of the column, a TC in the lower section is likely to respond slowly to changes in reflux and therefore may not control well during process disturbances. If the less preferred version of the material balance scheme is in place, a TC in the lower section might be thought to respond quickly since it would manipulate bottoms flow. However, this will not be the case because of the lag introduced by the reboiler responding to the change in column level.

Maintaining R/F, V/F, D/F (or possibly B/F) constant are feedforward strategies that we will address later in this chapter. While the Ryskamp scheme would appear to include some feedforward (in that it keeps R/D constant) it should be used with care. Figure 12.88 shows that the Ryskamp scheme commissioned without a TC in the lower section of the column gives poorer control of distillate composition than the energy balance scheme. With the TC in place it performs extremely well. In designing the scheme thought should be given to automatically disabling the R/D ratio controller if the TC is switched to manual (and re-commissioning it when the TC is returned to auto).

12.10 Pressure Compensated Temperature

We have dealt with many of the problems that can arise with tray temperature control by selecting the optimum trays. There remain some other issues that we need to address with

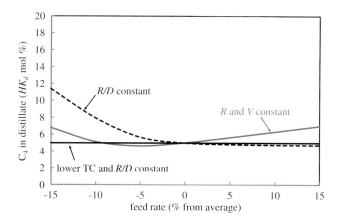

Figure 12.88 Impact of lower temperature control (Ryskamp scheme)

the controller design. The first of these is the effect of column pressure. The relationship between composition and bubble point depends on pressure, as shown in Figure 12.89. To maintain a composition of 5 % C_4 in distillate, when operating at normal (*reference*) pressure, the tray temperature would be controlled at T. If the column pressure is reduced by ΔP then the bubble point will reduce and the correlation between composition and bubble point changes. Maintaining the tray temperature constant will result in the C_4 in distillate increasing to around 6.7 %. To maintain a constant composition the tray temperature should be reduced by ΔT.

Figure 12.90 shows similar behaviour in the lower section of the column, where the reduction in pressure causes the C_3 in bottoms to fall to about 2.4 % – requiring a reduction in tray temperature larger than that required in the upper section.

The technique for dealing with this problem is known as *pressure compensated temperature* (*PCT*). This was covered briefly as an example of signal conditioning in Chapter 5. In its simplest form a linear correction is applied to the measured tray

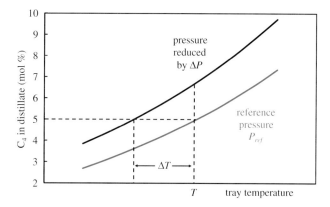

Figure 12.89 Effect of pressure on distillate composition

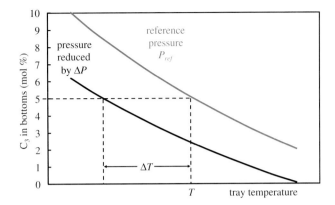

Figure 12.90 Effect of pressure on bottoms composition

temperature (*T*) based on the measured pressure (*P*).

$$PCT = T - \frac{dT}{dP}(P - P_{ref}) \qquad (12.52)$$

Taking our example of a reduction in pressure, the value of *PCT* will be higher than *T*. If we make *PCT* the PV of the tray temperature controller then it will compensate for the increase by reducing *T*. Provided we have quantified *dT/dP* correctly for each temperature controller (as $\Delta T/\Delta P$) then the product compositions will remain constant.

It is possible, although unusual, to apply the pressure correction to the SP of the tray temperature controller (T_{SP}). Some operators prefer this since it emulates what they would do when the pressure changes. Further the controller still displays the real tray temperature.

$$SP = T_{SP} + \frac{dT}{dP}(P - P_{ref}) \qquad (12.53)$$

We need to be cautious about the value that we use for *P*. Consider the situation where the column is operating at the condenser duty limit and the pressure controller is close to saturation. If we use the PV of the controller in the calculation of *PCT* then a transient increase in pressure will cause the lower tray temperature controller to increase the reboiler duty. This will overload the condenser more, causing a further increase in pressure and a further increase in reboiler duty. In these circumstances it would be better to use the SP of the column pressure controller in the calculation of *PCT*. This also has the advantage of being noise free. If the pressure controller is switched to manual, and PV tracking is implemented, the PCT calculation will automatically use the PV. If this is permitted then it may be necessary to filter the measurement and add some logic to prevent the compensation driving the unit towards the condenser limit. This situation can also arise if a column has multiple pressure measurements at different locations – one of which will be used for control but we may wish to use another for calculating *PCT*.

Theoretically the pressure transmitter should be at the same location in the column as the temperature transmitter. Fortunately on most columns the pressure difference, between the tray and the point at which column pressure is measured, is small compared to the operating pressure. Secondly we base PCT on the **change** in pressure which, provided the column pressure drop remains constant, will be the same throughout the column. We need to be cautious however on vacuum columns where changes in pressure drop can be of a similar order of magnitude to the column pressure. We will see later that the value of dT/dP in this type of column is extremely large and will so amplify any error in determining the change in pressure.

There are several ways in which dT/dP might be determined. The first assumes that sufficient good quality historical data exists so that an inferential property can be developed in the form

$$Q = a_0 + a_1 P + a_2 T \qquad (12.54)$$

Differentiating

$$dQ = a_1 dP + a_2 dT \qquad (12.55)$$

As P changes we want Q to remain constant, or dQ to be zero, and so

$$\frac{dT}{dP} = -\frac{a_1}{a_2} \qquad (12.56)$$

If historical data are not available then another approach is plant testing. This entails operating at different pressures and either allowing any composition controllers to take corrective action or manually adjusting conditions to bring the compositions back to target. Collecting data at several different pressures will allow tray temperature(s) to be plotted against pressure. The gradient of the line is then dT/dP.

There are several theoretical approaches to determining dT/dP. We have already used the Antoine Equation (12.1) to estimate relative volatility. Differentiating we get

$$\frac{dP}{P} = \frac{B.dT}{(T+C)^2} \qquad (12.57)$$

By substituting for T, again using the Antoine Equation, dT/dP can be defined as a function of either T and P or only P.

$$\frac{dT}{dP} = \frac{(T+C)^2}{B.P} = \frac{B}{P(A - \ln(P))^2} \qquad (12.58)$$

Note that, if the version of the Antoine Equation being used is based on $\log_{10}(P)$ rather than $\ln(P)$ then this can be replaced before differentiation.

$$\log_{10}(P) = \log_{10}(e) \times \ln(P) = \frac{\ln(P)}{\ln(10)} = 0.4343 \times \ln(P) \qquad (12.59)$$

Assuming a typical operating pressure of 12 barg (or an absolute pressure of 13.01325 bara or 12.84308 atm) for a LPG splitter, and the Antoine coefficients given in Table 12.2, we obtain the values for dT/dP shown in Table 12.6.

An alternative to the Antoine Equation is the Clausius-Clapeyron Equation which predicts dT/dP directly from the heat of vaporisation (H_v).

Table 12.6 Pressure compensation factors from Antoine Equation

Component		dT/dP	
		°C/bar	°F/atm
propene	C_3H_6	3.27	5.97
propane	C_3H_8	3.39	6.18
n-butane	C_4H_{10}	3.93	7.17

$$\frac{dT}{dP} = \frac{RT^2}{H_v.MW.P} \qquad (12.60)$$

R is the Universal Gas Constant and has a value of 8.314 kJ/kg-mole/K (1.9859 BTU/lb-mole/°F). Unlike the Antoine Equation the temperature T is on an **absolute** basis. H_v should have units of kJ/kg (or BTU/lb). The pressure (P), like Antoine, is on an absolute basis.

To apply Clausius-Clapeyron, we would normally use the actual tray temperature as the value for T. In the absence of this, its value is estimated by applying the Antoine Equation and converting to an absolute basis. Indeed this comparison between predicted and measured tray temperature can normally be used to confirm that the coefficients used in the Antoine Equation are correct. Table 12.7 shows that the values of dT/dP derived are very close to those generated from the Antoine Equation.

However the Antoine and Clausius-Clapeyron Equations are restricted to **pure** components. Antoine coefficients are published for pure components and the Clausius-Clapeyron Equation includes the heat of vaporisation – also only known for pure components. While we could assume that the estimates for dT/dP blend linearly with composition, we would need to know the composition on the tray on which the temperature is controlled. A simple approximation is to assume that dT/dP varies linearly up the column between the value for the HK component and that for the LK. So, for example, based on the Antoine results the value for tray 3 is given by

$$\frac{dT}{dP} = 3.93 - \frac{3}{20}(3.93 - 3.39) = 3.85 \qquad (12.61)$$

Table 12.7 Pressure compensation factors from Clausius-Clapeyron Equation

Component		T (from Antoine)		MW	H_v		dT/dP	
		K	°R		kJ/kg	BTU/lb	°C/bar	°F/atm
propene	C_3H_6	303.5	546.3	42.08	439.43	188.92	3.18	5.81
propane	C_3H_8	311.8	561.2	44.10	425.68	183.01	3.31	6.03
n-butane	C_4H_{10}	366.3	659.4	58.12	385.95	165.93	3.82	6.97

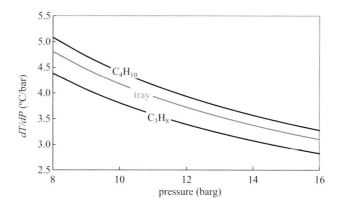

Figure 12.91 Variation of pressure compensation factor with pressure (Antoine)

and that for tray 18 is given by

$$\frac{dT}{dP} = 3.93 - \frac{18}{20}(3.93-3.39) = 3.44 \qquad (12.62)$$

While clearly a very simplistic approach, since dT/dP varies relatively little between the pure key components, the error introduced by the approximation is small. However the underlying assumption that dT/dP is constant is likely to introduce a much larger error. Equations (12.58) and (12.60) both show that dT/dP varies with pressure. Figure 12.91 shows, for values derived from the Antoine Equation, what error is potentially introduced.

On most columns it would not be possible for the pressure to be varied over a wide range before reaching equipment limits. For example, on the case study column, a 1 bar variation in column pressure would change the true value of dT/dP by about 0.2 °C/bar (around 0.4 °F/atm) – introducing an error of about 5 % into the temperature correction. Under these circumstances a more rigorous approach is probably not justified. However there are columns which have multiple operating modes between which pressure may change greatly.

The Antoine and Clausius-Clapeyron Equations can be adapted to determine PCT directly. Rearranging the Antoine Equation (12.1)

$$T = \frac{B}{A - \ln(P)} - C \qquad (12.63)$$

Writing it for reference conditions,

$$PCT = \frac{B}{A - \ln(P_{ref})} - C \qquad (12.64)$$

Combining Equations (12.63) and (12.64),

$$PCT = T - B\left[\frac{1}{A - \ln(P)} - \frac{1}{A - \ln(P_{ref})}\right] \qquad (12.65)$$

Rearranging the Clausius-Clapeyron Equation (12.60),

$$\frac{dT}{T^2} = \frac{R}{H_v.MW} \times \frac{dP}{P} \quad (12.66)$$

Integrating

$$\left[\frac{-1}{T}\right]_T^{PCT} = \left[\frac{R}{H_v.MW}\ln(P)\right]_P^{P_{ref}} \quad (12.67)$$

Remembering that both T and PCT are **absolute** temperatures,

$$PCT = \frac{1}{\frac{R}{H_v.MW}\ln\left[\frac{P}{P_{ref}}\right] + \frac{1}{T}} \quad (12.68)$$

while addressing the issue of the variability of dT/dP with pressure, Equations (12.65) and (12.68) still assume pure components. To apply them we have to make two assumptions. The first is to make some estimate of the composition of the liquid on the tray. The second is that, for the Antoine approach, the coefficients A and B can be derived by taking a weighted average of the values for the LK and HK components. With the Clausius-Clapeyron approach the assumption that H_v and MW can be derived in this way is perhaps more correct.

A method which needs make no assumption about composition is based on charts developed by Maxwell and Bonnell (Reference 2). The method is published in imperial units and so pressure (P) must be converted to an absolute basis in units of mm Hg and the temperature (T) to an absolute basis in °R. The method will at first appear complex. However the end result is effective and simple to implement. The calculation can readily be implemented in a spreadsheet. This helps avoid the almost inevitable errors that will arise from using a calculator and provides a method that can quickly be applied to any column.

Values for the coefficients A, B, C and D are selected from Table 12.8.

The values are used to calculate χ.

$$\chi = \frac{A - B\log_{10}(P)}{C - D\log_{10}(P)} \quad (12.69)$$

Differentiating

$$\frac{d\chi}{dP} = \frac{D(A - B\log_{10}(P)) - B(C - D\log_{10}(P))}{P(C - D\log_{10}(P))^2 \ln(10)} \quad (12.70)$$

Table 12.8 Coefficients for Maxwell and Bonnell calculations

Pressure	A	B	C	D
$P < 2$	6.761560	0.987672	3000.538	43.00
$2 < P < 760$	5.994296	0.972546	2663.129	95.76
$P > 760$	6.412631	0.989679	2770.085	36.00

The normal boiling point of the liquid is calculated from

$$T'_b = \frac{748.1\chi}{\frac{1}{T} - 0.0002867 + 0.2145\chi} \qquad (12.71)$$

Differentiating

$$\frac{d\chi}{dT} = \frac{-T'_b}{(748.1 - 0.2145T'_b)T^2} \qquad (12.72)$$

Combining Equations (12.70) and (12.72)

$$\frac{dT}{dP} = \frac{\left(\frac{d\chi}{dP}\right)}{\left(\frac{d\chi}{dT}\right)} \qquad (12.73)$$

The result for dT/dP will have units °R/mm Hg and should be converted as required. While the method is arithmetically complex it only requires the typical operating pressure (P_{ref}) and the normal tray temperature. The method can be extended to accommodate the variation in dT/dP with pressure. Figure 12.92 shows this variation. To incorporate this we calculate dT/dP at the normal tray temperature and a number of different pressures within the expected operating range (shows as individual points in Figure 12.92).

The values for dT/dP are plotted against the **reciprocal** of the **absolute** pressure, as shown in Figure 12.93.

The resulting points will lie close to a straight line. The gradient (m) and it intercept (c) on the vertical axis can be used to predict dT/dP from the operating pressure (P) in absolute units.

$$\frac{dT}{dP} = \frac{m}{P} + c \qquad (12.74)$$

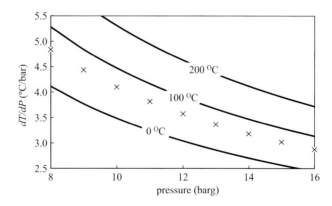

Figure 12.92 Variation of pressure compensation factor with pressure (Maxwell)

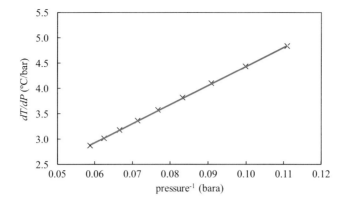

Figure 12.93 Adjusting pressure compensation factor for pressure

Integrating

$$[T]_T^{PCT} = [m.\ln(P) + cP]_P^{P_{ref}} \qquad (12.75)$$

Thus

$$PCT = T - c(P - P_{ref}) - m.\ln\left[\frac{P}{P_{ref}}\right] \qquad (12.76)$$

Those checking the reference from where this technique was developed will find mention that it should only be applied to pure hydrocarbons and narrow-boiling range petroleum fractions. However this restriction applies primarily to the conversion between the true normal boiling point (T_b) and the boiling point corrected to a *Watson K* of 12 (T'_b). The procedure for calculating *PCT* does not involve this conversion and experience shows that the resulting formula works well.

While not of importance here, for background information, the Watson K factor is used in the oil industry as means of characterising the paraffinicity of a mixture of hydrocarbons and is derived from the *molar average boiling point (MABP)* (in °R) and the specific gravity at 60 °F (*SG*).

$$K = \frac{\sqrt[3]{MABP}}{SG} \qquad (12.77)$$

If the true normal boiling point is required, for example to validate the method by comparison with a known boiling point, then the procedure is as follows. If *P* is less than 760 mm Hg or the normal boiling point is above 400 °F then

$$T_b = T'_b + 2.5(K-12)\log_{10}\left(\frac{P}{760}\right) \qquad (12.78)$$

If *P* is greater than 760 mm Hg and the normal boiling point is below 200 °F then

$$T_b = T'_b \qquad (12.79)$$

If P is greater than 760 mm Hg and the normal boiling point is between 200 °F and 400 °F, then

$$T_b = \frac{T_b' - 8.24625(K-12)\log_{10}\left(\frac{P}{760}\right)}{1 - 0.0125(K-12)\log_{10}\left(\frac{P}{760}\right)} \qquad (12.80)$$

The one remaining limitation of the method based on Maxwell-Bonnell, as in those based on Antoine and Clausius-Clapeyron, is that it only accounts for the change in bubble point as pressure changes. None of the theoretical methods takes account of the change in relative volatility. Thus, on increasing column pressure, there will be a decrease in separation and the purity of the products will worsen. This effect is secondary to the change in bubble point and will result in only a minor change but, if accuracy is required, then *dT/dP* should be determined empirically, as described by Equation (12.56), or a more rigorous definition of *PCT* be developed from simulation.

The value of *dT/dP* increases as the bubble point of the product increases. We have seen that for LPG, it is typically 3 to 4 °C/bar. For a vacuum column operating at an absolute pressure of 30 mm Hg, with a tray temperature of 400 °C, *dT/dP* will be in excess of 800 °C/bar. Under these circumstances it is important that pressure changes are measured accurately. The pressure transmitter should be therefore located close to the temperature measurement so that changes in pressure drop across the column have no effect. The impulse line should be self-draining back to the column to avoid introducing noise into the pressure measurement due to liquid boiling in the line.

There are columns which do not lend themselves to tray temperature control, for example because temperature is insensitive to composition. Under these circumstances pressure compensation may be applied directly to what would otherwise be the MV of the temperature controller. The pressure compensation factor can be determined empirically. The approach is similar to that described for quantifying *dT/dP*. It is based on the assumption that an inferential can be developed based on pressure (*P*), the manipulated flow (*F*) and other independent variables. The manipulated flow may be reboiler duty, reflux, distillate or bottoms – depending on the choice of level control strategy.

$$Q = a_0 + a_1 P + a_2 F + \ldots \qquad (12.81)$$

Differentiating

$$dQ = a_1 dP + a_2 dF \qquad (12.82)$$

For no change in composition as pressure is changed,

$$\frac{dF}{dP} = -\frac{a_1}{a_2} \qquad (12.83)$$

Alternatively a linear (or nonlinear) function can be developed from process simulation.

12.11 Inferentials

While there are good reasons for locating tray temperature controllers away from the point at which the products are withdrawn, this does have a disadvantage. Because fractionation takes place between the temperature controller and the product then any change in the liquid and vapour traffic in this section of the column will change the correlation between product composition and tray temperature.

For example, if the temperature is controlled on tray 3, the bottoms composition will vary slightly as reflux is changed. This is illustrated in Figure 12.94. With no temperature control in place, reflux has a huge effect on bottoms composition. With tray 3 under temperature control the bottoms composition is kept close to target but not exactly so. The controller is effectively keeping constant the composition of the liquid on tray 3. As reflux is increased more separation takes places between tray 3 and the base of the column, so improving the product purity.

By considering what else is taking place in the column, it is possible to develop an inferential which takes account of this effect. Figure 12.95 shows how the tray 3 temperature controller increases the reboiler duty to compensate for the cooling effect of the additional reflux.

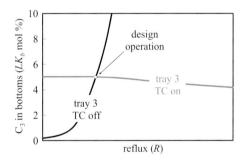

Figure 12.94 Effect of reflux on bottoms quality

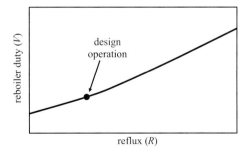

Figure 12.95 Effect of reflux on reboiler duty

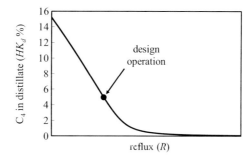

Figure 12.96   Effect of reflux on distillate composition

Figure 12.96 shows the effect on distillate composition. With tray 3 under temperature control the bottoms purity remains approximately constant. The additional reflux and reboiler duty increase separation by improving the purity of the distillate.

At the high reflux/reboil operation virtually all the C_4 is recovered from the distillate. As a result the bottoms yield increases as shown in Figure 12.97.

Figure 12.98 uses the data collected during this test to develop an inferential property calculation for bottoms composition. Despite the highly nonlinear behaviour displayed in the previous figures, a strong linear relationship exists between the C_3 content of bottoms and the reciprocal reboil ratio.

This relationship can be expressed as

$$LK_b = a_0 + a_1 \frac{B}{V} \tag{12.84}$$

V can be any measure of reboiler duty. It might be the steam flow to a condensing reboiler. If heat integrated it might be the flow of heating fluid multiplied by its temperature change. For a fired reboiler it might be the fuel flow.

There are a number of ways in which this inferential may be applied. At this stage, replacing the temperature controller with a virtual composition controller is not an option since the inferential is developed on the assumption that the tray temperature is held constant.

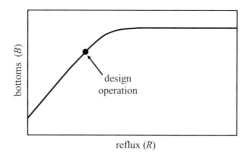

Figure 12.97   Effect of reflux on bottoms

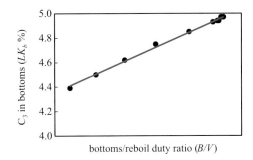

Figure 12.98 Inferential for bottoms composition

It could be configured as a virtual analyser controller cascaded to the temperature controller. The approach needs some care since the input to the inferential is effectively the output of the temperature controller. It is only because we use the actual reboiler duty rather than its SP that introduces a lag that makes the cascade stable.

Another approach would be to incorporate a correction term into the temperature measurement in much the same way as linear pressure compensation.

$$PV = T - \frac{dT}{dP}(P - P_{ref}) - K\left[\left(\frac{B}{V}\right) - \left(\frac{B}{V}\right)_{ref}\right] \quad (12.85)$$

K could be derived from historical process data, in much the same way as dT/dP was from Equation (12.56), or from simulation.

Another option is to include the reciprocal reboil ratio with other variables, such as column pressure and tray temperature, in regression analysis to develop an inferential of the form

$$LK_b = a_0 + a_1 P + a_2 T + a_3 \left(\frac{B}{V}\right) \quad (12.86)$$

This effectively incorporates *PCT* into the inferential. However it might be preferable to retain this in the tray temperature controller – for example to permit its use if the inferential is out of service. Arithmetically, if a **linear** *PCT* is used, the two approaches are identical. From Equations (12.52) and (12.56)

$$PCT = T + \frac{a_1}{a_2}(P - P_{ref}) \quad (12.87)$$

The inferential would then comprise

$$LK_b = b_0 + b_1 PCT + b_2 \left(\frac{B}{V}\right) \quad (12.88)$$

Substituting for *PCT*

$$LK_b = b_0 + b_1 \left[T + \frac{a_1}{a_2}(P - P_{ref})\right] + b_2 \left(\frac{B}{V}\right) \quad (12.89)$$

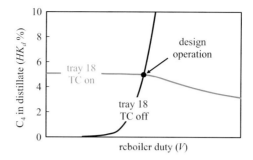

Figure 12.99 Effect of reboiler duty on distillate quality

Equating coefficients with Equation (12.86)

$$b_0 = a_0 + a_1 P_{ref} \tag{12.90}$$

$$b_1 = a_2 \tag{12.91}$$

$$b_2 = a_3 \tag{12.92}$$

If however a nonlinear *PCT* is required it may be easier to understand if retained. The inferential in Equation (12.88) would then be developed by regression.

A very similar approach can be applied to the distillate composition. If the temperature is controlled on tray 18, the composition will change slightly as reboiler duty is changed. This is illustrated in Figure 12.99. With no temperature control in place, reboiler duty has a huge effect on distillate composition. With tray 18 under temperature control the distillate composition is kept close to target but not exactly so. The controller is effectively keeping constant the composition of the liquid on tray 18. As reboil duty is increased more separation takes places between tray 18 and the top of the column, so improving the product purity.

Again it is possible to develop an inferential which takes account of this effect. Figure 12.100 shows how the tray 18 temperature controller increases the reflux to compensate for the heating effect of the additional reboiler duty.

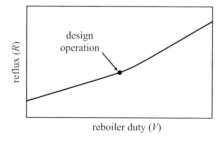

Figure 12.100 Effect of reboiler duty on reflux

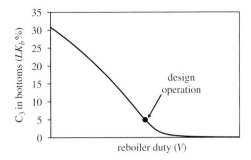

Figure 12.101 Effect of reboiler duty on bottoms composition

Figure 12.101 shows the effect on bottoms composition. With tray 18 under temperature control the distillate purity remains approximately constant. The additional reboil duty and reflux increase separation by improving the purity of the bottoms.

At the high reboil/reflux operation virtually all the C_3 is recovered from the bottoms. As a result the distillate yield increases as shown in Figure 12.102.

Figure 12.103 uses the data collected during this test to develop an inferential property calculation for distillate composition. Again a strong linear relationship exists, this time between the C_4 content of distillate and the reciprocal reflux ratio.

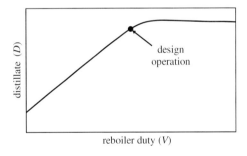

Figure 12.102 Effect of reboiler duty on distillate

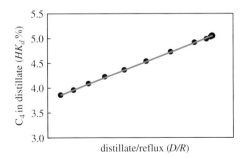

Figure 12.103 Inferential for bottoms composition

This relationship can be expressed as

$$HK_d = a_0 + a_1 \frac{D}{R} \tag{12.93}$$

The options for applying this correlation are equivalent to those for the bottoms composition controller. With care, it could be set up as composition controller cascaded to the temperature controller. Or a correction term could be incorporated into the temperature measurement.

$$PV = T - \frac{dT}{dP}(P - P_{ref}) - K\left[\left(\frac{D}{R}\right) - \left(\frac{D}{R}\right)_{ref}\right] \tag{12.94}$$

Or the reciprocal reflux ratio could be included with other variables, such as column pressure and tray temperature, in regression analysis to develop an inferential of the form

$$HK_d = a_0 + a_1 P + a_2 T + a_3 \left(\frac{D}{R}\right) \tag{12.95}$$

Or the *PCT* could be retained and the distillate inferential could be based on this.

$$HK_d = b_0 + b_1 PCT + b_2 \left(\frac{D}{R}\right) \tag{12.96}$$

If a nonlinear *PCT* is required, for example because of large variations in pressure, but for some reason cannot be developed from Antoine, Clausius-Clapeyron or Maxwell-Bonnell then regression may be an alternative. One approach would be to retain the *PCT* equation suggested by each method and apply a least squares technique to adjust the coefficients to obtain the best predicted composition. Another approach would be to steer a more general regression approach to consider nonlinear functions. For example all three methods indicate that ln(*P*) might be considered as an input.

As described in Chapter 9, care must be taken to include in the regression analysis only those process parameters that make engineering sense. But the form that those parameters take is not always immediately obvious. Figure 12.104 shows the true relationship between product composition and tray temperature at three operating pressures. The inclusion of only *P* and *T* would imply that the relationship is a series of parallel straight lines as shown (dashed).

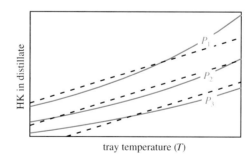

Figure 12.104 Predicting composition from pressure and tray temperature (homogenic)

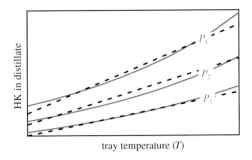

Figure 12.105 Predicting composition from pressure and tray temperature (heterogenic)

The influence that pressure and temperature have on composition is described by a *heterogenic* rather than a *homogenic* function. Terms derived from independent variables must also be included – such as cross-products, ratios, etc. For example, the additional inclusion of $P.T$ as an independent variable results in a closer match to the nonparallel lines, as shown in Figure 12.105.

Alternatively, the use of a nonlinear function, for example $\log_{10}(HK_d)$, reduces the error by fitting curves rather than straight lines as shown in Figure 12.106.

Examination of the inferentials developed shows that they each include a measure of cut (the *PCT*) and fractionation (reflux or reboil ratio). While *PCT* is an effective measure of cut, there are a number of ways of including fractionation. For example Equation (12.97) shows part of an inferential developed using two tray temperature measurements where T_1 is higher up the column than T_2.

$$HK_d = a_0 + a_1 T_1 - a_2 T_2 \ldots \tag{12.97}$$

The coefficients a_1 and a_2 are positive numbers. Since we expect HK_d to increase with tray temperature we might be tempted to exclude T_2 and repeat the regression. However this equation can be rewritten as

$$HK_d = a_0 + (a_1 - a_2)T_1 - a_2(T_2 - T_1) \ldots \tag{12.98}$$

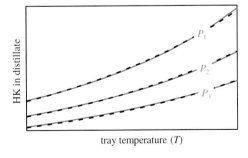

Figure 12.106 Predicting composition from pressure and tray temperature (nonlinear)

Provided $(a_1 - a_2)$ is positive, the correlation makes sense by including a measure of cut (T_1) with a coefficient of the correct sign. The term $(T_2 - T_1)$ is a measure of the slope of the temperature profile and so of separation. Increasing separation will reduce HK_d and so the sign of a_2 is also correct.

If pressure compensation is to be included it should not be necessary to apply it to both T_1 and T_2 since both values will be compensated by approximately the same amount. In Equation (12.98) T_1 could be replaced by its equivalent *PCT*. The inferential may exhibit unusual dynamic behaviour if the two trays are far apart. Steptesting will reveal if this is a problem. If so, it may be necessary to include an alternative measure of fractionation or omit it completely.

On high purity columns it is usually better to develop regressed inferentials for $\log(HK_d)$ and $\log(LK_b)$. This helps accommodate the nonlinear relationships common on these columns but also has the advantage that the predicted values for HK_d and LK_b will never be negative.

A function (f) based on the bubble points of the light and heavy key components (T_{LK} and T_{HK}) would also be worth considering if other approaches fail. The bubble points (at normal operating pressure) are derived from the Antoine Equation. For the HK_d inferential T is the temperature on the tray selected in the upper section of the column.

$$HK_d = f\left(\frac{T - T_{LK}}{T_{HK} - T}\right) \qquad (12.99)$$

For the LK_b inferential T is the temperature on the tray selected in the lower section of the column.

$$LK_b = f\left(\frac{T_{HK} - T}{T - T_{LK}}\right) \qquad (12.100)$$

In addition to regressed inferentials there are a range of commercially available first-principle models. These use conventional heat and mass balances plus published correlations. Since much of this technology is proprietary, the content here is restricted to a summary of the published correlations.

12.12 First-Principle Inferentials

The Fenske Equation has already been covered as Equations (12.30) and (12.33). Knowing the relative volatility of the components and the target separation, the minimum number of theoretical stages (N_{min}) can be calculated.

Underwood's Method comprises two equations. The first includes all n components in the feed, where α is the volatility with respect to the least volatile component and x_f the mole fraction. Knowing the feed quality (q), the value of ϕ can be calculated.

$$1 - q = \sum_{i=1}^{n} \frac{\alpha_i (x_f)_i}{\alpha_i - \phi} \qquad (12.101)$$

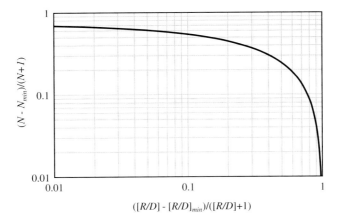

Figure 12.107 Gilliland's correlation

This value is then used to determine the minimum reflux ratio $(R/D)_{min}$. The equation includes the m components in the distillate, where x_d is the mole fraction.

$$\left(\frac{R}{D}\right)_{min} + 1 = \sum_{i=1}^{m} \frac{\alpha_i (x_d)_i}{\alpha_i - \phi} \tag{12.102}$$

Gilliland's Correlation can then be used with the value of N_{min} from Fenske and $(R/D)_{min}$ from Underwood to determine the actual number of stages (N) from the actual reflux ratio (R/D).

$$\frac{N - N_{min}}{N + 1} = f\left(\frac{\left(\frac{R}{D}\right) - \left(\frac{R}{D}\right)_{min}}{\left(\frac{R}{D}\right) + 1}\right) \tag{12.103}$$

The correlation was published as a chart, as shown in Figure 12.107.

There are a range of equations that have been devised to describe Gilliland's correlation. The simplest of these is the *Eduljee Equation*.

$$\frac{N - N_{min}}{N + 1} = 0.75 \left[1 - \left(\frac{\left(\frac{R}{D}\right) - \left(\frac{R}{D}\right)_{min}}{\left(\frac{R}{D}\right) + 1}\right)^{0.5668}\right] \tag{12.104}$$

Other equations include those developed by Chung and by Molkanov.

Once N is known it can be used in a number of other correlations. For example the *Jafarey, Douglas and McAvoy Correlation* was developed by simplifying *Smoker's Equation* (Reference 4)

$$\log(S) = N \log\left[\alpha \sqrt{1 - \frac{R + qD}{(R + D)(RLK_f + qD)}}\right] \tag{12.105}$$

Alternatively the *Colburn Equation* (Reference 5) gives the mole fraction of a chosen component in the vapour leaving the nth tray (y_n) based on the mole fraction of the component in the bottoms (x_b).

$$\frac{y_n}{x_b} = \frac{(U^n-1)(\alpha-1)}{(U-1)} + 1 \qquad (12.106)$$

U is derived from the vapour-to-liquid molar flow ratio on the tray. For example, in a two product column, above the feed tray

$$U = \alpha \frac{R+D}{R} \qquad (12.107)$$

Below the feed tray

$$U = \alpha \frac{V}{V+B} \qquad (12.108)$$

A refinement is to add to $R + D$ the rate of change of reflux drum inventory. This helps overcome the problem caused by the process not yet reaching steady state. This is particularly useful if averaging level control has been installed on the drum in order to minimise flow disturbances to the downstream process. Similarly the rate of change of the liquid inventory in the base of the column can be added to $V + B$.

The number of trays (N) can also be determined empirically by adjusting it to fit sets of known conditions obtained from plant test runs. N need not be an integer and so can incorporate tray efficiency.

12.13 Feedforward on Feed Rate

Change in feed rate is a common source of process disturbance. We have seen that commissioning a single temperature controller (in the lower section of the column) manipulating fractionation may control bottoms composition well but makes worse control of distillate composition. The same effect applies to feedforward controllers. Figures 12.108 and 12.109 show the effect, with the energy scheme in place, of maintaining one of the MVs in proportion to feed rate, but not the other. While keeping both in proportion to feed rate will result in virtually no change to product compositions, feeding forward to only one causes both product compositions to vary by substantially more than they would with no feedforward in place!

Figures 12.110 and 12.111 show the same test with the material balance scheme in place. Again the distillate quality would be better controlled with no feedforward, rather than with just one of the ratios kept constant. While keeping either ratio constant marginally improves control of bottoms quality, keeping both constant will result in virtually no change.

While these figures show that both MVs need to change with feed rate this does not necessarily imply that **feedforward** control should be configured. It would be quite common for one variable to be under tray temperature control and the other ratioed to

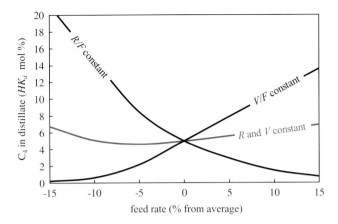

Figure 12.108 Effect of feed rate feedforward on distillate (energy balance scheme)

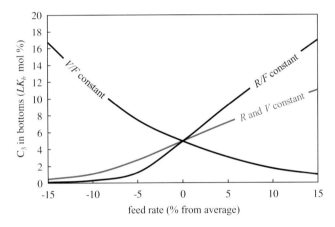

Figure 12.109 Effect of feed rate feedforward on bottoms (energy balance scheme)

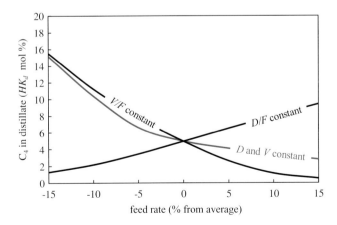

Figure 12.110 Effect of feed rate feedforward on distillate (material balance scheme)

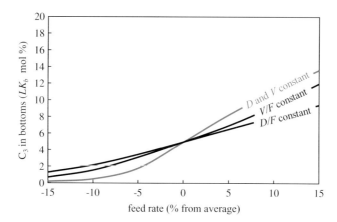

Figure 12.111 Effect of feed rate feedforward on bottoms (material balance scheme)

feed rate. Indeed this is similar to the configuration suggested by Ryskamp. However, ratio feedforward offers three other potential benefits. Firstly it gives a dynamic advantage. If tray temperatures respond slowly to changes in feed rate, then feedforward would permit tighter control of temperature. Secondly it removes the need to retune the tray temperature controllers to compensate for the change in process gains caused by the change in feed rate. Finally, if tray temperature control is not practical or temporarily out of service, then feedforward would keep tray temperatures approximately constant.

Figure 12.112 shows the application of full feed rate feedforward on a column with the energy balance scheme. Either or both of the reflux-to-feed and the steam-to-feed ratios can remain with operator entered SPs or their SP can be adjusted by a higher level control (such as tray temperature, inferential or on-stream analyser).

Dynamic compensation is likely to be necessary to ensure that the reflux and steam flows are adjusted at the right time. The method for tuning these deadtime/lead-lag algorithms is described in Chapter 6. Part of this procedure involves steptesting the DV, in this case feed rate, to obtain the dynamic response of the PV, in this case tray temperature. This can present a problem on some columns.

Figure 12.113 shows the variation of some selected tray temperatures, in our 20 tray column, as feed rate is varied. The gradient of the appropriate line is the process gain between the PV and the DV, i.e. $(K_p)_d$. So, if we had chosen tray 18 for temperature control, we might experience difficulty obtaining the process dynamics – since the gradient of this line changes sign. Since the value of $(K_p)_d$ is used in none of the tuning calculations in **ratio** feedforward this will not cause a problem with controller performance. However, if steptesting is performed in the region where the gain changes sign then model identification is likely to fail. While we are not concerned about the value of $(K_p)_d$, we do need values for θ_d and τ_d. It may be necessary to perform separate steptests, each staying within a region where the gradient is approximately constant, and then average the values obtained for deadtime and lag.

In the same way that we showed that two composition controllers interact, so will feedforward control. While it is correct at steady state to maintain a constant ratio between MVs and feed rate it is likely to be the case that the two MVs should not be changed at the

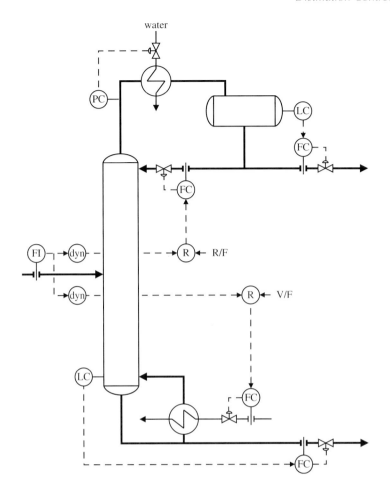

Figure 12.112 Feed rate feedforward (energy balance scheme)

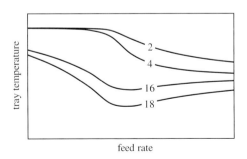

Figure 12.113 Variation of tray temperatures with feed rate (energy balance scheme)

348   Process Control

same time. Indeed our tuning of the dynamic compensation is designed to keep the product compositions constant. If, because of testing errors or the process being far from first order, we see transient disturbances in one of the compositions then it will be tempting to adjust the dynamic compensation by trial-and-error. However if we alter the timing of changes to, for example, reflux this will affect both product compositions – making it appear that the other feedforward compensation needs retuning. It is therefore unlikely that we would ever achieve perfect compensation and retuning should only be attempted if performance is particularly poor.

12.14 Feed Composition Feedforward

Feedforward on feed composition can be a valuable enhancement but may not be practical. Firstly it requires an on-stream analyser on feed. Few plant owners would install this as standard and there may not be sufficient economic justification to add it later. Secondly it may not be possible to acquire an analyser that responds quickly enough. If the change in feed composition affects tray temperatures and/or inferentials before being reported by the analyser then the feedback controller(s) will take corrective action. A delayed measurement of feed composition would then be less valuable than having no measurement.

However, if the feed is produced by an upstream unit and routed directly to the column (i.e. not via storage), it may be possible to develop an inferential based on the operating conditions in that unit.

Another difficulty may be in the determination of the feedforward gain (K) that should be used. Unlike feed rate feedforward, feed composition feedforward requires a bias not a ratio algorithm and so K is not 1 (see Chapter 6 for explanation). Figure 12.114 shows how each of the possible MVs should be adjusted as feed composition changes. The shape of these lines was explained earlier in the chapter (see Figure 12.68). K is the gradient of the line.

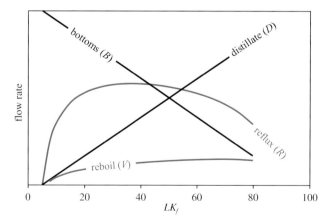

Figure 12.114   Effect of feed composition

If we have either version of the material balance scheme then feedforward to distillate flow or bottoms flow is straightforward. Since these are linearly related to LK_f. Differentiating Equation (12.25)

$$\frac{dD}{d(LK_f)} = \frac{F}{LK_d - LK_b} \quad (12.109)$$

Similarly

$$\frac{dB}{d(LK_f)} = \frac{-F}{LK_d - LK_b} \quad (12.110)$$

K is a function of feed rate (F). If feed rate is not constant this can be dealt with by adaptive tuning or feeding forward to the D/F (or B/F) ratio implemented for feed rate feedforward. The values of LK_d and LK_b are the product specifications.

However feeding forward to either distillate or bottoms flow, although beneficial, does not adjust fractionation – in this case the reboiler duty. As Figure 12.114 shows, the gradient of the reboiler duty line changes sign. If the average feed composition is close to this point then it is unlikely that feedforward would be effective. Indeed a better approach might to assume that the line is horizontal and therefore exclude this part of the feedforward strategy.

The situation is more complex if the energy balance scheme is installed. The sign of the gradient reverses for both MVs – reboil and reflux. And the reversal does not occur at the same feed composition. Care needs to be taken therefore that, for the range of possible feed compositions, it is reasonable to assume a linear relationship.

12.15 Feed Enthalpy Feedforward

Feed enthalpy feedforward is, rather fortunately, not often of great benefit. Changes in enthalpy are usually small compared to reboiler duty. If feedforward is justified then measuring enthalpy may present a problem. Provided the feed is below its bubble point or above its dew point then, provided its composition and pressure are reasonably constant, it is sufficient to use temperature as the DV. However if the feed is partially vaporised a measurement of wetness is not possible. Since change in wetness represents a large change in enthalpy it is probably not realistic to make any assumption about its value.

If we can successfully measure enthalpy we next need to determine what action to take when it changes. We could simply maintain the heat balance by adjusting the reboiler duty to compensate for the change. For this we need to be able to measure reboiler duty in consistent units. Further since energy is entering the column at a different point, the liquid and vapour traffic in part of the column will change. So maintaining the heat balance is not sufficient to maintain product compositions.

Alternatively, we could attempt to obtain the feedforward gains (K) empirically by plant testing, providing that we can introduce a disturbance into feed enthalpy. We may be able to determine K from analysis of historical data but if these were collected while tray temperature (or some other composition) control was in service then it will only be possible to model steady state behaviour. Similarly we could identify K from steady state simulation. Dynamic compensation would then have to be tuned by trial and error.

12.16 Decoupling

So far we have considered schemes that control either the distillate composition or the bottoms composition but not both simultaneously. On most columns the two controllers would interact to the point of instability. There are a number of techniques available which help alleviate this problem.

We have seen that the Ryskamp scheme largely breaks the interaction in one direction so that corrections made to the bottoms composition have little impact on the distillate. Although the converse is not true; an adjustment to the reflux ratio will affect the bottoms composition, but when its controller takes corrective action it will not 'fight' the distillate composition controller.

Another similar approach is to ratio, to feed rate, the MV not being used for control of composition. While not providing feedback control it does reduce variation in the uncontrolled composition and permits operator adjustment of the feedforward ratio target if necessary. If control of this composition is not important then some deviation might be a worthwhile price to pay to retain simplicity.

If dual composition control is required then relative gain analysis, as described in Chapter 8, will help assess the level of interaction. While not an entirely accurate tool, because it only considers steady state interactions, it is indicative of the severity of the problem. Since we need to perform steptests to tune the composition controllers the additional effort involved is minor.

If one composition is considerably less important than the other, then it may possible to commission both controllers as normal PID controllers. The principle is to detune the less important controller, so that while it acts very slowly, the other controller can take corrective action to deal with any interaction. The more important controller should manipulate cut, as the dominant MV, or it may be impossible for it to reach its SP until the less important controller has set the cut to a feasible value. This may be in conflict with dynamic considerations and result in bottoms quality being controlled by a MV at the top of the column and distillate quality being set by a MV at the bottom. The technique is to steptest the more important controller as normal but put this into automatic mode before steptesting the other. When tuning the less important controller we would typically reduce the controller gain to about 25 % of what would be used if there was no interaction. However it may be necessary to reduce it further.

The $\Sigma T/\Delta T$ method is occasionally referred to texts on the subject of distillation control. It should only be considered on columns with the material balance control strategy. With this strategy in place a change in the distillate flow causes the column temperature profile to shift horizontally (see Figure 12.74). Because the profile largely maintains is shape there is no change in the difference between two tray temperatures. The sum, however, will change. A change in reboiler duty however causes the profile to rotate (see Figure 12.75), thus the temperature in the upper section of the column will change in the opposite direction to one in the lower section. Thus the sum of the two temperatures will remain approximately constant. The difference however will change. In principle therefore ΣT responds only to changes in cut – i.e. distillate flow (or bottoms flow if the less preferred material balance scheme is in place). And ΔT responds only to changes in fractionation – i.e. reboil duty (or reflux).

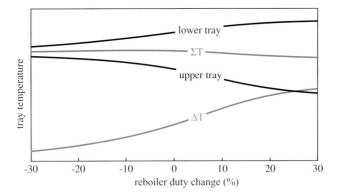

Figure 12.115 Response of ΣT and ΔT to reboiler duty

Figure 12.115 shows how well the scheme would perform on our case study column. To keep the measurement in roughly the same range, ΣT is the average (half the sum) of the two temperatures. The principle appears to work well with very little change in ΣT over a wide range of reboiler duty and ΔT varying reasonably linearly. This would suggest that ΔT could be well controlled by manipulating reboiler duty.

Figure 12.116 shows the effect of varying distillate flow. Here the principle is not working well, with ΣT and ΔT both changing by similar amounts. Thus controlling ΣT by manipulating distillate flow will cause changes to ΔT. However the tray temperatures were chosen to perform well as standalone controllers with the energy balance scheme in place. Selection based on the revised requirement would improve the decoupling. Performance might be further improved by the inclusion of coefficients such that ΣT is calculated as $(a_1 T_1 + a_2 T_2)$ and ΔT as $(b_1 T_1 - b_2 T_2)$. Nevertheless, the scheme as it stands has decoupled the controllers in one direction, which is sufficient to prevent the controllers fighting to the point of instability.

However the decoupling takes no account of process dynamics. With the trays far apart one temperature will change at a different time to the other. Thus, when reboiler duty is

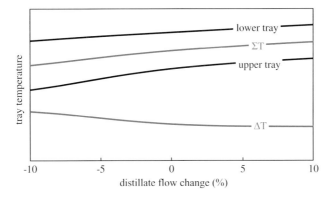

Figure 12.116 Response of ΣT and ΔT to distillate flow

changed, there will be a transient disturbance to ΣT. This will cause the controller to take corrective action and may trigger instability. It is not practical to dynamically compensate the temperatures since for disturbances at the top of the column the upper temperature will change first while, for disturbances at the bottom of the column, the lower temperature will change first. The scheme should therefore only be considered if the dynamics of the two temperatures are similar – no matter what the source of the disturbance.

Success was claimed for a similar scheme based on on-stream analysers (Reference 3). Here the PV controlled by reboiler duty was defined as $(HK_d - LK_b)$ while that controlled by distillate flow was defined as the average of HK_d and LK_b. It is likely that analyser deadtime was large compared to the dynamics of the process and masked any differences between top and bottom of the column.

In Chapter 8 a more rigorous approach to decoupling is described. Configured using the DCS block structure it is a complex implementation and prone to a number of problems. For the reasons given it should only be considered if, for some reason, a proprietary MVC cannot be applied.

12.17 Multivariable Control

One of the common applications of MVC is providing dual composition control on distillation columns. It is often the most practical way of resolving interactions. Here we work through a simple example of its design. Figure 12.117 shows the addition of MVC to a column with an energy balance scheme.

Notable by their omission are the tray temperature controllers. Some implementers believe this to be the better approach. Since two tray temperature controllers will interact they argue that they should be decoupled within the MVC. But often the starting point for the MVC implementation is a working temperature controller on one tray. Others argue that if a basic controller is working well then it should be retained. It helps linearise the process and provides graceful degradation when the MVC is out of service. A third view is that the temperature controller should be retained as a back-up scheme but this means the operator needs training in both the normal control configuration and the back-up scheme. There is no universally correct approach; each should be considered on a per case basis.

In our example the two CVs are both compositions, either inferentials or analysers. Our MVs are reflux and reboil. Plant testing gave the process gain matrix shown in Table 12.9.
The MVC will therefore predict the process behaviour

$$CV_1 = -0.962MV_1 + 4.17MV_2 + bias_1 \qquad (12.111)$$

$$CV_2 = 0.806MV_1 - 5.32MV_2 + bias_2 \qquad (12.112)$$

The MVC continuously updates the bias terms as required so that the predicted CV matches the actual CV. It does this using the full dynamic model whereas our example is written on a steady-state basis.

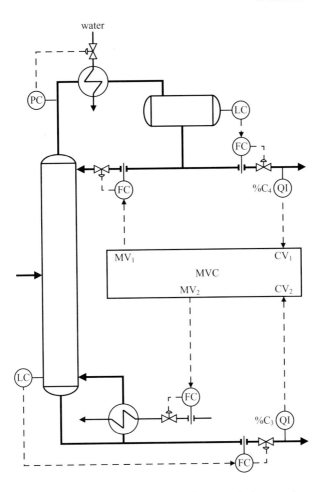

Figure 12.117 MVC on column with energy balance scheme

If we assume that when both CVs are at their targets of 5 %, the reflux is 56.5 and the reboiler steam rate 14. We can therefore write Equations (12.111) and (12.112) as

$$-0.962MV_1 + 4.17MV_2 = 4.027 \qquad (12.113)$$
$$0.806MV_1 - 5.32MV_2 = -28.941 \qquad (12.114)$$

Figure 12.118 shows a plot of these lines of constant composition.

Table 12.9 Process gain matrix

		MV_1 Reflux	MV_2 Reboil
CV_1	% C_4 in distillate	−0.962	4.17
CV_2	% C_3 in bottoms	0.806	−5.32

354   Process Control

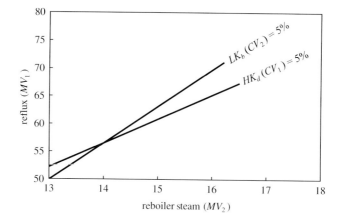

Figure 12.118 Contours of constant composition

Figure 12.119 shows the operating constraints, to which has been added a maximum reboiler duty.

Figure 12.120 shows the feasible operating area. Most MVC include a linear program (LP) for optimisation. This technology can only find the most profitable *node*, where constraints cross. There may be a more profitable operating point within the feasible operating area. A nonlinear optimisation technique would be required to identify this point; this is covered later in this section.

So the MVC will select one of the three nodes:

- the 'minimum energy' case, where both products are exactly at their specifications
- the 'maximum propane' case, where the propane is at specification and the maximum permitted reboiler duty is used to recover as much C_3 as possible from bottoms
- the 'maximum butane' case, where the butane is at specification and the maximum permitted reboiler duty is used to recover as much C_4 as possible from distillate.

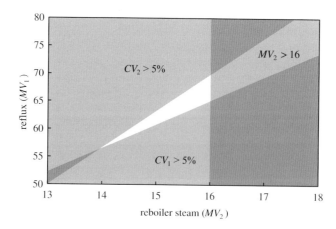

Figure 12.119 Operating constraints

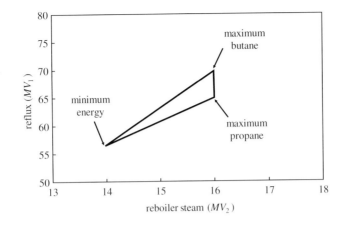

Figure 12.120 Feasible operating region

What drives the MVC to select a node are the *objective coefficients* (or *cost coefficients*). These are applied to each MV (*P*) and, in some MVC packages, also to each CV (*Q*).

If *m* is the number of CVs and *n* the number of MVs then the MVC *objective function* or *cost function* (*C*) is given by

$$C = \sum_{i=1}^{m} P_i CV_i + \sum_{j=1}^{n} Q_j MV_j \qquad (12.115)$$

As we shall show later, there are several advantages to using real economics in the controller. However with our controller in its current form this is not immediately obvious. What economic value would be placed on reflux, or on product compositions? To facilitate this we include additional variables in the controller. In this case we include the two products flows as additional CVs. No constraints are placed on these variables so they do affect the feasible operating region. The product flows simply provide a more convenient way of defining the controller economics.

Table 12.10 shows the extended gain matrix.

In adopting this approach it is important that the gain matrix is consistent with the rules of mass balance. In our example, at constant feed rate, any change in the yield of one product must be reflected by an equal and opposite change in the yield of the other. Thus, in this case, $(K_p)_{31}$ must be equal and opposite to $(K_p)_{41}$ and $(K_p)_{32}$ must be equal and opposite to $(K_p)_{42}$. Since plant testing is unlikely to produce exactly this result some adjustment will be necessary.

Similarly, if feed rate were included as an MV, as shown in Table 12.11, mass balance requires that $(K_p)_{33}$ and $(K_p)_{43}$ sum to 1.

If the material balance scheme were implemented on the column, then the MVs would be distillate flow and reboiler steam. Plant testing gave the results shown in Table 12.12.

In this case, for the mass balance to be correct, $(K_p)_{31}$ must be -1, $(K_p)_{32}$ must be 0 and $(K_p)_{33}$ must be 1. There is no need to include propane flow as a CV, since its inclusion as an MV already permits an objective coefficient to be assigned.

Table 12.10 Extended process gain matrix

		MV_1 Reflux	MV_2 Reboil
CV_1	% C_4 in distillate	−0.962	4.17
CV_2	% C_3 in bottoms	0.806	−5.32
CV_3	propane flow	−1.000	5.40
CV_4	butane flow	1.000	−5.40

Table 12.11 Inclusion of feed rate as a MV

		MV_1 Reflux	MV_2 Reboil	MV_3 Feed rate
CV_1	% C_4 in distillate	−0.962	4.17	−0.098
CV_2	% C_3 in bottoms	0.806	−5.32	−0.446
CV_3	propane flow	−1.000	5.40	0.310
CV_4	butane flow	1.000	−5.40	0.690

If the MVC package does not permit CVs to be given objective coefficients, then the coefficients for the MVs should be modified to take account of the effect that changing each MV has on each CV.

$$Q'_j = Q_j + \sum_{i=1}^{m} (K_p)_{ij} P_i \qquad (12.116)$$

Using the results in Table 12.10, if we consider first the 'maximum butane' case then we see from Figure 12.120 that at this point the reboil steam is 16 and the reflux 69.7. Thus, in moving from the 'minimum energy' case, the change in butane production is given by

$$\Delta CV_4 = 1.000 \times (69.7 - 56.5) - 5.40(16 - 14) = 2.4 \qquad (12.117)$$

This is matched by an equal loss of propane production. For this change to be economic the value of the additional butane must be greater than the cost of the additional steam.

$$2.4 P_4 - 2.4 P_3 > 2 Q_2 \quad \text{or} \quad P_4 - P_3 > 0.833 Q_2 \qquad (12.118)$$

Similarly, for the 'maximum propane' case, the reboil steam is 16 and the reflux 65.2.

Table 12.12 Extended process gain matrix (with material balance scheme)

		MV_1 Distillate	MV_2 Reboil	MV_3 Feed rate
CV_1	% C_4 in distillate	1.045	−2.17	0.253
CV_2	% C_3 in bottoms	−0.760	−2.17	−0.647
CV_3	butane flow	−1.000	0	1.000

Thus

$$\Delta CV_3 = -1.000 \times (65.2 - 56.5) + 5.40(16 - 14) = 2.1 \quad (12.119)$$

For this to be profitable

$$2.1P_3 - 2.1P_4 > 2Q_2 \quad \text{or} \quad P_4 - P_3 < -0.952Q_2 \quad (12.120)$$

The purpose of this analysis is to demonstrate the impact of process economics on MVC. In this case, if the price difference between the two products approaches or exceeds the unit cost of steam, it would incorrect to operate at the 'minimum energy' point. It is common practice to treat objective coefficients as weighting factors that are adjusted by trial and error to force the controller to drive the process to what is believed to be the optimum operation. Doing so risks better achieving the wrong objective and so losing money when the MVC is commissioned.

If real process economics are used then, during the testing phase, it will become apparent that the operating strategy suggested by the MVC is different from the established strategy. Rather than simply adjusting the objective coefficients, the difference should be reconciled. Either the MVC is not a faithful model of the process, for example because the process gains are incorrect or a key model is missing, or the choice of economics is wrong. If neither of these problems exists then the inevitable conclusion is that the current operating strategy is wrong. Given the complexity of many processes this should not be unexpected. It is likely that never before has such a detailed analysis of process optimisation been undertaken.

There are several other advantages to the use of true process economics. It is common for process operators to artificially constrain the MVs. In our example, consider the impact of the operator placing an upper limit on reflux of 60. As Figure 12.121 shows, this would severely constrain the maximum butane or maximum propane operating modes. With the use of real economics it is possible to quantify the cost of this.

For the 'maximum butane' case, Figure 12.121 shows the steam rate reduced to 14.5. The change in butane yield caused by constraining the reflux is given by

$$\Delta CV_4 = 1.000 \times (60.0 - 69.7) - 5.40(14.5 - 16) = -1.6 \quad (12.121)$$

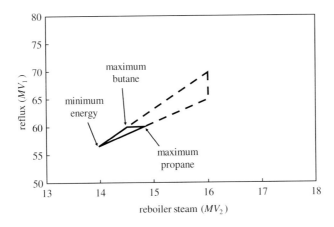

Figure 12.121 Over-constrained reflux

358 Process Control

Taking account of the steam saving, the change in the value of the cost function is

$$\Delta C = -1.6 \times (P_4 - P_3) - 1.5 Q_2 \qquad (12.122)$$

Expressed per unit change in the CV, this value is known as the *shadow price* or *Lagrange multiplier*. The amount by which the CV can change before another constraint is reached is the *allowable increase*. Some MVC packages make these values available to the engineer. If not, then a spreadsheet package will permit the engineer to build a steady state simulation of the MVC. Most spreadsheet packages then automatically generate this information.

Performing a similar calculation for the 'maximum propane' case, Figure 12.121 shows the steam reduced to 14.8. The change in propane yield is therefore

$$\Delta CV_3 = -1.000 \times (60.0 - 65.2) + 5.40(14.8 - 16) = -1.3 \qquad (12.123)$$

Taking account of the steam saving, the change in the value of the cost function is

$$\Delta C = -1.3 \times (P_3 - P_4) - 1.2 Q_2 \qquad (12.124)$$

A further advantage of the use of real economics is that they readily provide the economic incentive for process debottlenecking projects. If we assume that the constraint placed on reboiler duty is real then, if we want to maximise the yield of one of the products, there is benefit in relaxing this constraint. For example relaxing it to 17 would result in expanding the feasible area as shown in Figure 12.122.

For the 'maximum butane' case, Figure 12.122 shows the reflux increased to 76.3. The change in butane yield resulting from increasing the reboiler duty is given by

$$\Delta CV_4 = 1.000 \times (76.3 - 69.7) - 5.40(17 - 16) = 1.2 \qquad (12.125)$$

Taking account of the steam cost, the change in the value of the cost function is

$$\Delta C = Q_2 - 1.2 \times (P_4 - P_3) \qquad (12.126)$$

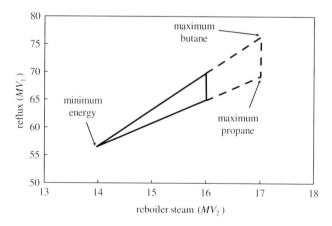

Figure 12.122 Relaxing the reboiler constraint

This value, expressed per unit change in the MV, is known as the *reduced cost* or *reduced gradient*. Again the MVC may make this available, along with the allowable increase, or it may be calculated by the spreadsheet.

Repeating the calculation for the 'maximum propane' case, Figure 12.122 shows the reflux increased to 69.5. The change in propane yield is given by

$$\Delta CV_3 = -1.000 \times (69.5 - 65.2) + 5.40(17 - 16) = 1.1 \qquad (12.127)$$

Taking account of the steam cost, the change in the value of the cost function is

$$\Delta C = Q_2 - 1.1 \times (P_3 - P_4) \qquad (12.128)$$

Finally, the use of real economics means the MVC objective function is a real measure of process profitability. This can then be used to assess the value of the application. While the minute-to-minute value is probably too noisy and subject to change if the objective coefficients are changed, it can be used offline. For example, the average C_4 content of propane before implementation might have been 4.1 %. With the MVC in place it was increased to 4.5 % and so the improvement is 0.4 %. We repeat this calculation for the other limiting constraints such as the C_3 content of bottoms and the reboiler steam flow.

We then run a simulation of the MVC first with the normal constraints and then with the constraints tightened by the improvement. For example the constraint on C_4 in propane is moved from 5 % to 4.6 %. The difference in objective function between the two simulation runs will be profit improvement (per hour, if flows are measured per hour).

While using real economics is an ideal, there are sometimes practical difficulties in achieving this. For example, it is common to include column pressure as an MV. The resulting gain matrix is shown in Table 12.13.

The relationship between product composition and pressure is relatively straightforward to identify. Indeed this is done in developing the PCT. But the impact that pressure has on product yields is often difficult to quantify during step tests. As a result $(K_p)_{33}$ and $(K_p)_{43}$ are often omitted. The controller therefore will only manipulate pressure to relieve a constraint; it sees no economic incentive to adjust it otherwise. The solution often adopted is to define an objective coefficient for the pressure MV, usually a small positive value in the belief that pressure should be minimised to reduce energy requirements. However this may result in an opportunity to significantly improve profitability being overlooked. A better approach would be to find some other means of identifying the process gains, for example by regression of historical data or from process simulation.

Table 12.13 Inclusion of pressure as a MV

		MV_1 Reflux	MV_2 Reboil	MV_3 Pressure
CV_1	% C_4 in distillate	−0.962	4.17	−1.43
CV_2	% C_3 in bottoms	0.806	−5.32	2.78
CV_3	propane flow	−1.000	5.40	−2.30
CV_4	butane flow	1.000	−5.40	2.30

Table 12.14 Retention of tray TC

		MV_1 Reflux	MV_2 Tray TC
CV_1	% C_4 in distillate	−0.600	0.418
CV_2	% C_3 in bottoms	−0.006	−0.750
CV_3	propane flow	−0.330	0.650
CV_4	butane flow	0.330	−0.650
CV_5	steam flow	0.130	0.125

At the beginning of this section we considered the retention of the tray temperature controller. Assuming this is in the lower section of the column and manipulates reboiler steam then the SP of this controller becomes an MV, instead of that of the steam flow controller. However we may wish to include steam flow as a CV so that we can apply an objective coefficient. The control matrix then becomes that shown in Table 12.14.

With the controller configured in this way, when the reflux was steptested the tray TC took corrective action to maintain the temperature at SP. This affects the response of all five CVs. Thus, if the TC is switched to manual once the MVC is operating, the process gains will revert to those in Table 12.11. Of particular concern would be $(K_p)_{21}$ since its value changes sign. It is unlikely that the MVC could continue to operate with such an error. MV_2 would have to be configured as a *critical* MV so that the MVC is automatically disabled if the TC is switched to manual.

12.18 On-stream Analysers

Much of the application of on-stream analysers has been covered in general in Chapter 9. Here we focus on those issues specific to their use on distillation columns.

One of our objectives is to minimise the sample delay; this can be achieved by locating the analyser as far up-stream as possible. For example, it is common to withdraw the distillate sample from the discharge of the product pump. But consideration should be given to withdrawing it from the overhead vapour line. This avoids the delay caused by the condenser and reflux drum. Plus the velocity of the vapour in the sample line will be far greater than liquid velocity. Some care should be taken with design of the sample system. It would be unwise to simply tee into the overhead line. Vapour can begin to condense on the internal surface of the line and any sample taken could then contain liquid containing more HK than the distillate. A better approach is to insert a probe into the top section of the column, through the elbow in the overhead line. The sample line to the analyser should be heated and insulated to ensure that no condensation takes place.

The vapour sample would only be representative of distillate composition if the condenser is total. However, if a vapour product is only produced intermittently then it may be practical to detect production and temporarily disable the use of the analyser measurement in any control scheme.

On trains of distillation columns there may be other possible methods of reducing sample delay. Consider the arrangement in Figure 12.123 in which our case study LPG splitter is the last column.

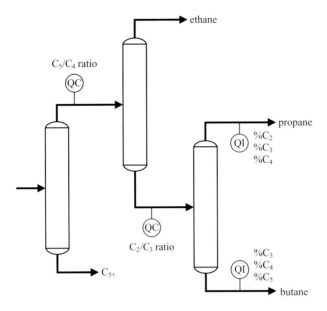

Figure 12.123 Reducing analyser sample delay in distillation trains

It is common to have specifications on the C_2 and C_4 content of propane. We have considered schemes that might control the latter by manipulation of the reflux on the LPG splitter. But to control the C_2 content we manipulate the reboiler duty on the preceding de-ethaniser. A composition controller based on a C_2 analyser on the propane product will respond very slowly to changes in its MV. A better approach would be to locate an analyser on the de-ethaniser bottoms. The C_2/C_3 ratio will be close to the C_2 content of propane but provides a much earlier indication and hence much faster disturbance rejection.

Similarly it is common to have C_3 and C_5 specifications for butane. Control of C_3 would likely be by manipulation of the LPG splitter reboiler duty. However control of C_5 would be manipulation of the reflux on the debutaniser – two columns upstream! Measuring the C_5/C_4 ratio of debutaniser distillate would provide a huge dynamic advantage.

Of course costs need to be taken into account, but the **incremental** cost of an additional analyser, if installed in the same housing, is much smaller than the cost of a standalone installation. And further use may be made of such analysers – for example as part of feedforward strategies.

12.19 Towers with Sidestreams

The logic followed in designing control strategies for two-product columns can be extended to those with one or more sidestreams. The pressure controller, as before, is the first designed – using much the same approach as that described. The next step, the level control strategy, first depends on the process configuration. Figure 12.124 shows a column with no liquid distillate. There is little choice but control reflux drum level with the reflux. Assuming feed rate is not available as a MV, column level may be controlled conventionally

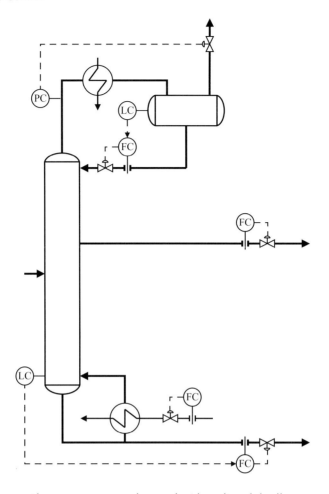

Figure 12.124   Level control with no liquid distillate

by manipulating bottoms flow as shown. If the bottoms flow is too small to provide sufficient control than manipulation of the sidestream flow is an option.

With three liquid products we can elect to reserve the sidestream flow as a MV for composition control. There are three options, each of which is simply the addition of the sidestream to schemes we have already described. Figure 12.125 shows its addition to the energy balance scheme.

Figure 12.126 shows its addition to the preferred material balance scheme.

Figure 12.127 shows its addition to the less preferred material balance scheme.

The column LC can, if the more usual MVs cannot be used for some reason, be configured to manipulate the sidestream flow. Figure 12.128 is the result of making this change to the scheme shown in Figure 12.125. By adjusting the sidestream, the internal reflux is changed and so also the liquid that accumulates in the base of the column.

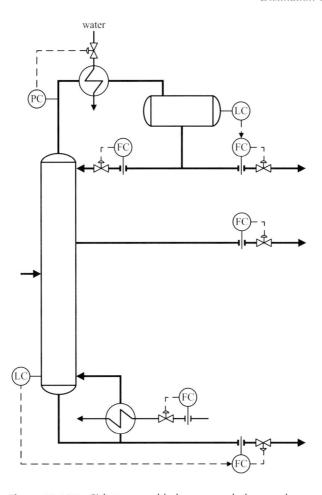

Figure 12.125   Sidestream added to energy balance scheme

We have to give some additional consideration to the composition controllers. By adding the sidestream we have increased by one the number of MVs – i.e. the sidestream draw flow. But we potentially add two composition targets. We add a third key component – *middle key* (*MK*). We have our existing composition targets now expressed as %MK in distillate and % MK in bottoms, but we may also have specifications for both %LK and %HK in the sidestream. Since we now have more PVs than MVs then we cannot meet all four product composition specifications. One of the compositions will always be in giveaway, i.e. better than specification. If it is always the same specification then we can safely ignore it and design a scheme to control the remaining three. If, however, the operation and/or the targets cause the three limiting specifications to change then all four must be included in the controller and some logic added to automatically select which will be controlled. This function is readily provided by MVC packages.

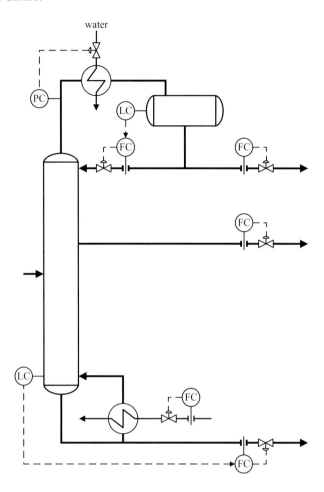

Figure 12.126   Sidestream added to preferred material balance scheme

12.20   Column Optimisation

Once the column has schemes which provide effective energy and material balance and composition control there may still remain a number of variables which can be manipulated to improve profitability. It may be possible to adjust feed rate, feed composition or feed enthalpy. There is usually scope to adjust column pressure. And, if there is a large difference in product prices, compositions can be adjusted to be better than specification.

In adjusting the operation several of a wide range of equipment constraints may be encountered. These include condenser duty which may limit because of high coolant inlet temperature (e.g. on air-fins in hot weather) or because there is a maximum permitted coolant exit temperature (e.g. corrosion by salt water). The condenser limit might be approached because the column pressure is too low such that the dew point at the top of the column approaches the coolant temperature. High feed enthalpy can similarly overload the condenser, as can fouling on either the tube or shell side.

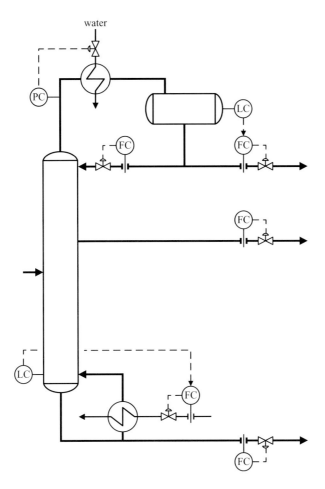

Figure 12.127 Sidestream added to less preferred material balance scheme

Similarly the reboiler may constrain. If duty is provided by heat integration then a low heating medium inlet temperature may not give sufficient LMTD. Hot oil based heating systems are subject to a minimum outlet temperature, below which high viscosity gives pumping problems. Too high a column pressure will increase the bubble point so that again the LMTD is insufficient. On fired reboilers a maximum fuel rate may be reached, constrained by combustion air availability, tube skin temperature, burner pressure etc. There may be metallurgical constraints on temperature. And, like condensers, reboilers can be subject to fouling.

Parts of the unit may have hydraulic limits. These might apply to any vapour product and to pumps on feed, products and reflux. They can also apply to the flow of coolant through the condenser and heating fluid through the reboiler. As we saw at the beginning of this chapter the column may blow, weep or flood.

In selecting the optimisation technology we need to determine what form the problem takes. If the number of available MVs exceeds the number of **active** constraints then there

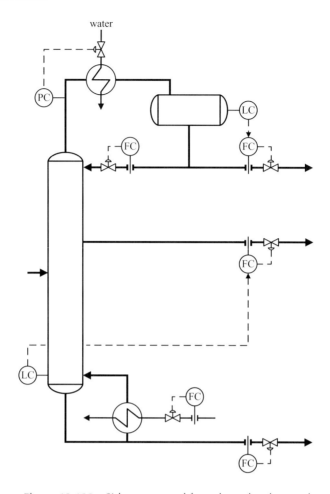

Figure 12.128   Sidestream used for column level control

is at least one degree of freedom and some form of nonlinear real-time optimiser will be required. If there are no degrees of freedom then MVC is sufficient.

12.21   Optimisation of Column Pressure

We have seen that reducing pressure reduces the temperature at which the column operates and hence improves relative volatility. Figure 12.129 shows the effect this has on separation (at constant cut).

If, instead of keeping the reboiler duty constant as we reduce pressure, we manipulate the duty to keep the product compositions constant then we get the result shown in Figure 12.130.

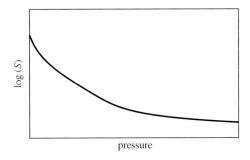

Figure 12.129 Effect of pressure on separation

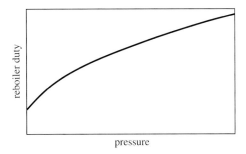

Figure 12.130 Effect of pressure on reboiler duty

While designing the column to operate at minimum pressure would minimise its energy requirements, construction cost must also be considered. Figure 12.131 shows that, as pressure is reduced, the condenser inlet temperature falls. This could require a greater condenser area and hence larger equipment plus the increased cost of the support structure. It could also result in the inlet temperature falling below that of the available cooling fluid and hence require the costly installation of a refrigeration system.

While these decisions are outside the scope of the control engineer, there still remains an opportunity to adjust pressure within the constraints of the installed equipment. Again it is

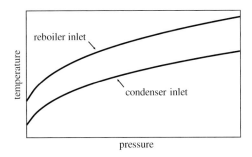

Figure 12.131 Effect of pressure on reboiler and condenser limits

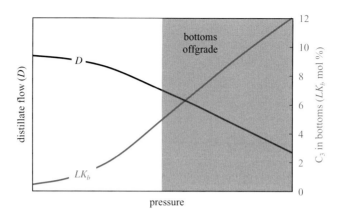

Figure 12.132   Effect of pressure on product yield

tempting to believe that pressure should be minimised to exploit any condensing capacity and so minimise energy consumption. While this may be true on some columns, it does impose an artificial constraint on feed rate. It may be more economic to raise the pressure to increase the unit capacity. Or recovery of the more valuable product may be increased.

Figure 12.131 also shows the impact on the reboiler inlet temperature. If increasing pressure is economically attractive then doing so will reduce the temperature difference between the heating medium and the bottoms draw-off. This could result in the reboiler constraining the process.

For example, if in our case study LPG splitter, we fix the reboiler at its limit and vary pressure we obtain the result shown in Figure 12.132. With the energy balance scheme in place, increasing the pressure reduces distillate yield and increases the C_3 content of bottoms. However, there is a maximum pressure above which the bottoms product will be off-grade. Since the energy cost is constant, changing pressure simply shifts yield between distillate and bottoms. If the bottoms product is more valuable then we should maximise the pressure within the composition target. If the distillate product is more valuable then we should minimise pressure until some other constraint is reached, such as condenser duty.

The economic value of optimising pressure can vary greatly. On our LPG splitter, exploiting the improvement in relative volatility to reduce energy consumption (as in Figure 12.130) will bring relatively little profit. However if there is a significant difference in product prices then exploiting it to improve yield (as in Figure 12.132) can be very lucrative. Of course, on large energy users such as our C3 splitter, even a small percentage energy saving is very attractive.

12.22   Energy/Yield Optimisation

In the same way that pressure can be manipulated to exploit the difference in product prices, so can target compositions. We saw in the first section of this chapter how cut and separation can be adjusted to maximise revenue. The feasible operating region developed for these variables can be converted to one based on the actual MVs. For our case study LPG splitter,

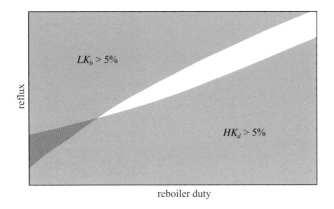

Figure 12.133 Feasible operating region

with the energy balance scheme in place, the MVs available for composition control are reboil and reflux. Figure 12.133 shows the feasible operating region.

If both products are of equal value then there is no point in producing one at the expense of the other. The object should be to minimise energy consumption by making both products exactly at their purity specifications, i.e. operate where the constraints in Figure 12.133 intersect.

If however propane is more valuable than butane we still want to produce propane exactly on specification (to maximise the C_4 sold at propane prices) but may wish to operate with giveaway against the butane specification. If the additional C_3 recovered from bottoms is more valuable than the additional energy cost we want to move away for the intersection of the constraints along the line of constant HK_d. The decision we have to make is how far we move along this line.

Figure 12.134 shows the effect of adjusting reboiler duty. A composition controller on propane, adjusting reflux, keeps the distillate composition constant but there is no control of

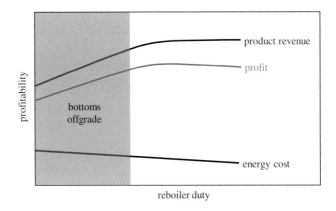

Figure 12.134 Effect of reboiler duty on profitability

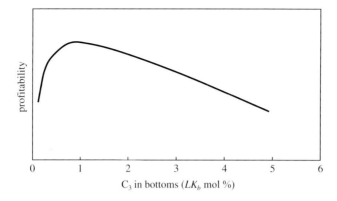

Figure 12.135 Effect of butane giveaway on profitability

bottoms composition. As duty is increased the energy cost increases linearly. However, the line of product revenue initially rises but soon flattens. As the bottoms purity increases there is little C_3 left to recover and so increasing reboiler duty has little effect on yields. There will be a point at which the additional propane yield does not justify the additional cost of steam. This, theoretically, is the point of maximum profitability. Taking the energy cost from the product revenue gives the profit curve, which reaches a maximum at this point.

This point of maximum profit exists in theory on all distillation columns. However it may not be an attainable point. For example, as shown in Figure 12.134, reducing reboiler duty will eventually cause the bottoms product to go off-grade. In our case the maximum is not in this region, but that will not be the case for all columns. Indeed if we were to reduce the price difference between propane and butane, the peak would move into this off-spec region. We may also reach other constraints that place an upper limit on reboiler duty – such as limit on the reboiler itself, the condenser or the column. Under these circumstances the most profitable operation would be at this constraint.

Figure 12.135 plots the profitability curve plotted against the bottoms composition to illustrate how much giveaway there is, on the bottoms specification of 5 % C_3, when at optimum operation.

References

1. Ryskamp, C.J. (1980) New strategy improves dual composition control. *Hydrocarbon Processing*, **59**(6) 51–59.
2. Prediction of vapour pressure of pure hydrocarbons and narrow-boiling petroleum fractions, in API *Technical Data* Book, procedure 5A1.19.
3. Weber, R. and Gaitonde, N.Y. (1982) Non-interactive distillation tower analyzer control. Procedures of American Control Conference, Arlington, VA, p. 87.
4. Jafarey, A., Douglas, J.M.and McAvoy, T.J. (1979) Short-cut techniques for distillation column design and control. Number. *Industrial Engineering Chemistry, Process Design and Development*, **18**(2), 197.
5. Colburn, A.P. (1941) The simplified calculation of diffusional processes – II. Number of transfer units or theoretical plates. *Industrial Engineering Chemistry*, **33**, 459–467.

13
APC Project Execution

13.1 Benefits Study

The cost of process control, as a fraction of the total construction cost of the process, has risen substantially since the early 1960s. Then is it was around 5 %, now it is closer to 25 %. In the 1960s the view was that some instrumentation was necessary but costs should be kept low. As a result plants had the minimum of measurements – just enough for safety and operability. Much of the instrumentation was local to the process, not repeated in the control room, and most of the controllers were single loop with the occasional cascade controller.

Processes are now much more extensively instrumented, with most of it in the control room. The instrumentation has become 'smarter' supporting a wide range of features such as linearisation, alarms, self-diagnostics, networking and so on. The control buildings have become far more sophisticated with blastproofing, climate control, ergonomic design, specialist lighting etc. The control systems have progressed from local and panel mounted controllers to DCS with operator consoles and links to supervisory control computers and data collection systems. Data from other sources such as the laboratory, product storage, scheduling department etc. are increasingly integrated with the control system. Sophisticated control applications based in the DCS and using MVC packages are now commonplace. Rigorous equation-based closed loop real-time optimisers are installed on a wide range of processes.

While some of the increased investment was driven by higher safety standards, increased environmental concerns and greater awareness of the value of process data, a large proportion of the justification derives from improved process control. In this section we focus on how those benefits can be quantified.

The management of some manufacturing companies are so convinced of the value of improved control that they require only the most cursory examination of the benefits. While control engineers clearly welcome this approach, it is not without risk. A detailed study would ensure that reality matches expectation; if it does not then it is far better to disappoint before large costs are incurred. It also supports decisions as to whether more costly options

---

Process Control: A Practical Approach   Myke King
© 2011 John Wiley & Sons, Ltd

should be included. For example does the **incremental** benefit, of the use of an on-stream analyser versus an inferential, justify the **incremental** cost?

An unfortunate fact is that managers move jobs. By the time it is underway the manager that fully supported the project as a 'no-brainer' may be replaced by one more sceptical. While the decision to progress the project is unlikely to be reversed, there is a danger that the necessary ongoing costly support will be allowed to wane. The study is an opportunity for everyone involved to 'buy in' to the benefits; this is important for the **long-term** success of the project.

A reasonable level of accuracy in benefit estimate should be sought. Calculations should be based on actual process performance rather than global statements such as '2 % increase in capacity' or '1 % energy saving'. Underestimating benefits could result in an attractive project not being sanctioned. Overestimating them will get the project approved but if the actual benefits fall well short of what was claimed, it is unlikely that the appropriation request for next project will be given much credibility. The best tactic is to slightly understate the true value – both as a contingency against possible implementation problems and to generate kudos when the claimed benefits are exceeded.

It can be tempting to delay benefit studies if process revamps are being explored or in progress. In most sites this is often a permanent state and so the study could be delayed by many years and miss opportunities to capture benefits. Experience shows that the total benefits are reasonably robust to changes in not only the process, but also the site economics. The operating constraints may change but the process operator's ability to reach each constraint remains unchanged. The benefits will still be captured but perhaps not in the way initially envisaged. The advantage of MVC is that it can readily be reconfigured to deal with such changes in circumstances.

It is important to select the right team to execute the study. Resourcing it entirely in-house carries risk. As described in Chapter 12, it is common for controllers to be configured to better achieve the current operating strategy – even if that strategy is incorrect. An outsider, properly examining the process economics, is more likely to challenge existing operating strategies. If the argument to change is convincing there is no need to wait for the project. Early manual implementation is likely to capture benefits larger than those attributable to improved control. An outside specialist is also more likely to not only know what newer technologies are available but, more importantly, how successful they have been elsewhere.

One might consider getting the MVC implementer to conduct the benefits study. The argument often presented for this approach is that whoever calculates the benefits should be able to guarantee them if they are also responsible for their capture. Indeed this is often the argument presented by implementers, suggesting they might not be held accountable to capture benefits quantified by others. The statistical techniques used to determine the benefits are common to all organisations. It is extremely unlikely that an implementer would decline a project if a requirement was to deliver the process improvements quantified by another using the same methods as his own.

Further, the single-supplier approach is likely to force the owner to compromise on the best choice of provider. It is often not the case that the best implementer is also the best at performing benefits studies. Given the relative size of the pieces of work, the owner will choose in favour of implementation skills and risk overlooking potentially valuable profit improvement opportunities.

The risk of having the implementer conduct the benefit study is that there is no guarantee that it will be executed impartially. Deliberately steering the project definition to favour the implementer rather than the owner is not in the implementer's long-term interest; it will eventually harm their reputation. But the implementer will look for benefits that can be captured by the technology he offers, rather than those which might require technology from a competitor or one with which the implementer is unfamiliar. While some implementers will claim to be 'independent', in that they are not tied to a particular technology supplier, they still have a very strong vested interest in maximising the profit they will make from the overall project.

13.2 Benefit Estimation for Improved Regulatory Control

At the regulatory control level there are the obvious benefits of keeping the process stable, safe and operable by a reasonable number of process operators. The process design team will most likely have already met these requirements, so we take this as our base case above which we have to economically justify further costs.

Addressing first the basic regulatory control layer, we should recognise that if higher levels of control are economically justified then, even if the basic controls capture no benefit in their own right, they provide the necessary foundation for these higher levels. Indeed it was commonplace for APC to be included in a re-instrumentation project to provide the justification for the improved basic control. The problem is that this can create the culture that basic control is a 'necessary evil' and that its cost should be minimised. This is in conflict with a culture of maximum profit. There remain many sites where basic instrumentation is neglected, despite large investments in APC. Good basic control should not only be recognised as necessary for APC, but that it can be valuable standalone.

Good regulatory control permits faster changes of operating conditions. Processes will often have different operating modes. They may process different feeds or produce different grades of product. On many processes the change of mode is the **only** significant disturbance. Mode changes can be costly. When switching between product grades, the material produced before the new specification is reached will have a lower value. It may require storage, reprocessing or is downgraded to a lower value product. If we can shorten the mode change then less of this material is produced. Further, because we produce more of the target product in the same time period, we have effectively debottlenecked the process. This is particularly true if off-spec material has to be reprocessed – using up valuable capacity and increasing operating costs. We can reduce storage costs by moving closer to a make-to-order process. And we may be able to gain market share by being a more reliable and flexible supplier.

Improved regulatory control can reduce maintenance costs. It is self-evident, if pumps and compressors are subject to fewer pressure surges, temperature deviations and so on, then bearings, seals and gearboxes are less likely to fail. While virtually impossible to predict these savings there have been cases where the *mean time between failures* (*MTBF*) for machines in difficult services have increased by a factor of three. If the equipment is critical, in that its failure requires a process shutdown or turndown, then not only are maintenance costs reduced but process capacity is increased.

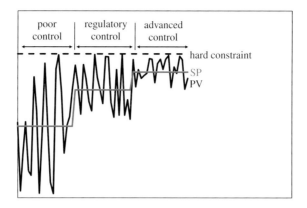

Figure 13.1 Benefit of improved control versus a hard constraint

Improved control makes better use of the process operator. Gone are the days where process control projects were justified in terms of a reduction in manpower. Most processes now have manning levels close to the minimum required for routine start-ups and shut-downs, and to deal with emergencies. Nowadays improved control makes the operator's role less mundane. Instead of being fully occupied keeping basic variables at desired conditions, more time can be spent on improving the operation. This of course has an impact on the quality of the personnel required and their level of training, but a good operator is wasted if employed in a mundane role.

A common way of estimating benefits for regulatory control is to quantify how much closer a process constraint may be approached. Figure 13.1 shows the classic drawing. With poor control the process operator will enter a controller SP such that worst deviation does not violate the constraint. Improving the regulatory control and reducing the deviations increases operator confidence that the controller can better handle process upsets and so he will move the SP closer to the constraint.

This type of analysis should only be applied to a hard constraint - one that should only be approached from one side. An example is maximising pressure against a trip or relief valve setting. It should not be universally applied to assessing the benefit for more closely approaching a specification of a liquid product. If the product is likely to be routed to storage before being sold or processed further then its specification is a **soft** constraint. It can be temporarily violated and, provided corrective action is taken, the finished product can still be produced exactly on grade, no matter how large the deviations from SP.

Indeed a frequent oversight in benefit estimation is to compare the limiting property of a product being sold against its specification. It is tempting to believe that if the product is exactly on specification then there is no benefit to be had. However, if the product has been produced on a process not well controlled so that the **rundown** property is highly variable, large savings may still be possible.

Most processes are fundamentally nonlinear. If we examine some of the key equations governing process behaviour, this quickly becomes apparent. Heat transfer is fundamental to almost every process. Whichever way this is achieved involves nonlinearity. For example, in Chapter 10 we applied Stefan's Law to estimating benefits on a fired heater.

The law states that the rate of heat transfer (Q), in the radiant section of a fired heater, varies highly nonlinearly with the temperature (T).

$$Q = \sigma T^4 \tag{13.1}$$

In the convection section the *Five Fourths Power Law* applies.

$$Q \propto T^{5/4} \tag{13.2}$$

While linear with respect to heat transfer coefficient (U) and area (A), the rate of heat transfer in a heat exchanger is highly nonlinear with respect to temperature. This is governed by the log mean temperature difference (LMTD). So, for example, if the flow though one side of an exchanger is doubled we do not achieve double the heat transfer.

$$Q = \frac{UA(\Delta T_{in} - \Delta T_{out})}{\ln\left(\frac{\Delta T_{in}}{\Delta T_{out}}\right)} \tag{13.3}$$

Chemical reactions are governed by Arrhenius's Law which states that the rate of reaction (k) is proportional to an exponential function of absolute temperature.

$$k \propto Ae^{-E/RT} \tag{13.4}$$

Key to the distillation process, Antoine's Law relates vapour pressure (P_0) to an exponential function of temperature.

$$\ln(P_0) = A - \frac{B}{T+C} \tag{13.5}$$

Even mixing is not immune to nonlinearity. If blending on a **weight** basis, SG does not blend linearly. Blending two streams of flows F_1 and F_2, with specific gravities SG_1 and SG_2 gives a combined *SG* of

$$SG = \frac{(F_1 + F_2)SG_1 SG_2}{F_1 SG_2 + F_2 SG_1} \tag{13.6}$$

Combining streams of different viscosities is governed by the *Refutas Equation* which converts viscosity (v) to a linearised *viscosity blending number (VBN)*.

$$VBN = 14.534.\ln(\ln(v+0.8)) + 10.975 \tag{13.7}$$

The VBN is determined for each stream, blended on a **weight** basis, and the result converted back to viscosity.

Figure 13.2 shows the upshot of all the sources of nonlinearity. As almost any process parameter is varied the effect that it has on operating cost is nonlinear. Thus the cost of manufacturing a product **always** exactly on specification is less than that of producing it when **on average** it is at specification.

Figure 13.3 shows an example of the calculation that might be performed on a very high purity LPG splitter. The distillate specification is 0.05 mol% C_4 in propane. The material sold from storage has a concentration very close to this specification and so there would appear little benefit to be obtained from better controlling it. However the rundown

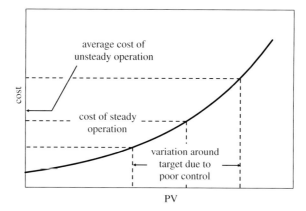

Figure 13.2 Benefit of improved control versus a soft constraint

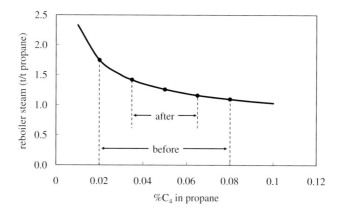

Figure 13.3 Benefit of improved control of product purity

composition of the stream to storage varies between 0.02 and 0.08 %. Figure 13.3 shows the reboiler steam consumption per unit of propane product. This relationship could be obtained from historically collected process data or from process simulation. It shows that steam usage varies from 1.10 to 1.74, giving an average of 1.42 t/t propane. By halving the variation in C_4 content to the range 0.035 to 0.065 %, the steam variation reduces to 1.16 to 1.42, giving an average of 1.29 t/t propane. The saving of 0.13 t/t propane, on an average size column, would easily justify the cost of an on-stream analyser and the control application.

With the advent of MVC the split of benefits between regulatory control and constraint control became less clear. Traditional APC applications, such as those controlling product composition, became part of the MVC. But both exploit the 'comfort zone' left by process operators. They achieve this by taking action far more frequently than even the most attentive process operator could achieve. So they will rapidly exploit even a transient opportunity to more closely approach a target. And they will avoid costly violation of constraints. Constraint control, as it is now defined, captures by far the largest portion of the benefits available on continuous processes. On batch processes, constraint control usually has only limited application; the larger benefits are more likely to be attributable to scheduling techniques.

While MVC will capture additional benefits associated with its choosing the optimum constraints to approach, like its predecessors it will approach these more closely by reducing the variation in the deviation from target. The assumption that the variability or standard deviation (σ) is halved by the implementation of improved regulatory control has become a de facto standard in the process industry. It has no theoretical background; indeed it would difficult to develop a value theoretically that is any more credible. Post-implementation audits usually confirm that this assumption is realistic.

There are a variety of ways in which the assumption can be applied. The *Same Percentage Rule* (Reference 1) is based on the principle that if a certain percentage of results already violate a specification, then after improving the regulatory control, it is acceptable that the percentage violation is the same. Figure 13.4 shows six months of data collected for a benefits study. The results have been normalised to % of specification, so that the maximum acceptable value is 100 % (shown as the solid line). There is an economic incentive to approach this target as close as possible. The chart shows that, in doing so, several results violated this target. The average value was 97.5 (shown as the dashed line) and the standard deviation 2.3.

Figure 13.5 shows the same data as a cumulative frequency plot. The solid black line is the best fit normal distribution. If improved control halves the standard deviation (coloured line) then the mean can be increased by half the current average deviation from target, i.e.

$$\Delta \bar{x} = 0.5(x_{\text{target}} - \bar{x}) = 0.5(100 - 97.5) = 1.25 \tag{13.8}$$

In our example, improved control will increase the average result from 97.5 to 98.75. Figure 13.5 shows that this gives the same percentage violation of the specification.

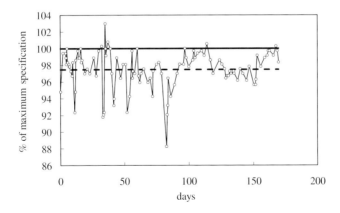

Figure 13.4 Process performance against a maximum specification

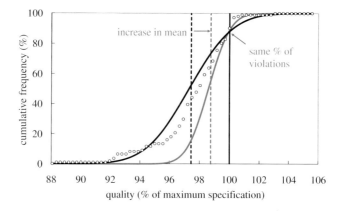

Figure 13.5 Same percentage rule

The *Same Limit Rule* is used when a large number of results violate the specification. In this case we choose a new target which is violated by a reasonable proportion of results. The resulting change in the mean is given by

$$\Delta \bar{x} = \frac{\sigma}{2} P(m) \qquad (13.9)$$

The value for $P(m)$ is taken from Figure 13.6. For the normal 5 % violation the value is 1.65; for the more demanding 1 % it is 2.33.

Figure 13.7 shows data for a different product quality from the same study. This time the specification is a **minimum**. In this case the majority of the results violate the limit. Indeed the average was below the limit at 97.9. The standard deviation was 9.1.

Figure 13.8 shows the same data plotted as a cumulative frequency plot. The solid black line is the best fit normal distribution. Having chosen a value for $P(m)$ to give 5 % violation of the new minimum limit, the average is reduced to 90.4.

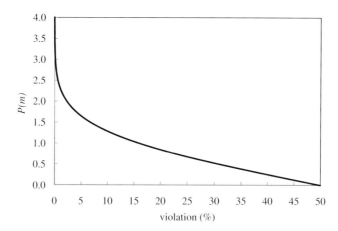

Figure 13.6 Cumulative probability function of the standard normal distribution

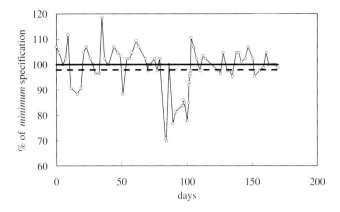

Figure 13.7 Process performance against a minimum specification

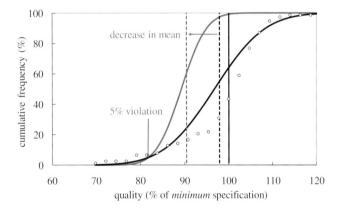

Figure 13.8 Same limit rule

Once the process improvement is quantified in engineering units, it has to be converted to a profit increase. Care has to be taken in identifying the effect of such marginal changes. For example, on our case study LPG splitter, increasing propane yield by recovering more C_3 from the butane product might appear to involve quite simple economics. This is the case if the additional propane is sold to a customer and another customer receives less butane. We need only to consider the selling price of both streams and perhaps any change in reboiler energy consumption. But it may be that we have some internal use for butane. For example, butane is often a blend component in gasoline. If less is available then we may sell less gasoline so it might be tempting to use the gasoline selling price in our calculation of profit. However, we also have to consider the change that we have made to the butane composition. By reducing its C_3 content we have reduced its RVP (Reid vapour pressure), which reduces the RVP of the blended gasoline. We might choose to inject **more** butane into gasoline and, as a result sell more gasoline and even less butane to our customer. The situation is complicated by the fact that butane is traded on a weight basis and gasoline on a volume basis. Our calculations therefore need to take into account the change in the density of not only the additional gasoline sold, but of **all** the gasoline sold.

Describing all but an example of the complexity of process economics is beyond the scope of this book. The example makes the point that the engineer should consider the full impact of the change being made. Obtaining marginal prices from the planning and economics section is usually too simplistic. A more rigorous approach will occasionally reveal that an assumed operating objective is incorrect and that meeting such an objective more closely with improved control will **lose** money.

13.3 Benefits of Closed-Loop Real-Time Optimisation

Estimating benefits for closed-loop real-time optimisation (CLRTO) requires particular attention. We have to distinguish CLRTO from what can be achieved with MVC. MVC is based on simple process models developed empirically. They support simple process economics and, if properly configured, will locate a constrained optimum. CLRTO is much more rigorous and is based on first-principle nonlinear engineering models. Its implementation can be extremely costly. The process simulation is often equation-based. For complex processes it is not unusual for many thousands of equations to be configured. While many come pre-packaged from a library of equipment types they have to be checked and calibrated against real process data. This incurs large technology and engineering fees.

In addition to the process simulation and optimiser, other modules are required. The optimiser is likely to be steady-state and therefore should only be permitted to execute if the process is at steady state. Further, if the optimiser takes any significant time to identify optimum conditions, the steady-state detection must also check that the process conditions have not changed since the start of the optimiser cycle. Tuning steady-state detection can be demanding; processes rarely attain exact steady state so sufficient leeway has to be given so that the optimiser runs reasonably frequently but without jeopardising the accuracy of the result.

Data reconciliation is required to handle suspect or missing measurements; if mass and energy balances do not close then the optimiser may fail.

Costly expertise is required to maintain the system, requiring an engineer strong in the technology itself and very experienced with the process and its economics.

There is also the potential for the optimiser to generate a nonoptimum solution. An economic value has to be placed on any stream crossing the process simulation's boundary. Often such streams are feeds or products that do not have a clear purchase or sales price. Valuing intermediate products and energy sources can be extremely complex on integrated sites. Choosing a set of economics consistent with more global optimisation requirements is challenging and prone to error.

Optimum operation is likely to be a function of feed composition. On many processes, susceptible to feed changes, the necessary on-stream analyser technology may not exist. The use of laboratory data introduces a delay which could cause the optimisation to be in error for many hours.

The market for rigorous CLRTO applications has peaked; the initial hype has been largely replaced with some level of scepticism, based on the experience of the users. In many cases the optimiser moved the process to a set of constraints and then did nothing

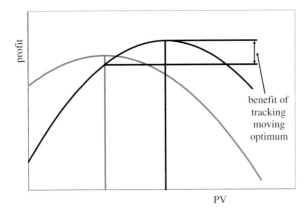

Figure 13.9 Quantifying benefits for CLRTO

more. While capturing substantial benefits, most of these could have been obtained from a considerably less costly offline process study. In other cases there are no independent means of determining whether the process has been truly optimised. On many sites the technology has fallen into disuse due to a combination of scepticism about its value and the difficulty in retaining high-quality technical support.

Having stated the problems there are, however, a number of very successful applications. The key to achieving this is to first properly consider the technical feasibility, the costs and the benefits. Figure 13.9 shows a simplified example of optimising a single PV. As the PV is adjusted the profit passes through a maximum. This is clearly the first condition for CLRTO to be considered; if the maximum is beyond a constraint then it can be located much more cost effectively by MVC. The second condition is that the optimum is not too 'flat'; the current operation may be far from optimum but the economic impact may be small. Thirdly the optimum must move. There must be changes in key parameters such as feed price, availability and composition; product price, demand and quality specification; or process changes (such as degradation of catalyst, coking etc.). The changes must be frequent and significant. If this is not the case then the majority of the benefits can be captured by offline studies, perhaps using the same simulation tool, to develop sets of recommended operating conditions for each scenario. The benefit of CLRTO comes from the incremental value of tracking a moving optimum compared to controlling at an optimum previously defined offline.

If the benefits are insufficient to justify rigorous CLRTO, or if the plant owner would rather capture less of the benefit and not have to support the complex technology, there are a growing number of less rigorous intermediate solutions. These are effectively extensions to MVC but apply *quadratic programming (QP)* rather than *linear programming (LP)*. The fundamental difference is that objective coefficients are applied to a quadratic function of process variables. This permits an unconstrained optimum to be identified. Like the MVC, these packages use dynamic models of the process, determined empirically, enhanced as necessary with simple engineering models. Because they do not require steady state they execute far more frequently.

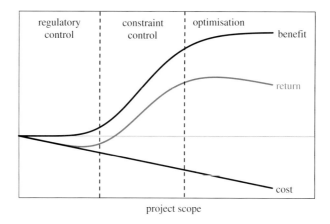

Figure 13.10 Optimising project scope

Figure 13.10 shows the effect of varying the project scope. Simple regulatory controls tend to be costly without capturing large benefits in their own right. On existing processes it is common for the instrumentation to need considerable attention, for many of the control algorithms to be changed and tuning constants to more be accurately determined. This can involve significant hardware costs, particularly if it includes upgrading the DCS. It will also involve substantial manpower. It is unlikely that a project of such limited scope would show a strong economic return.

A means of justifying the investment in regulatory control is to enlarge the scope of the project to include constraint control. This is likely to substantially increase the benefits but with usually a more modest impact on cost. The temptation is to progress constraint control without paying much attention to regulatory control on the basis that, if we can reduce the cost of this layer, then the overall return would improve. But this approach will reduce the benefits captured by the constraint controllers and this lost opportunity will be perpetuated. The cost of re-engineering the constraint controllers to take account of any **later** improvements to the regulatory controls will probably be prohibitive.

The benefits captured by optimisation are very process-specific. They can match or exceed those captured by constraint control or they can be a very minor addition. The case illustrated in Figure 13.10 is one where a rigorous simulation-based optimiser would not be justified but perhaps a much lower cost empirical technique, capturing a portion of the available benefits, might be worthwhile. Since the choice of technology and implementer are likely to be influenced by the scope of the project then it is important to decide upon this first.

13.4 Basic Controls

One of the main aims of this book is to draw attention to the improvements possible at the basic control level and to emphasise the importance of completing these before embarking on a MVC project. The choice of control algorithm, its tuning and any additions such as feedforward control or deadtime compensation, affect the overall process dynamics. Once step-testing for the MVC has been completed then any change to the basic controls would be very unwelcome and unlikely to be implemented until other circumstances, such as a process revamp, require steptesting to be repeated.

The first priority is to ensure that the instrumentation is operating correctly. Any controllers often out of service should be identified and the problems resolved. All control valves should be checked to ensure that they are correctly sized and the controller not liable to saturate. The type of valve (equal percentage or linear) should be checked. Calibration of valve positioners should be checked and any mechanical defect, such as excessive stiction or hysteresis, be rectified. Any excessively noisy measurements should be dealt with, if possible, by resolving the problem at source – rather than by the use of filtering. Resolution of some instrument problems can be delayed waiting for delivery of replacement parts or for a plant shutdown. It is therefore wise to identify any problems as soon as possible so that these are not on the critical path of the project.

Once instrumentation is fully functional then control configuration changes should be completed and tuning optimised. If the schedule permits (or if the work is being completed for a new process) then, for the advantages of adopting a standard approach, all controllers should be addressed. If this is not the case then only those controllers having a significant impact should be checked. While it may be possible to substantially improve the time a flow controller takes to return to SP, a saving of a few seconds will not be noticeable on a process where dynamics are measured in minutes.

Those controllers that should be reviewed include all level controllers, most temperature controllers (such as those on fired heaters, distillation column trays etc.) and any other controller where the process dynamics are relatively slow. Any controller identified by the process operator as problematic should also be addressed – whether or not important to the performance of MVC. This will help greatly with operator acceptance of the whole project.

The preferred control algorithms have been described in earlier chapters but in summary they are:

- signal conditioning to help linearise process behaviour
- filtering to remove noise (and the removal of unnecessary filters), using the least squares filter if the lag introduced by the standard exponential filter is excessive
- proportional-on-PV, derivative-on-PV noninteracting PID algorithm tuned for SP changes using the method described in Chapter 3
- averaging level control where appropriate (using gap controller if flow disturbances are very variable), tight level control otherwise
- ratio feedforward on feed rate (particularly if feed rate changes by more than $\pm 20\,\%$).

As an alternative other algorithms, such as bias feedforward and deadtime compensation, can be implemented in the MVC – depending on which approach is better for operator understanding and what back-up scheme is necessary if the MVC is out of service. It is also possible to move averaging level control from the DCS to the MVC. This should only be considered if it is desirable to let the MVC select which flow to manipulate (i.e. vessel inlet or outlet) depending on where the process is constrained. The DCS controller will still be required as back-up.

If the MVC is to be implemented by a contracted specialist company it is unlikely that including, in their scope of work, a **detailed** check on basic controls will be successful. The implementers generally require a much lower standard of basic control performance. They will generally check in any case, during the *pre-test* phase of the project, that all the MVs operate on automatic and generally respond properly. They probably will not consider the performance of controllers not destined to be MVs, even though their tuning may still affect the overall process dynamics.

Even if asked to be more rigorous the implementer may not have the skills necessary for this work. Further, since they are usually under budget and schedule pressures, it is not in their interest to identify problems that delay implementation of the MVC or require a great deal of their attention. The plant owner should take on this work, long before placing the contract with the MVC vendor, if necessary by bringing in outside expertise.

13.5 Inferentials

While the effort in building inferentials is relatively small, the **elapsed** time can be very large. There are several questions that need to be addressed before the MVC project is started, these include:

- Do sufficient good quality data exist to support the development of the inferential?
- Should regression or a first-principle model be used?
- Should a specialist supplier be used?
- Should the inferential be built in the DCS or in a proprietary software package?
- Is the inferential sufficiently accurate?
- Should laboratory updating be applied?
- If the inferential proves infeasible, what additional instrumentation should be included?

Although MVC implementers also offer inferentials they will deliver the best that their technology offers using the installed instrumentation. They will have little interest in working alongside a potential competitor supplying inferentials, nor will they wish to delay their development while additional data are collected or new instrumentation installed.

Work should therefore begin well in advance of any controller design work and certainly before any MVC implementation project is awarded. Many of the questions can be answered by the plant owner, supported by a specialist if required, by first attempting to develop regression type inferentials. It will quickly become apparent whether further data collection is required. For example it may be necessary to operate under different test run conditions with accurate time-stamping of samples taken at steady state. Automatic laboratory updating can be explored to see if this significantly reduces bias error or worsens random error.

Accuracy against that required can be checked. If sufficient accuracy cannot be achieved then a specialist can be brought in on a no-win no-fee basis to see if more accurate first-principle models can be developed. Regression models are easily implemented in the DCS using standard blocks. Other technologies require custom code or the use of a proprietary package.

If neither approach is satisfactory then the installation of additional instrumentation can be explored – either to provide further inputs to an inferential or replace it altogether with an on-stream analyser. Such instrumentation can be long-delivery and possibly shutdown-critical. Step-testing cannot be completed until such instrumentation is in place.

13.6 Organisation

As with all projects the key to success of APC projects is commitment from senior management. Too often, process control is seen as a necessary evil – involving costly instrumentation, software and people. The only exposure most managers have had to process control is the theory they were taught at university (for which they probably still do not see the need!) and a tour of the hardware installed in the control building. Many are still not convinced of its importance in maximising process profitability. It is perceived as an option; the process seemed to run just fine for years before it was installed so why do we need it now?

A manager will authorise almost any necessary expense to reinstate a piece of failed equipment, without which the plant cannot operate. As a result process start-up may be advanced 24 hours and thus increase annual capacity by about 0.3 %. Would the same manager authorise similar expenditure to improve by 15 % the performance of an MVC that can achieve a 2 % increase in capacity utilisation? Both have the same effect on production.

For a project to have long-term success a management culture is needed which asks 'Why is APC not installed?' as opposed to 'Why should APC be installed?' While it may be possible to convince management to sanction the project, the risk is that this commitment is short term. When there is pressure to relocate key personnel to other areas of the business, this may be done to the detriment of APC. Performance will then slowly degrade and a major initiative, probably as costly as the original project, will be required to re-establish the capture of the benefits. Often such an initiative will not be forthcoming until there is a change in management.

A management 'champion' will ensure everything necessary for project success is put in place. There will a strong commitment to ongoing projects, executed as part of an agreed master plan. Approval of expenditure will be rapid. Staff of the highest competency will be assigned to the project(s) in sufficient numbers. Work outside the control of the APC team, such as instrumentation improvements, will be given the correct priority. The importance placed on APC will become apparent to the vendor who will then assign better staff to the project and pay greater attention to their client. The more senior the manager – the more successful will be the project. The benefits captured by APC are often said to be proportional to the salary of the manager sponsoring the project! Often the champion is only in place by chance and then may soon be replaced by someone less enthusiastic. Commitment to APC is rarely a criterion used in appointing a manager.

The implementation team can do much to nurture what interest there is. To maintain the momentum, rather than seeking approval separately for individual projects, commitment should be sought for all the projects conceived in the master plan. While the sanction process might be lengthier, it is likely the larger budget requires that more senior management are involved – further raising the profile. Given that implementation would then take several years it would survive the periods when less enthusiastic managers are in place.

Ongoing benefits are not as obvious as the same benefits first captured on commissioning APC. Even less obvious is the slow decline that neglect will cause. A new manager may not be aware of what was achieved by the project. Many of the monitoring tools described in this chapter offer the opportunity to regularly engage with senior management to help maintain a high profile for APC and to provide continuity during changes in management personnel.

It is a common problem that management mistakenly assume that less expertise is required for ongoing support compared to what was necessary for implementation. While it is possible to contract the APC vendor to take a major involvement in implementation, ongoing support requires in-house expertise. The accuracy of inferentials, the availability of on-stream analysers and the reliability of the MVC dynamics all need close monitoring. There needs to be a check at least daily the each MVC has not been over-constrained, is using the correct objectives coefficients and is driving the operation against the correct constraints. There needs to be frequent liaison between all the groups that can influence the success of the controller – including process supervision, process technical support, planning and economics section, instrument and system support personnel and process operators.

Regular (e.g. monthly) meetings, chaired by the APC engineer, should be attended by the plant manager, process technical support, process economics specialist, instrument technician and system support. The main agenda items include APC performance for the last month, problems encountered, solutions developed, forthcoming changes and an agreed action plan (including who, what and when).

An experienced APC engineer should be involved in the approval of the design of all proposed process modifications. Additions agreed at the process design stage can be implemented for almost no incremental cost, compared to the possibly unjustifiable costs of retrofitting the change. Examples include ensuring sufficient tray temperature thermowells are installed on distillation columns, ensuring sufficient meter runs and orifice flanges are installed for the later addition of flow meters and ensuring sufficient space is reserved for the future installation of an analyser house. Such involvement will also help avoid many poor process design practices that later cause process control difficulties. These include wrongly placed and wrongly ranged level gauges where surge capacity can be used, omission of instrumentation important to inferentials, selection of control strategies known to give problems with inverse response or have other dynamic problems.

While recruiting good control engineers is possible there is an ongoing need to train existing staff. Projects present an ideal opportunity for on-the-job training. Trainees should be selected firstly on their level of enthusiasm for the subject and, because they need to liaise with almost every part of the organisation, on their interpersonal skills. They need a strong process background acquired either through education (e.g. chemical engineering graduates) or though experience (e.g. ex process operators). They will need a general appreciation of process economics.

New engineers should attend training courses in DCS configuration, usually provided by the DCS vendor, and in the chosen MVC package, usually by its developer. What is often overlooked is training in the areas covered by this book. There are many courses offered in the academic world, the majority of which are unashamedly highly theoretical. Such courses are probably the biggest cause of potentially very competent control engineers choosing another branch of the engineering profession. A small minority of the academic institutions are beginning to appreciate the difference between theory and practice, but often the lecturers have not been in a position to accumulate the practical expertise to pass on.

In assigning a new engineer to a project team it is important that the APC vendor understands what is required in terms of training. Most vendors do not see training client staff as a threat to their business and most will enthusiastically take on the training role. They do however need to factor in what impact it might have on their man-hour budget and schedule.

Once trained, retaining the expertise presents another problem. APC vendors are always on the lookout for good staff. A newly trained engineer will be looking for the next challenge; if all that is offered is ongoing support, rather than another project, working permanently on projects with an APC vendor might appear attractive. Much depends on the value the employer places on technical expertise. Too often those following technical careers are perceived as doing so because they would fail as managers. Many companies prefer to train generalists rather than specialists. Ideally a career in APC in larger companies should offer development into some centralised engineering/consultancy role that carries the same kudos as the equivalent position in the management hierarchy.

Other moves which help retain expertise is to have an approved master plan in place so that staff can see firstly a commitment to the technology by the company and secondly a role for themselves on future projects. Promoting APC engineers into management positions can demonstrate that the experience gained is valued by the company; few technical positions in a manufacturing company require as much understanding of the business as that developed by APC engineers.

Rotating staff through the APC group is beneficial on two counts; it imports knowledge of the process operation or its control systems and it exports APC supporters into other parts of the organisation. Many successful APC projects are on plants now managed by former APC engineers. Some of the most successful APC engineers were once process operators. Similarly exchange of experienced personnel with others sites, or even with vendors, will help develop expertise. Importantly movement demonstrates that joining the APC group opens up career opportunities rather than closes them down. Successful rotation routes are shown in Figure 13.11.

The debate in many organisations is where to locate the process control team. Their function is astride the interface between process technical services and instrumentation support, so arguments are made for them to reside in one group or the other. The group also has strong links with operations department and the planning and economics group, so there is logic in locating them in either of these. In fact, all of these options have been explored somewhere and each has succeeded and each has failed! The important consideration is not where but who. A well-chosen engineer reporting to a manager well-versed in APC will be effective in just about any part of the organisation.

The structure which does have several reported failures is the plant-centric approach. Here, all the staff associated with a particular plant are located in the same section reporting to the plant manager. Often the plant may not be complex enough to require the full-time attention of the APC engineer, so he or she will be given other responsibilities. Further not all plant managers are equally enthusiastic about APC. The APC engineer, who in a technology-centric structure would work on other processes for multiple managers, can become disillusioned with the profession or his employer.

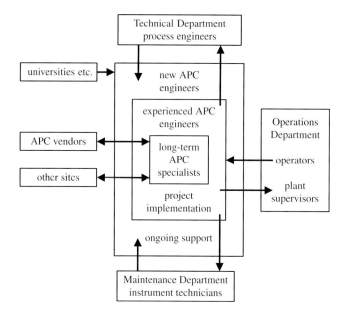

Figure 13.11 Staff rotation

MVC is a complex technology not always readily understood by all console operators. Technical support is usually only on site for 25 % of the time, for the other 75 % the console operator has the greatest influence over APC utilisation. If not properly trained, one operator can take an action which loses in one shift the profit that the controller has made in one year. There needs to be at least one operator per shift fully conversant with the APC. This is best achieved by assigning a respected leading operator to the implementation team. He will represent the process operators and, as importantly, be seen by the other operators as their representative. As a result there will be much greater confidence in the APC. Plus the lead operator, when returning to his normal shift position, will be an invaluable source or expertise outside of office hours.

APC expertise within a company can often be overstretched in trying to support existing APC applications while also being involved in a major implementation project. APC is treated quite differently from other additions to the process. For example, a new compressor, once accepted as operating correctly, becomes the responsibility of the plant manager. Any necessary maintenance will be coordinated by him or her, as will any additional training prompted by operator misunderstanding. Only in the event of more complex problems will the specialist be again involved.

There is no reason why much of APC monitoring should not be delegated in this way. Once accepted by the plant manager then his or her role should include routine checks that it is not being over-constrained and is operating against the right constraints. Tools developed for monitoring on-stream analyser performance can be passed to the analyser technicians and/or engineer. Similarly, inferential performance monitoring can be passed to the process support engineer, as can the responsibility for reporting APC performance. The APC engineer would only be called in the event of problems that fall outside of such routine support.

Prior to commissioning the MVC there is a need to involve the planning and economics group in a review of the process gain matrix, the economics used and the result of any off-line MVC simulations. While at first they might feel that they might struggle understanding the technology, it should quickly become apparent that the steady state part of the controller is very similar to LP-based planning tools. They will recognise process gains as *vectors*, objective coefficients as *reduced costs* etc. The operating constraints as understood by the planning group will be conservative and so any operating plan is likely to be suboptimal. Recognising that some of these constraints can be exceeded if required will influence the way they specify the operating plan. For example an instruction to run at a particular feed rate might be replaced with 'maximise feed rate, but not more than...'. They should also be given access to the offline version of the MVC so that they can explore the impact of changing economics before passing them to operations department. They should also be able to view the online version to check that new strategies are being adhered to.

13.7 Vendor Selection

There are a growing number of APC technology suppliers and implementers. Some only implement technology that they have developed. Others have exclusive licences to work with just one technology developed by others. And there are others that theoretically offer a choice of technologies – although in practice they will often favour one over the others. The situation is further complicated by specialist suppliers. For example there are those that specialise in inferential properties, those that have developed CLRTO packages, those that offer process-specific technology and so on.

Choice of vendor and implementer can be bewildering to an engineer, especially for one doing it for the first time. The first step is the production of an *invitation to bid (ITB)*. In addition to the plant owner's normal commercial terms, this should include:

- brief description of the process; simplified process flowsheets showing the basic controls will help the bidder assess the number of MVs and hence the work involved in step-testing and commissioning
- brief description of the control system, including its network(s), modules and how they are split per process, data acquisition and historian
- history of any previous APC implementations
- the key results of the benefits study listing, in engineering units, the anticipated process improvements
- list of any existing inferentials and their accuracy
- list of personnel that the client plans to assign to the team, their experience, what their role is and how the implementer is expected to involve them
- any specialist organisations that the successful bidder will be expected to work with and whether they will be subcontracted to the implementer or contracted to the plant owner
- list of documentation expected and what language should be used
- what performance guarantees are required
- any special procedures that the vendor must follow, for example, safety reviews
- pricing basis required, i.e. lump sum or reimbursable, travel and living expenses included or rechargeable, staged payments etc.

The ITB should also describe what is required of the *bidder*, including:

- list of reference sites with contact details for similar work completed by the bidder for other clients
- list (with résumés) of those staff shortlisted for each of the project roles
- proposed project schedule barchart
- description of involvement of plant owner's personnel
- contents of any training courses offered
- examples of specimen documentation.

Once the ITB has been issued, but before the bid due date, meetings should be held with all the bidders. These serve two purposes, an opportunity to clarify the ITB as required and assess the bidders' capabilities. Ideally this should be in their headquarters since this usually gives an opportunity to meet with a wide range of the bidders' technical staff and see demonstrations of their products. However, many of the smaller implementation companies do not have offices; their personnel largely work from home or on site. In which case the meeting can be at any convenient location, providing that the bidder brings to the meeting likely members of the project team.

The plant owner should take the opportunity to visit a selection of reference sites. This helps a little with vendor selection but the vendor is unlikely to suggest a site where his reputation is poor. It is more an opportunity to benefit from the experience of others that have completed similar projects.

Bid analysis is important to check that each bidder has complied with the requirements; clarification should be sought where necessary. How much it contributes to vendor selection can vary greatly. Sometimes, from the contact the plant owner has had with each bidder, selection may well have been made before the bids were submitted. Clarifying the bid is only necessary since it will become part of the contract. On other occasions, vendor selection can become quite complex. This can arise where one bidder has strengths not matched by others but has weaknesses compared to the others. If there are a large number of owner's personnel involved in the bid selection there can often be very different views on which is the preferred bidder. Under these circumstances a more methodical approach is required.

Kepner-Trago Analysis is a decision support methodology. It is occasionally discredited and can have the reputation that it only confirms a decision already made. This can arise if it is not applied impartially. The technique is to first brainstorm, involving all the staff involved in the decision, and list all the criteria on which selection should be based. For each criterion, a numerical value should be placed on its importance. These values should not be just arbitrary weighting factors. They should as far as possible represent the **financial** impact of meeting, or failing to meet, each criterion. It is useful to generate this table before finalising the ITB, so that any additional information required from the bidders can be included. Analysis of the bids then includes scoring each bidder against each criterion and totalling the financial impact. Table 13.1 shows a much shortened example for one bidder.

Once the preliminary analysis is complete then, to aid selection:

- delete from all of the analyses any issue on which the bidders score equally, so the focus is on differences
- delete any issue valued so low that it cannot have any impact on the decision, so the focus is on important differences

Table 13.1   Kepner-Trago analysis

Key issue	Potential benefit	$k	Rating	Value
Training of owner's staff	50 % cost saving on next project	150	50 %	75
Inferential technology	cost of analyser + 9 months benefits	175	70 %	123
Lead engineer quality	20 % of total benefits for 3 years	600	85 %	510
Location of support	3 faults/year fixed 1 day earlier	25	100 %	25
Overall		950	77 %	733

- explore how sensitive the result is to the ratings given to determine if the most pessimistic or optimistic view would alter the result
- discuss with the bidder any items where he might be able to improve his rating.

The fees quoted can vary widely from bid to bid. Bidders may be bidding tactically, for example to win a client already working with a competitor or to add an important site to its references. Bidders can also inflate their quotation if not enthusiastic about winning the work but think it may harm their relationship with the owner if they decline to bid. However it is important to reconcile any major differences to ensure that the bidder has properly understood the scope of work.

Clearly cost is an issue in selecting the bid but needs to be balanced by consideration of the benefits. The result of the Kepner-Trago analysis permits exactly this, financially justifying the decision to select a higher quotation.

No bid should be rejected until negotiations are completed with the leading bidder. Indeed, if a single tender approach has been adopted, the sole bidder should still be allowed to believe that he is bidding competitively. There are a wide range of techniques for doing this, for example revising the bid submission date at the request of 'others'. While negotiation on price is a possibility, more important is to get the bidder to agree to assign the best engineers to the project. The owner should identify what makes the project attractive to the bidder. For example, the bidder might see particular value in being able to bring future potential clients to the site once the project is complete and therefore be prepared to offer something in return. The owner is in an even stronger negotiating position if previous contracts were awarded to the bidder's competitor. A reference site where the bidder has displaced its competition is a valuable marketing advantage.

13.8   Safety in APC Design

While safety should be uppermost in any design decisions there is wide variation in how this is managed. The attitude of some engineers is that safety should be handled by the basic controls and, if anything, the APC will directionally improve safety. However there are many examples of an APC application causing a major process upset. Further it is often the case that the basic controls are modified to support the addition of APC.

Some operating companies have adopted a formal approach to APC safety, treating it in much the same way as any other plant modification. The well-documented *Hazard and Operability (HAZOP)* approach is generally too rigorous for APC, since it is difficult to prevent it from becoming a full process review. However if HAZOP is being applied to a major process revamp, APC should be included in the review. The *Control, Hazards and Operability (CHAZOP)* study is more suited to identifying the risks associated with the control system itself. *Failure Mode Event Analysis (FMEA)* has been used successfully, although in addition to the risks arising from APC it will identify problems with the basic controls already in place. The key to success is to ensure the review meeting has a strong leader whose role is to prevent the discussion moving outside the area of process control and also ensure every point raised is properly followed up.

Some companies have extended their permit-to-work system to include APC. Such permits require the signature of key personnel such as the plant manager, the head of the APC team etc. Permits are required for step-testing as well as commissioning.

Once safety procedures are agreed it is also necessary to agree criteria about when they are applied. What level of modification to a control strategy requires a formal review and/or a permit? Most would allow changes to operator graphics without a permit. Many allow changes to tuning. But would a change to a different version of the PID algorithm justify the paperwork? Would the addition of signal conditioning, etc. etc?

13.9 Alarms

A problem that can arise with the use of DCS is a proliferation of alarms. Each controller can be configured to give alarms for HI/LO PV, high rate of change and excessive deviation from SP. Other messages can be generated to draw attention to the change of auto/man mode, configurations changes etc. Left unchecked the large number of alarms can cause the process operator to miss those that are particularly serious. There are several major incidents on record where the deluge of alarms was later identified as a contributory factor.

While not strictly in the area of process control, a project and its associated budget does give the opportunity to address the issue. Experiences of those having performed such exercises are remarkably consistent:

- It is essential that operations department are strongly committed to the work; indeed without this it is probably not worth progressing.
- A strong project team, including a very experienced process operator and process engineer, is required. The design, support and monitoring of any resulting alarm management system usually falls to the control engineer.
- An alarm reduction study, possibly assisted by external consultants, should be progressed before the decision made to install any alarm management package.
- Resolving the problem is not a one-off piece of work; there is a need for a regular review of all alarm actuation and revision of alarm management configuration.

The team are set the objective of meeting the guidelines (Reference 2) published by the Engineering Equipment & Materials Users' Association. These recommend the following upper limits per operator console:

- No more than 10 *standing alarms*, i.e. alarms which have been acknowledged
- No more than 10 *background alarms* per hour, i.e. alarms for information purposes that may not require urgent attention

- No more than 10 alarms in the first 10 minutes after a major process problem develops.

There are a number of commercial alarm management systems available. Some features are often already built into the DCS as standard. Others require the addition of a package – maybe from a third party. These are not a substitute for an ongoing review of alarm statistics, but they do add useful functionality. They can be particularly useful in identifying repeating *nuisance* and long-standing alarms.

- Alarms can be categorised so that the greatest attention is drawn to the most important alarms.
- Alarms can be suppressed; this can be applied to alarms known to be the result of an upset that has already generated a *first-up* alarm. Or it can be used in predictable situations such as startups and routine shutdowns.
- Retrospective analysis of incidents is supported by tools that enable the alarm database to be searched for unnecessary duplicate alarms, multiple occurrences of the same alarm, exact timing and sequence of alarms, alarms associated with the same equipment or plant area etc.

References

1. Martin, G.D., Turpin, L.E. and Cline, R.P. (1991) Estimating control function benefits. *Hydrocarbon Processing*, **70**(6), 68–73 and Martin, G.D. (2004) Understand Control Benefits Estimates. *Hydrocarbon Processing*, **83**(01), 43–46.
2. Alarm Systems, a Guide to Design, Management and Procurement, EEMUA Publication No. 191 (1999).

Index

Acid, 119–122, 228
Adaptive tuning, 79, 166, 224, 349
Adiabatic index, 244
Adiabatic, 244
Air register, 228–229
Air-fin condenser, 285, 288–292, 364
Alarm, 93, 97, 100–107, 111, 213, 240, 303, 372, 392–393
Anti-reset windup, 81, 173, 256
Anti-surge, 252, 255
Antoine Equation, 263–264, 328–331, 334, 340, 342, 375
APC, x, 2, 371–388
Apparent deadtime, 7, 22, 26
ARC, x, 1–2
Arrhenius' Law, 238, 375
Automatic mode, 10, 30, 56, 80, 174, 186, 350
Averaging filter, 131–132, 134
Averaging level control, 26, 60, 92, 97, 100–105, 111, 152, 153, 301, 304, 305, 344, 383
Axial compressor, 243
Azeotrope, 270

Bang-bang control, 81
Base (alkali), 80, 119–120
BDC, 251
BFOE, 223
Bias algorithm, 115, 151, 172–173, 224
Bias error, 150, 206, 209–210, 212, 384
Bias feedforward, 151–152, 158–161, 181, 186, 210, 222–224, 383
Bias update, 209, 211
Blowing, 259, 261, 286, 365
Boyle's Law, 243
BPD, 94–99
Bubble point, 124, 126, 265, 268–269, 286, 287, 289, 306, 315–318, 326, 334, 342, 349, 365

Bumpless initialisation/transfer, 30, 173, 184,
Butane, 262–269, 280, 329, 354–361, 369–370
Butterworth filter, 130

Calibration, 10, 81, 118, 119, 125, 140–145, 171, 199, 201, 202, 213, 215–216, 222, 239, 315, 380, 383
Callendar-van Dusen Equation, 117
Calorific value, see GHV and NHV
Carbon dioxide (CO_2), 217–221, 229–230
Carbon monoxide (CO), 123, 217–220, 222, 231–233, 244
Carbon, 230
Cascade control, 9–10, 47, 49, 91–92, 123, 148, 301, 304, 337, 371
Centrifugal compressor, 244
Charles' Law, 243
CHAZOP, 392
Chromatograph, 38, 198, 213, 220, 222, 285
Chung Equation, 343
Clausius-Clapeyron Equation, 328–332, 334, 340
Cloud point, 208
CLRTO, 2, 380–381, 389
Cohen-Coon tuning, 56, 88
Colburn Equation, 344, 370
Cold property, 208
Constant molal overflow, 268
Constraint conditioning, 122–124
Constraint control, xii, 2, 122, 169–173, 376–377, 382
Control horizon, 185
Controller gain, 30, 45, 54, 61, 67, 70, 71, 73–75, 78–80, 98–101, 103, 105–108, 110–113, 115–116, 127, 150, 160–161, 164–165, 172–173, 183, 294, 350
Correlated steps, 15
Cost coefficient, xii, 355

Process Control: A Practical Approach   Myke King
© 2011 John Wiley & Sons, Ltd

Counter-rotation, 249
Critical MV, 189, 360
Critically damped, 51
Cross-limiting, 234–236
CUSUM, 209–210, 212
Cut, 271–282, 292, 301–309, 315, 318, 325, 341–342, 350, 366
Cylinder loading, 251
Cylinder, 95–96, 243, 251

Dahlin algorithm, 167–168
Dalton's Law, 264
Damping ratio, 51
Deadband, 82–83, 108–109, 113, 142, 144
Deadtime algorithm, 20, 153, 221
Deadtime, 5–7, 12, 17–18, 22, 25–27, 31, 34–39, 49, 55, 59, 61, 63, 65, 74, 82, 83, 86, 88, 147, 151–153, 155, 158–159, 162, 163–168, 180, 198–199, 210–211, 213, 285, 307, 346, 352, 383
Decay ratio, 52
Decoupling, 174, 177, 179–184, 350–352
Degrees of freedom, 170, 366, 367
Densitometer, 216, 220–222
Density, 114, 117, 139, 221, 254, 256, 261, 380
Derivative gain limit, 43, 84–85
Derivative spike, 37, 49, 167
Derivative time, 20, 35, 69, 71, 72, 76
Derivative-on-error, 37, 167
Derivative-on-PV, 37, 50, 65–66, 84–85
Diatomic, 244
Digital transmitter, 38
Direct synthesis, 63, 85
Direct-acting, 30, 172–173
Discharge coefficient, 117, 254
Discharge throttling, 248, 251, 297
Disturbance variable, 17, 147
DMC (dynamic matrix control), 185
Downcomer flooding, 261
Downcomer, 259–261
Dry basis, 230
Dual firing, 222–223
Dynamic compensation, 1, 153, 156–159, 161–162, 180–182, 207, 210–212, 221, 225, 308, 346, 348, 349
Dynamic decoupling, 180, 182

Economiser, 115
Eduljee Equation, 343
Ejector, 297–298
End point, 273
Energy balance, 100, 285–286, 292, 299–312, 318–319, 323–325, 345–349, 351–353, 363, 368–369, 380
Energy/yield optimisation, 368
Enriching section, 259
Equal percentage valve, 136–140, 247, 383
Equilibrium constant, 121
Error-squared, 105–112
Ethanol, 270
Exhausting section, 259
Exponential filter, 127–135
External reflux, 289, 292
External reset feedback, 81

Fast loop, 212–213
FBP, 273–274
Feed composition, 12, 262, 271, 284–285, 306–307, 312–322, 348–349, 364, 380
Feed enthalpy, 271, 284–286, 306, 349, 364
Feed quality, 267, 342
Feedback control, 115, 147, 148, 222, 225, 348, 350
Feedback, 23, 81, 115, 148, 182
Feedforward control, *xii*, 1, 79, 115–116, 148–162, 181, 186, 210, 222–226, 236, 257, 307, 314, 324–325, 344–350, 361, 383
Feedforward variable, 186
Feedforward-feedback, 150
Fenske Equation, 274–275, 342
Fenske tray efficiency, 282
FIR, 185
First-principle inferential, 199, 201–203, 342, 380, 384
First-up alarm, 393
Five Fourths Power law, 375
Flooded condenser, 290–293
Flooding, 198, 261–262, 286
Fluid hammer, 52
FMEA, 392
FOPDT, 27
Forced draught, 227–228, 232
Fractionation, 280, 285, 292, 298, 299, 325, 335, 341, 342, 344, 349, 350
Freeze point, 208
FSR, 185
Fuel gas flow, *xiii*, 215
Fuel pressure, *xiii*, 223, 225–227
Full position form, 30–31, 50, 80, 173

Gap controller, *xi*, 108–113, 383
Gas flow, *xiii*, 21, 125, 169, 215–217, 221, 256
GHV, 216
Gilliland correlation, 343
Guide-vanes, 249–250, 297

HAZOP, 391–392
Heat of combustion, see NHV and GHV
Heat of vaporisation, 22, 216, 267, 268, 287, 289, 290, 328, 329
Heavy key component, 262–263, 289
Henry's constant, 264
Henry's Law, 264
Heterogenic, 341
HHK, 263, 316
High signal selector, 81, 172, 227, 235
HK, 262, 278, 318, 329, 331, 340–341, 360, 363
Homogenic, 341
Hot gas bypass, 293–294
Hot vapour bypass, 293
Hydrogen (H_2), 120, 125, 216–222, 229–230
Hydrogen sulphide (H_2S), 217–222
Hysteresis, 16, 81, 232, 383

IAE, 57–58
IBP, 273–274
Ideal control algorithm, 42
Ideal Gas Law, 243, 265
IMC, 63, 85, 89, 166–168
Incremental form, 31, 173
Induced draught, 228
Inlet guide-vanes, 249–250, 297
Inlet temperature feedforward, 223–226, 236, 311
Instrument range, 4, 79, 94, 104, 125, 161, 187
Integral only, 49, 85
Integral time, 20, 33, 68, 70, 72, 74, 75, 17
Integrating process, 20–23, 51, 53, 55–56, 62
Interaction, 170, 174–178, 239, 257, 287, 291, 296, 310, 312, 350, 352
Interactive control algorithm, 41–43, 56, 63, 65, 72, 77, 84–85
Internal reflux, 289, 292, 293, 308, 362
Inverse response, 8, 13, 27–28, 66, 116, 203, 294, 302, 323, 386
ISE, 57, 59
Isentropic, 244–245
ITAE, 47–48, 57–60, 65, 66, 71–74

ITB, 389–390
ITSE, 57, 59–60

Jafarey, Douglas and McAvoy correlation, 343, 370
Jet flooding, 261

Kepner-Trago, 390–391
Key component, 262–263, 316, 330, 363
Kickback, 78

Laboratory update, *xii*, 208–210, 384
Lagrange multiplier, 358
Lambda method, 61–63, 167, 185
Laplace transform, *ix*, 13, 27, 40, 41, 83, 153, 158, 161, 184
Latent heat of vaporisation, 22, 216, 267, 268, 287, 289, 290, 328, 329
Lead, 13, 28, 154, 155, 158, 166, 211
Lead-lag algorithm, 152–155, 166, 180, 210–211, 234–235, 306, 346
Least squares filter, 132–135, 383
Level control, *xi*, 20, 26, 32, 33, 36, 60, 78, 82, 91–116, 152–153, 287, 291, 299–314, 318, 324, 334, 344, 346, 361–362, 366, 383
Light key component, 262–263, 268
LK, 262, 271, 277
LLK, 263, 271, 316
LMTD, 286, 290, 365, 375
Load change, 39, 45–50, 53, 57–58, 65–66, 70, 74, 75
Loop gain, 79–80, 160–161, 183
Low signal selector, 81, 172, 226
LP (linear program), 171, 186, 354, 381, 389
LPG splitter, 262, 283, 316, 328, 360, 361, 368, 375, 379
LPG, 334

MABP, 333
Manipulated variable, 3, 30, 60, 184
Manual mode, 10, 11, 17, 30, 80, 148, 151, 186, 225, 235
Manual reset, 30, 33
Master controller, 9, 239
Master plan, 385
Material balance, 199, 299–312, 318–319, 324–325, 344–346, 349–350, 355–356, 362, 364–365
Maxell-Bonnel, 331–334, 340

McCabe-Thiele diagram, 266, 269–270
Measured value, 29
Methane (CH_4), 229–230
Middle key, 363
MIMO, 170, 174
MISO, 170–173
MK, 363
Model identification, *xi*, 11, 15–16, 20, 22, 26, 51, 78, 80, 171, 225, 255, 346
Molecular weight, *xiii*, 125, 215–217, 220, 222, 230, 243, 246, 256, 268
Molkanov Equation, 343
Monatomic, 244
MTBF, 373
Murphree tray efficiency, 283
MV overshoot, 60–66, 71–73

Natural draught, 228, 232
Nested controllers, 300
NHV, 24–25, 216, 217, 219–222, 225
NIR, 285
Nitrogen (N_2), 217–222, 229–230, 244
NMR, 285
Node, 354–355
Noise, *xi*, 15, 17, 18, 26, 38, 43, 50, 73, 91, 99, 100, 111–113, 126–135, 159, 173–174, 247, 327, 334, 383
Non-correlated steps, 16
Non-interactive control algorithm, 41–44, 56, 64–66, 72, 84, 167
Nonlinear exponential filter, 130–131
Nonlinearity (in distillation), 275, 285, 374–375, 275, 281–282, 285, 287, 316, 317, 320, 334, 336, 338, 340–342, 354, 366
Nonlinearity (in level control), 91, 94–96, 104, 105, 107–109, 111–113
Nonlinearity, *xi*, 10, 16, 24–25, 79, 117–123, 129, 136–137, 149, 159, 161, 202, 232, 237, 256, 374–375, 380
Non-self-regulating process, 20, 23, 89
Nonstandard algorithm, 50

Objective coefficient, 2, 186–188, 355–360, 381, 389
Objective function, 187, 189, 191, 355, 359
On-off control, 81–83
Open-loop unstable, 23
Open-loop, 11, 39, 46, 51, 159, 164, 166, 210

Optimisation, 2, 38, 49, 50, 106, 142, 161, 170, 188, 194, 197, 208, 265, 280, 284, 285, 298, 354, 357, 364–368, 371, 380–383
Order (of process dynamics), 6–8, 12, 13–15, 17–20, 26, 27, 52, 55, 61, 63, 65, 78, 86, 130, 159, 164, 166, 167, 204, 211, 348
Orifice (flow meter), 21, 92, 117, 125, 215, 220, 239, 253, 386
Oscillation, 27, 33, 34, 38, 51–54, 58–60, 78, 106–107, 111–112, 140, 239
Output conditioning, 136, 139, 247
Overdamped, 51–52
Overheads, 260
Oxygen (O_2), 10–11, 123, 222, 229–237

Padé approximation, 13, 87, 128, 154
Paraffinicity, 208, 333
Parallel compressors, 257
Parallel control algorithm, 41
Parallel coordinates, 187
Parallel variables, 176
Partial condenser, 259
Partial pressure, 264–265
Pass balancing, *xiii*, 237–241
PB, 31
PCT (pressure compensated temperature), 125, 325–342, 359
Pearson R^2, 203
Penalty function, 57–59
Performance index, 207, 209, 212
pH, 79–80, 119–122
PI controller, 33–34, 38–39, 51, 54–58, 63, 66, 72–74, 77, 88, 103–104, 168, 171
Polytropic efficiency, 245
Polytropic head, 243, 246, 247, 253–256
Pour point, 208
Prediction horizon, 185
Predictor-corrector, 164
Preheater, 227–228
Pre-rotation, 249
Pressure compensation, 114, 124–126, 215, 263, 325–342
Pressure control, 52, 225–227, 236–237, 285–299
Pressure optimisation, 298, 366–368
Pre-test, 384
Primary controller, 9–10, 47–49, 79, 81, 92
Process gain (definition of), 3–4
Process lag, xii, 5–7, 17, 18, 46, 63, 74, 129, 132, 135, 163, 166, 198, 285, 289, 292, 308

Process variable, 3, 29, 202, 381
Propane, 262–276, 280, 287, 316, 329, 354–361, 369–370, 375–379,
Propene, 263–265, 270, 280, 316, 329
Proportional band, 31
Proportional kick, 31, 44–45
Proportional-on-error, *xi*, 44–49, 70, 78, 103, 161, 167
Proportional-only, 32–33, 35, 52–53, 58, 98, 101, 103–104
Proportional-on-PV, *xi*, 43–50, 53, 57, 65–66, 78, 84–85, 103, 111, 161, 383
Pseudo-code, 201
PV overshoot, 14, 28, 61
PV tracking, 80–81, 148, 255, 327

q, 267–268, 284, 342–343
QP (quadratic program), 381
Quick opening valve, 139–140, 256

Radar plot, 188
Random error, 200, 206, 208–212, 384
Raoult's Law, 264
Ratio algorithm, 115–116, 147–152, 186, 310, 348
Realisable, 159
Reboiler, *xiii*, 22, 97, 100, 113–114, 259–262, 267, 280–288, 300–327, 334–341, 349–361, 365–370, 377, 379
Reciprocating compressor, 243, 251
Rectangular rule, 40
Rectifying section, 259
Recycle, 21, 140, 169, 249–250, 252–257, 297
Reduced cost, 359, 389
Reduced gradient, 359
Reflux ratio, 267, 273, 305, 309, 310, 312–315, 339–340, 343, 350
Reflux, xiii, 97, 100, 114, 259–262, 267, 269–271, 273–274, 285, 289, 291–293, 299–325, 334–362, 365, 369–370
Refrigerant, 287–288
Refutas Equation, 375
Regression analysis, 13–14, 132, 184, 199–203, 337–341, 359, 384
Regulatory control, *x*, *xiii*, 1–2, 4, 11, 373–374, 376–377, 382, 387
Relative volatility, 263–267, 283, 287, 328, 334, 342, 366, 368
Relay method, 53–54
Repeatability, 15, 17, 96, 126, 208

Repeats, 33, 35
Reproducibility, 208
Reset action, 33, 81
Reset windup, 81, 173, 256
Resistance temperature detector, 118–119
Reverse-acting, 31, 172–173, 300
RGA (relative gain array), 174–179
Robustness, 24, 140, 165, 256, 372
RTD, 118–119
RTO, 2, 380–381, 389
Runaway process, 23
RVP (Reid vapour pressure), 263–264, 380
Ryskamp scheme, 309–315, 325–326, 346, 350, 370

Same limit rule, 378
Same percentage rule, 377–378
Sample conditioning, 213
Sample-and-hold, 38, 168, 213
Saturation (of controller), 23, 81, 122, 168, 173, 176, 180, 184, 256, 295, 298, 300, 307, 327, 383
Scan interval, 5, 33, 35, 37, 41, 58, 65, 66, 74–76, 93, 97–98, 104, 108, 111, 128, 167, 168
SCFM, 125
Secondary controller, 9–10, 47–49, 79, 81, 91–92, 98
Self-regulating process, 20–23, 27, 51, 53, 55, 62–65
Self-tuner, 80
Separation, 261–266, 269–285, 301–303, 307, 313, 318, 325, 334–339, 342–343, 367, 368
Series control algorithm, 41–42
Set point, *xi*, 49, 94
Shadow price, 358
Sigma-T delta-T, 350–352
Signal conditioning, *xii*, 1, 96, 117–145, 186, 219, 326, 383, 392
SISO, 169–171, 173, 177, 185
Skin temperature, 238–240, 365
Slave controller, 9
Smith predictor, 163–168
Smoker's Equation, 343
Soft sensor, *xii*, 197
SOPDT, 27
SP change, 37, 39, 44–53, 57–56, 61–63, 67, 70, 72–74, 78, 384
Specific gravity, 213, 216, 217, 219, 333
Specific heat, 24, 148–150, 178, 224, 244, 289

Speed control, 249, 251, 258
Sphere, 95–96
Spillback, 297–298
Split-range, 140–145, 298–299
Stage (distillation), 259
Standard conditions, 125, 215, 221
Standing alarm, 392
Steady state, 1, 3, 5, 8–9, 15–23, 27, 32, 37, 47, 51–52, 54–55, 61, 93, 98, 101, 103, 106, 109, 154, 156, 159, 166, 177, 180, 188, 199, 201, 207, 210–212, 344, 346, 348–350, 352, 358, 380–381, 384, 389
Steady-state decoupling, 177, 180
Steepest slope, 17–18, 55
Stefan's Law, 237, 374
Step-test, *xi*, 49, 157, 159, 172, 174, 183–184, 201, 203, 342, 346, 350, 360, 383, 385, 389, 392
Stiction, 16, 81, 232, 383
Stonewall, 246, 258
Stripping section, 259
Suction throttling, 247–249, 251
Sulphur dioxide (SO_2), 244
Surge (compressor), 65, 140, 169, 186, 246–258
Surge capacity, *xi*, 95, 97, 100, 101, 103–104, 106–107, 110–112, 304, 386
Swell, 114, 124

Tangent lines, 94
Tangent of steepest slope, 17–18
Taylor approximation, 12, 87–88, 128, 154
TBP curve, 272–274
TDC, 251
Temperature profile, 317–318, 322, 323, 342, 350
TFOE, 223
Theoretical stage, 259
Theoretical tray, 259, 266–267, 269–270, 274
Thermocouple, 5, 117–118
Thermowell, 5–6, 315, 386
Three-element level control, 114–115, 152
Tight level control, 91, 97, 99–101, 111, 113–115, 304, 383
Timeout, 213
Time-stamping, 199–200, 212, 384
Titration curve, 79, 120

Total condenser, 260, 268, 296
Transport delay, 5, 7, 25, 285
Trapezium rule, 40
Tray efficiency, 259, 261, 282–284, 344
Trial-and-error tuning, *ix*, *xi*, 3, 34, 50, 53, 61, 63, 66, 155, 308, 348, 349, 357
Triatomic, 244
Trouton's Law, 268
Tube metal temperature, 169, 237
Turbo-machine, 244, 246, 251, 252
Turndown ratio, 25, 161, 233
Two degrees of freedom controller, 49

Ultimate gain, 52, 54
Ultimate period, 52
Unbiased CV, 193–194
Underdamped, 51–52
Underwood's Method, 342–343
Unit reaction rate, 55
Universal Gas Constant, 238, 243, 329
USGPM, 94, 98, 99

Valve constant, 139
Valve position, 3–6, 21, 81, 97, 122–123, 136–138, 141–144, 150, 171–173, 238–241, 294, 383
Vapour pressure, 263–264, 370, 375, 379
VBN, 375
Vector, 389
Velocity form, 30–31, 33, 40, 42, 80
Velocity, 100, 225, 243, 259, 286, 360
Vertical drum, 94
VFD, 232, 249
Virtual analyser, 197, 337
Viscosity, 227, 365, 375,
Volatility, 263–267, 283, 287, 328, 334, 342, 366, 368,

Watson K, 333
Weeping, 261
Wet basis, 230
Wetness, 268, 349
Windup, 81, 173, 180, 256
Wobbe index, 221–222

Ziegler-Nichols method, 17–18, 28, 52, 54–56, 88
z-transform, 13, 40, 130, 154